POLYMERS FOR ELECTRICITY AND ELECTRONICS

POLYMERS FOR ELECTRICITY AND ELECTRONICS

Materials, Properties, and Applications

JIRI GEORGE DROBNY
Drobny Polymer Associates
Merrimack, New Hampshire

WILEY

A JOHN WILEY & SONS, INC., PUBLICATION

Published by John Wiley & Sons, Inc., Hoboken, New Jersey
Published simultaneously in Canada

Library of Congress Cataloging-in-Publication Data:

Drobny, Jiri George.
Polymers for electricity and electronics : materials, properties, and applications / Jiri George Drobny.
p. cm.
Includes bibliographical references and index.
ISBN 978-0-470-45553-1 (hardback)
1. Electronic apparatus and appliances—Materials. 2. Polymers. I. Title.
TK7871.15.P6D76 2012
621.381—dc23

2011026135

10 9 8 7 6 5 4 3 2 1

To Julia, Joshua, Daniel, Abigail, and Leah

CONTENTS

PREFACE

There have been several excellent books published recently that mainly concentrate on the theoretical aspects of polymers for electronics. Therefore an updated publication focusing on the practical side of this subject and complementing it would be very useful. This book has two goals. One is to provide a ready reference for professionals working in industrial practice and the other is as a textbook for advanced students of electrical and electronic engineering, material science, polymer science, and engineering. It will also be useful for educators, attorneys, and marketing specialists.

The introduction concentrates on the basic principles involved with the electrical properties of materials. Chapter 2 covers polymeric materials and focuses on structure–property relationships as well as standard process technologies. The main purpose of Chapter 2 is to cover the basic concepts of polymer science and the technology deemed necessary to understand polymers as valuable materials, which are used in electrical and electronic applications. Chapter 3 covers the specific electrical and electronic behavior of polymers, not only as well-established insulating materials, but also as conductive materials of equal importance. Chapter 4 provides essential information on special polymers, including intrinsically conductive and ferroelectric polymers, and so on. Chapters 5–7 deal with some examples of specific industrial fabrication methods. These methods include examples of formulations (essentially compounding concepts and starting formulations for different applications) and practical applications of polymers and polymer-based materials with their essential measuring methods. Chapter 8 covers current known developments and trends. There are numerous references at the end of each chapter, which may be used for additional information and/or an in-depth study. In addition, a list of recommended publications for further reading on the subjects covered in that chapter is provided. The appendices include a table of electrical properties of selected polymers and lists of additional ASTM testing standards and corresponding international standards. In addition, there is a list of pertinent acronyms and abbreviations, as well as a general bibliography and a rather comprehensive glossary.

Throughout the text, SI units are preferred, although occasionally, other units are also shown (e.g., degrees Fahrenheit for temperature, pounds for weight, pounds per square inch for pressure, and poise for viscosity).

My thanks go the John Wiley & Sons team for bringing this work to fruition: To Jonathan Rose for encouragement and coordination of this demanding project, to Kristen Parrish for production, and to Jeannette Stiefel for meticulous copyediting.

Thanks are also due to the American Institute of Physics, Cambridge University Press, Elsevier Books, Hanser Punlishers, Institute of Electrical and Electronic Engineers, Nature Publishing Group, Panasonic Corporation, Springer SBM, Springer Verlag, Taylor and Francis, and others for permissions to reprint their copyrighted material. The following companies provided valuable illustrations: Arkema Inc., Bayer Material Science, Belden Wire & Cable Co., Davis Standard, LLC, DuPont Company, Energy Sciences, Inc., IBA Industrial, The Okonite Company, Royle Systems Group, RTP Company, and Teknor Apex Company.

Special thanks go to my friend Professor Ivan Chodak from the Polymer Institute of the Slovak Academy of Sciences for providing valuable materials and for his cooperation on many projects related to this publication.

Merrimack, NH and
Prague, The Czech Republic,
April 2011

JIRI G. DROBNY

CHAPTER 1

INTRODUCTION

Electricity is one of the most important parts of modern everyday life. It is used to run product machinery, transportation, communications, medical procedures, military operations, research, and so on. Most of the principles that explain electricity as a form of energy have been known since the 19th century. Scientists like Faraday, Ampère, and Maxwell did most of the fundamental work. Since then, every decade has brought some refinements, new discoveries, and new applications. Polymers (i.e., plastics and elastomers) are very essential materials in electrical applications.

The first known use of a polymeric substance (a natural one at that) as insulation of the first Transatlantic telephone cable was gutta-percha (*trans*-polyisoprene) in the 1860s. Natural rubber (*cis*-polyisoprene) (NR) became a very important insulation material after the invention of vulcanization in the late 1840s.

With the development of synthetic polymers, the selection of insulating materials has been steadily growing (Table 1.1). Today, there are dozens of polymers serving as insulators at extremely low and high temperatures, at a wide range of frequencies, in adverse environments. There are special polymers or their compounds that are conductors or semiconductors and exhibit other specialized properties.

Polymers for Electricity and Electronics: Materials, Properties, and Applications, First Edition. Jiri George Drobny.

TABLE 1.1 Milestones in the Development of Synthetic Polymers for Electrical and Electronic Applications[a]

Year/Years	Polymer
1907	Phenolic resin
1926	Aniline–formaldehyde resins
1928	Urea–formaldehyde resins
1929	Styrene–butadiene rubber (SBR)
1930	Nitrile–butadiene rubber (NBR)
1931	Polychloroprene rubber (Neoprene)
1937	Butyl rubber (IIR)
	Polyurethanes
1938	Epoxy resins
	PTFE (Teflon)
	Poly(vinyl chloride) (PVC)
	Polyethylene (PE)
1940	Polyamide (Nylon)
1944	Silicones
1950	Linear polyethylene
1951	Chlorosulfonated PE (Hypalon)
1950s (late)	Ethylene–propylene rubbers (EPM, EPDM) elastomers
	Fluorocarbon elastomers (Kel-F)
1964	Carbon (graphite) fibers
1960s (late)	Styrenic block copolymers (Kraton)
	Polyvinylidene fluoride (PVDF)
	Ethylene–tetrafluoroethylene copolymer (ETFE)
	Fluorinated ethylene propylene (FEP)
	Tetrafluoroethylene–perfluoroalkyl vinyl ether copolymer (PFA)
1970s (late)	Conductive polymers
	Polyamide-based thermoplastic elastomers (Pebax)
	Copolyesters (Hytrel)
	Thermoplastic vulcanizates (TPV, Santoprene),
1980s	Thermoplastic polyolefins (TPO)
	Polyaniline
	Iodine-doped polyaniline (2000 Nobel Prize)
1990s to early 2000s	Conducting electroactive polymers (CEPs), including polypyrrole, polythiophene, sulfonated polyaniline
	Organic (polymeric) light-emitting diodes (OLEDs)
2001 to date	Organic (polymer) solar cells

[a] Due to the current multitude of projects involving new products and technologies, it is difficult to show exact years for the individual items.

1.1 ELECTRIC CHARGE

Elementary quantity of charge is ± 1 e, with the electron carrying a charge of -1 e and the proton $+1$ e. The unit of charge is 1 C (coulomb); in some cases this unit is extremely large. A more practical unit is the microcoulomb (μC), which is one millionth of a coulomb. The magnitude of the electron charge is 1.6021×10^{-19} C or 1.6021×10^{-13} μC.

1.1.1 Static Electricity

Static electricity refers to the build up of electric charge on the surface of objects. The static charges remain on an object until they either bleed off to ground or are quickly neutralized by a discharge. Although charge exchange can happen whenever any two surfaces come into contact and separate, a static charge only remains when at least one of the surfaces has a high resistance to electrical flow (as an electrical insulator). The effects of static electricity are familiar to most people because we can feel, hear, and even see the spark as the excess charge is neutralized when brought close to a large electrical conductor (e.g., a path to the ground), or a region with an excess charge of the opposite polarity (positive or negative). The familiar phenomenon of a static "shock" is caused by the neutralization of charge.

Electric charge is an intrinsic characteristic of the fundamental particles making up those objects; that is, it is a characteristic that automatically accompanies those particles wherever they exist. Essentially, many objects contain equal amounts of two kinds of charge: *positive* and *negative* charge and they are said to be *electrically neutral.* If the two types of charge are not in balance, then there is a net charge either positive or negative. Charged objects interact by exerting forces on one another. Charges with the same electrical sign repel each other, while charges with opposite sign attract each other.

The attraction and repulsion between charged bodies have many industrial applications, including electrostatic paint spraying and powder coating, fly-ash collection in chimneys, nonimpact inkjet printing, and photocopying.

1.1.2 Electric Fields

When forces exist at a distance, it is said that force field exists in space. In this case, the force field is referred to as the *electric* or *E field.* This

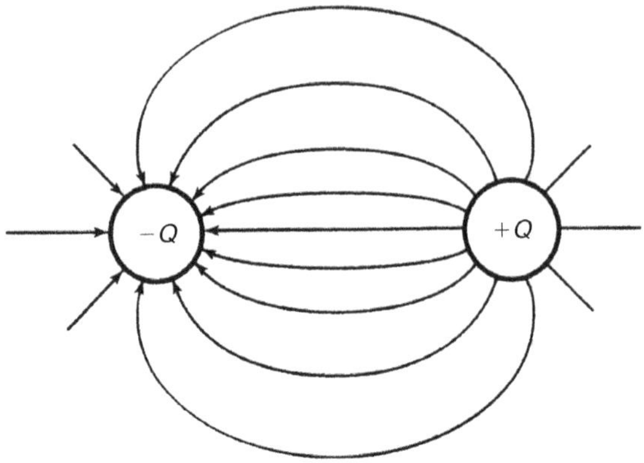

FIGURE 1.1 Electric field between oppositely charged objects. (Ku, C.C. and Liepins, R., *Electrical Properties of Polymers: Chemical Principles*, Hanser Publishers, Munich, 1987, p. 107)

field is represented by field lines represented by lines drawn between the charged objects. These symbolic lines connect units of positive charge (the absence of electrons) with units of negative charge. When more charges are involved, convention says that there are more lines (see Fig. 1.1). The E field exists through entire space; the lines are only representation. At every point in space, the force has a magnitude and direction (i.e., it is a *vector*).

1.1.3 The Triboelectric Effect

The *triboelectric effect* (also known as *triboelectric charging*) is a type of contact electrification in which certain materials become electrically charged after they come into contact with another different material and are then separated (e.g., through rubbing). The polarity and strength of the charges produced differ according to the materials, surface roughness, temperature, strain, and other properties.

Thus, it is not very predictable, and only broad generalizations can be made. Amber, for example, can acquire an electric charge by contact and separation (or friction) with a material like wool. This property, first recorded by Thales of Miletus, suggested the word "electricity" (from William Gilbert's initial coinage, "electra"), from the Greek word for amber, *ēlektron*. Other examples of materials that can acquire a significant charge when rubbed together include glass rubbed with silk, and hard rubber rubbed with fur.

TABLE 1.2 Triboelectric Series

Material	Polarity
	Positive ↑
Silicone elastomer with silica filler	
Borosilicate glass, fire polished	
Window glass	
Aniline–formaldehyde resin (acid catalyzed)	
Polyformaldehyde	
Poly(methyl methacrylate)	
Ethyl cellulose	
Polyamide 11	
Polyamide 66	
Rock salt (NaCl)	
Melamine–formaldehyde resin	
Wool, knitted	
Silica, fire polished	
Silk, woven	
Poly(ethylene glycol succinate)	
Cellulose acetate	
Poly(ethylene glycol adipate)	
Poly(diallyl phthalate)	
Cellulose (regenerated) sponge	
Cotton, woven	
Polyurethane elastomer	
Styrene–acrylonitrile copolymer	
Styrene–butadiene copolymer	
Polystyrene	
Polyisobutylene	
Polyurethane flexible sponge	
Borosilicate glass, ground surface	
Poly(ethylene glycol terephtalate)	
Polyvinylbutyral	
Formaldehyde–phenol hardened epoxy resin	
Polychlorobutadiene	
Butadiene–acrylonitrile copolymer	
Natural rubber	
Polyacrylonitrile	
Sulfur	
Polyethylene	
Poly(diphenylol propane carbonate)	
Chlorinated polyether	
Poly(vinyl chloride plasticized with 25% dioctyl phthalate	
Poly(vinyl chloride) without plasticizer	
Poly(trifluorochloroethylene)	
Polytetrafluoroethylene	↓ Negative

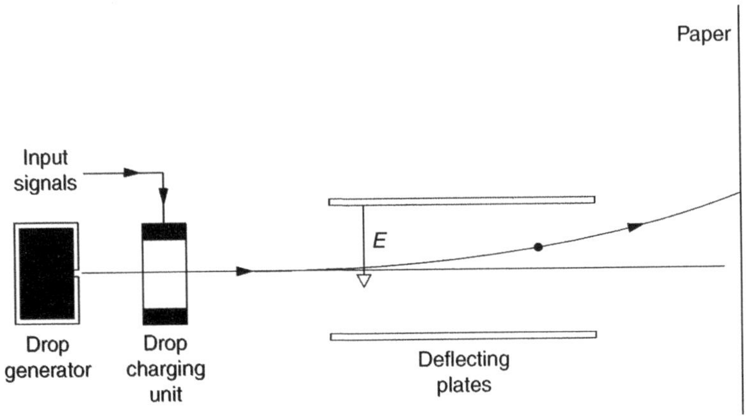

FIGURE 1.2 Essential features of the ink-jet printer. (Halliday, D., Resnick, R. and Walker, J., *Fundamental of Physics Extended, Fifth Edition*, John Wiley & Sons, New York, 1997, p. 547)

Materials are often listed in order of the polarity of charge separation when they are touched with another object in so-called *triboelectric series* (see Table 1.2). A material toward the bottom of the series, when touched to a material near the top of the series, will attain a more negative charge, and vice versa. The further away two materials are from each other on the series, the greater the charge transferred. Materials near each other on the series may not exchange any charge, or may exchange the opposite of what is implied by the list. This depends more on the presence of rubbing, the presence of contaminants or oxides, or upon properties other than on the type of material. Lists vary somewhat as to the exact order of some materials, since the charge also varies for nearby materials.

The phenomenon of acquiring charge through contact accounts for the attraction of dust particles by plastic items, the accumulation of electric charge in carpets in dry weather, and radio frequency (rf) and electromagnetic interference effects of plastics enclosures. One well-established application of this principle is ink-jet printing; the principle is shown in Fig. 1.2. In this application, an input signal from a computer controls the charge given to each drop and thus the position on the paper at which it lands. The accumulation of charge is also a key component in all *xerographic* processes.

1.2 COULOMB'S LAW

Coulomb's law expresses the interaction force between two charges at rest. It is valid for point charges only. This force is

$$F = \left[\frac{1}{4\pi\varepsilon_o}\frac{q_1 q_2}{r^2}\right]\boldsymbol{I}_r \quad (1.1)$$

where q_1, q_2 = magnitude of point charges (+ or −); r = distance between charges; $\boldsymbol{I}_r$ = unit vector along r; and ε_o = permittivity constant (= $8.85 \times 10^{-12}\,C^2 N^{-1} \cdot m^{-2}$). The force of attraction or repulsion between point charges at rest acts along the line joining the charges.

1.3 CONDUCTORS AND INSULATORS

Insulators or nonconductors (or dielectrics) are materials containing essentially no charges that are free to move. If such a material is subjected to an electric field, each electron and nucleus within the material experiences a force due to the electric field. However, these charges are bound tightly within the material and are not free to move under this force. Thus the external field does not cause charges to flow in the material, and therefore these materials do not conduct electric charges. Examples here are glass, ceramics, and polyethylene.

A conductor is an opposite of an insulator. In conductors, electrons exist that are free to move through the material. When an electric field is impressed upon such a material, the charges move to produce a current under the action of the field.

This classification is not sharply defined, however. There are materials that contain some charges that can move under the action of the field. If the proportion is significant, these are referred to as *semiconductors*.

1.4 CAPACITORS

Capacitors are devices capable of storing charge and there are several possible arrangements, the most common being parallel conductive plates or concentric cylinders or other shapes (see Fig. 1.3). The principle of a capacitor with parallel conducting plates is shown in Fig. 1.4.

The capability of storing charge is called capacitance, C, and is expressed as:

$$C = \frac{q}{V} \quad (1.2)$$

where q = charge in coulombs (C) and V = potential difference in volts (V).

The unit of capacitance is 1 farad, $1\,F = 1\,CV^{-1}$, but two practical units, 1 microfarad (1 μF) and 1 picofarad (1 pF) are much smaller;

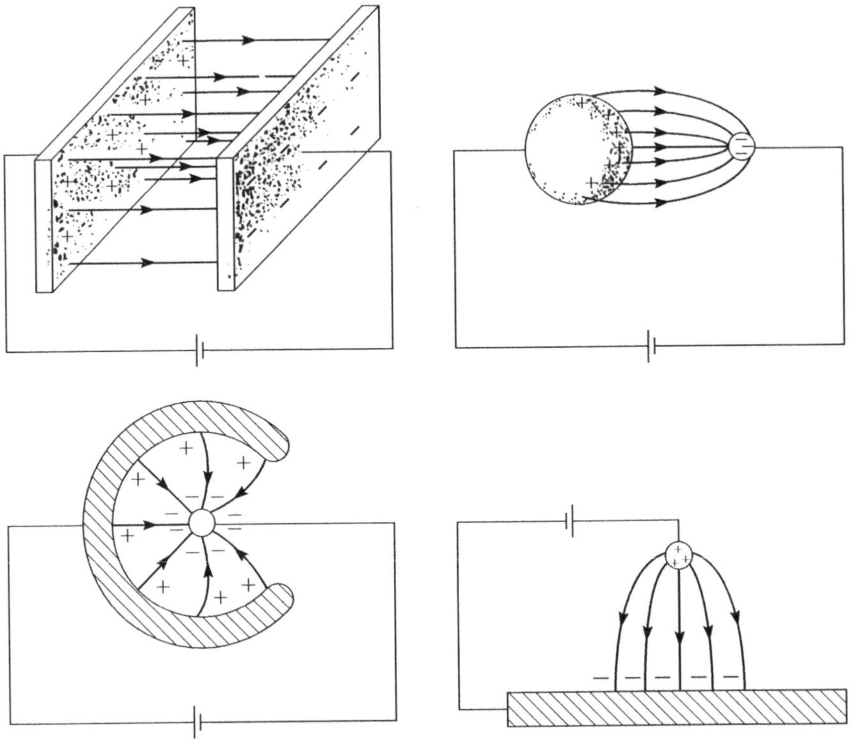

FIGURE 1.3 Examples of devices having capacitance.

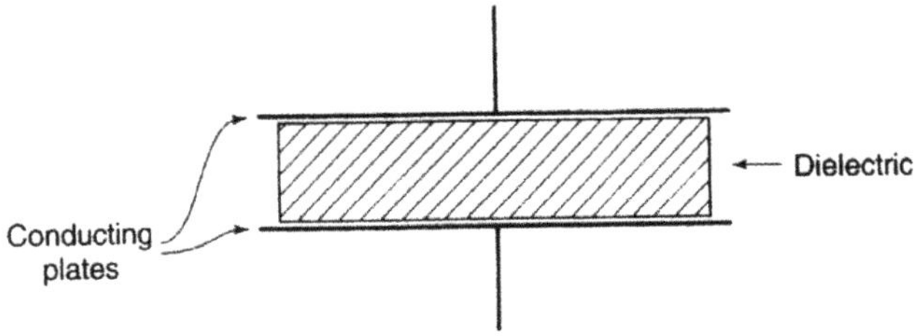

FIGURE 1.4 Principle of a capacitor with parallel conducting plates and a dielectric.

$1\,\mu F = 10^{-6}\,F$, $1\,pF = 10^{-12}\,F$. One farad of capacitance means that a charge of 1 coulomb can be stored for one volt of potential difference.

1.5 DIELECTRICS AND CAPACITANCE

Dielectrics (nonconducting materials) affect the capacitance of a capacitor. This factor is called a dielectric constant ε and is defined as:

$$\varepsilon = \frac{C}{C_0} \tag{1.3}$$

where C_0 = the capacitance of the capacitor in a vacuum and C = the capacitance with a dielectric between the plates.

The capacitance of a condenser measures the extent to which it is able to store charge. For a nonpolar material the increase in capacitance is due to the charges on the capacitor polarizing the molecules (attracting the positive charges in the molecules to one end and the negative charges to the other) with the result that increased charge appears on the surface, as shown in Fig. 1.5. When the insulating sheet is inserted between the metal plates the voltmeter reading drops from V_0 to V as shown in

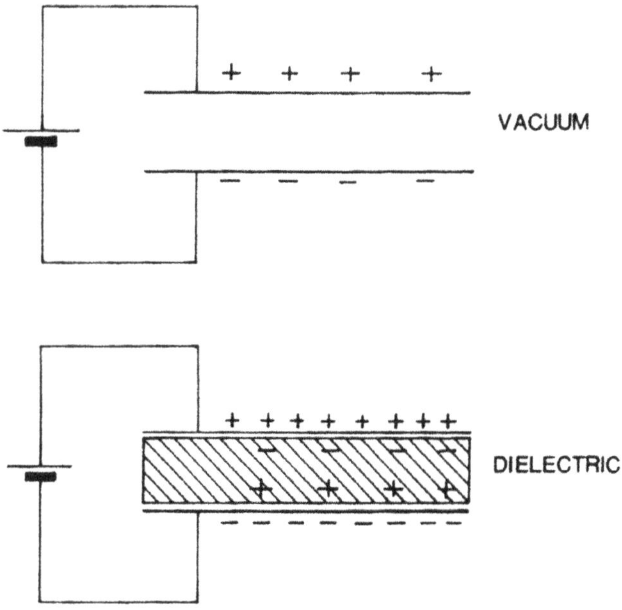

FIGURE 1.5 Surface charge on a condenser.

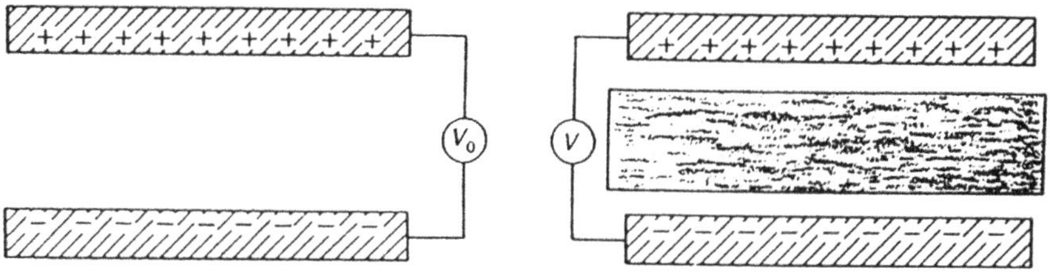

FIGURE 1.6 Effect of an insulator on voltage reading.

Fig. 1.6. Physical meaning of the dielectric constant is that the induced charge decreases the potential between the plates thus increasing the capacitance of the system. The value of dielectric constant alone addresses how an insulator stores the electrical energy so as to isolate electrical elements from each other and the ground. In other words, the higher the value of the dielectric constant, the more charge is stored on the plates for a given voltage. By definition the value of the dielectric constant of air is 1.0. Values of dielectric constants of some materials are in Table 1.3.

The electric potential energy of a charged capacitor, U, is given by

$$U = \frac{q}{2C} = \frac{1}{2}CV^2 \tag{1.4}$$

It is the work required to charge it. This energy can be associated with the capacitor's electric field, E. In vacuum, the *energy density* u or potential per unit volume, is given by

$$u = \frac{1}{2}\varepsilon_0 \cdot E^2 \tag{1.5}$$

TABLE 1.3 Values of Dielectric Constants of Some Materials (From Halliday, D., Resnick, R., and Walker, J., *Fundamentals of Physics Extended, 5th ed.*, John Wiley & Sons, Inc., New York, 1997, p. 639. With permission of John Wiley & Sons.)

Material	Dielectric Constant
Vacuum	1.00000
Air (1 atm)	1.00054
Polystyrene	2.6
Paper	3.5
Transformer oil	4.5
Pyrex	4.7
Ruby mica	5.4
Porcelain	6.5
Silicon	12
Germanium	16
Ethanol	25
Water (20°C)	80.4
Water (25°C)[a]	78.5
Titania ceramic	130
Strontium titanate	310

[a]All other measurements done at 20°C.

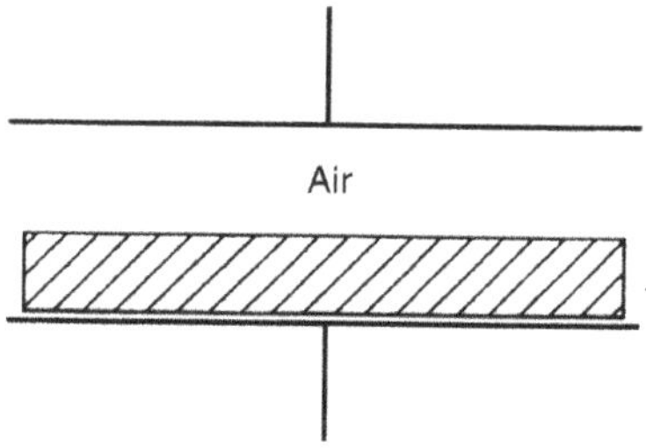

FIGURE 1.7 Air and dielectric between parallel plates. (Ku, C.C. and Liepins, R., *Electrical Properties of Polymers: Chemical Principles,* Hanser Publishers, Munich, 1987, p. 25)

If a dielectric occupies part of the space between the two parallel conducting plates (as shown in Fig. 1.7) the equipotential surfaces are no longer uniformly spaced across the space. Most of the potential difference appears across the air space. The ratio of the E field in the air to the field in the dielectric is the relative dielectric constant ε_r. It is convenient to consider a new field, one that is not dependent on the dielectric. This field is called the *displacement* or D field. This D field is generated by charges and is not a function of the dielectric. To make the units turn out correctly, the D field in air is the E field times the dielectric constant of free space. This constant ε_0 is also known as the *permittivity of free space.* This constant has the value 8.854×10^{-12}. The D field has units of charge per unit area (C m^{-2}). At the interface between the air and the dielectric, the D field has the same intensity on both sides of the interface. The energy stored in a field is proportional to the E, not the D field. In the system shown in Fig. 1.7, most of the energy is stored in the air space in contrast to a system without the air space. When the dielectric is introduced to an existing uniform E field, the field reconfigures itself to store the least possible energy [1].

Another effect of the introduction of a dielectric is to limit the potential difference that can be applied between the plates to a certain value V_{max}, called the *breakdown potential.* If the value is substantially exceeded, the dielectric material will break down and form a conducting path between the plates. Every dielectric material has a characteristic *dielectric strength*, which is the maximum value of the electric field that it can tolerate without breakdown. The values of the dielectric strength of several dielectrics measured at room temperature are in Table 1.4.

TABLE 1.4 Dielectric Strength Values of Selected Dielectrics at Room Temperature

Substance	Dielectric Strength (MV m^{-1})
Air	3
Alumina	13
Window glass	10–14
Pyrex	14
Silicone oil, mineral oil	10–15
Polyethylene	19–22
Polystyrene	16–24
Polyamide 66	14
Polychloroprene rubber	16–28
Paper	16
Bakelite	24
Ultrapure water	30
Waxed paper	40–60
Fused silica	25–40
Mica	20–70
PTFE	60

1.6 DIELECTRIC POLARIZABILITY

If an atom is placed into an electric field, the positive nucleus experiences a force in the direction of the field, while the negative electron cloud is pulled in the opposite direction. As a result, the centers of the positive and negative charge in the atom are separated. Now the atom has become polarized and acts as a dipole. The same is true about a molecule.

Evidently, the greater the polarizability of the dielectric, the higher the dielectric constant. The polarizability *of nonpolar* molecules arises from two effects: (1) *Electronic polarization*, in which the applied electric field causes a displacement of the electrons relative to the nucleus in each atom. (2) *Atomic polarization*, in which the applied field causes a displacement of the atomic nuclei relative to one another. In polar molecules, a third process contributes to the total polarizability: (3) *Orientation polarization*, in which the applied field causes a net orientation of the dipoles to the field.

A *polar* molecule has a permanent dielectric dipole moment; that is, although the molecule is electrically neutral, a distribution charge exists such that the centers of positive and negative charge are separated by a distance of molecular dimensions [2]. The dipole moment, which is equal to charge times distance, is measured in coulomb-meter (Cm). The old electrostatic debye unit of molecular dipoles (D) was equivalent to 3.335×10^{-30} Cm [3]. The overall polarizability of a molecule, α_T,

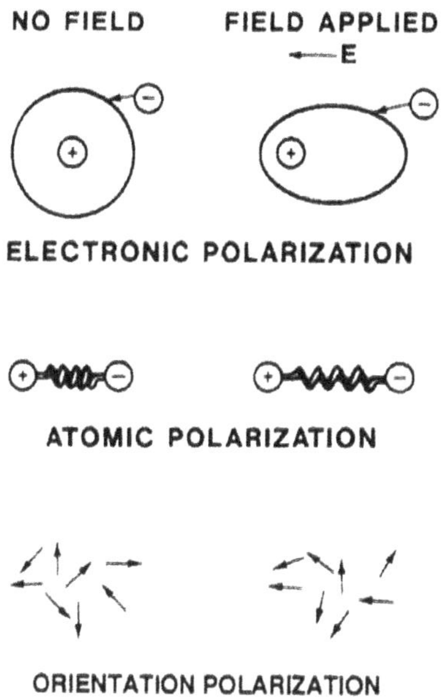

FIGURE 1.8 Mechanisms of polarization.

is the sum of three terms: $\alpha_T = \alpha_e + \alpha_a + \alpha_o$, where α_e, α_a, and α_o are the electronic atomic, and orientation polarizations, respectively [2]. The mechanisms of the above types of polarization are depicted by Fig. 1.8. Molecular dipole moments are listed in Table 1.5 and group dipole moments are in Table 1.6.

If an external electric field is present, the molecular dipoles experience a torque because of it. The torque tends to align the dipoles with the field in such a way that the positive end is moved forward in the direction of the field and the negative end is displaced in a direction antiparallel to it (see Fig. 1.9). If the electric field is alternating, the dipoles will rotate to adjust their positions accordingly. The process generates thermal energy and is the basis of microwave heating.

The correlation of polarizability α with dielectric constant ε is expressed by the Clausius–Mossotti equation:

$$\frac{N\alpha}{3\varepsilon_0} = \frac{\varepsilon - 1}{\varepsilon + 2} \tag{1.6}$$

where ε_0 is the dielectric constant (permittivity of the vacuum) and N is the number density (number of atoms or molecules per volume). The dielectric constant (relative permittivity) is a bulk (macroscopic) property and polarizability is a microscopic property of matter, thus the

TABLE 1.5 Molecular Dipole Moments

Compound	Dipole Moment (10^{-30} Cm)
H_2O	6.1
HF	6.4
HCl	3.6
HBr	2.6
CH_4	0
CCl_4	0
CO_2	0
NH_3	4.9
CH_3Cl	6.2
CH_2C_{12}	5.4
$CHCl_3$	4.0
CH_3F	6.0
C_2H_5OH	5.7
$(C_2H_5)_2O$	3.8
$C_6H_5CH_3$	1.2
C_6H_5Br	5.8
$C_6H_5NO_2$	13.1
$C_6H_4(CH_3)NO_2$ (*para*)	14.6
C_6H_5OH	5.1
$C_6H_4(CH_3)OH$ (*para*)	5.2

TABLE 1.6 Group Dipole Moments

Group	Aromatic Compounds, Moment (10^{-10} Cm)	Aliphatic Compounds, Moment (10^{-30} Cm)
$-CH_3$	1.3	0
Cl	5.3	6.3
–F	5.7	6.0
–Br	5.7	6.7
–I	5.7	6.3
–OH	4.7	5.7
$-NH_2$	5.0	12.3
$-NO_2$	8.5	7.7
$-CF_3$	5.3	13.4
$-CCl_3$	6.8	5.2
–CN	13.5	11.6
–CHO	9.9	8.3
$-OCH_3$	4.3	4.3
–COOH	6.1	5.4
$-COOCH_3$	6.1	6.0
$-COOC_2H_5$	6.3	6.0

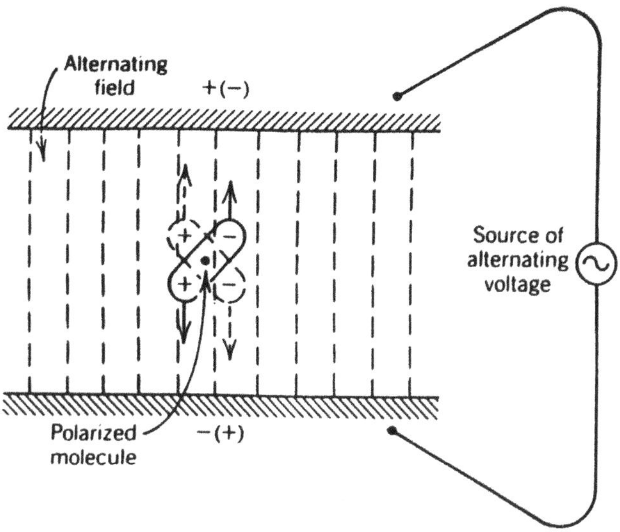

FIGURE 1.9 A molecular dipole immersed in a high frequency field.

relation bridges the gap between a directly observable macroscopic property and a microscopic molecular property.

The equation may be also written in terms of molecular weight M_w and density ρ:

$$\frac{N_A \alpha}{3\varepsilon_0} = \frac{\varepsilon - 1}{\varepsilon + 2} \frac{M_w}{\rho} \tag{1.7}$$

where N_A is Avogadro's number. The quantity $N_A \alpha / 3\varepsilon_0$ is called the *molar polarization* P_M and has the dimension of volume. As long as the theory leading to the equation remains valid, the molar polarization remains constant for a particular material irrespective of the temperature and pressure [4]. From the relationship between ε and α it is clear that the way to reduce ε is to lower α and to increase ε is to raise α. Note also that the Clausius–Mossotti equation applies only to nonpolar or weakly polar dielectrics.

1.7 VOLTAGE

The fundamental definition of voltage relates to the work required to move a unit of charge between two points. By definition, the unit of charge is positive. The amount of work does not require a reference level. The work required is measured by the potential difference. It is correct to say that the work per unit charge is the voltage difference. Thus the words *voltage* and *potential* are used interchangeably. The unit of voltage is a volt (V). Large voltages are expressed in kilovolt or 10^3 V (kV), megavolt, 10^6 V (MV), smaller ones are millivolt, 10^{-3} V (mV), and microvolt, 10^{-6} V (μV).

Any point can be selected as zero potential. If a remote point is selected, work might be required to get the given charge to the first body. If this work is 10 V, then the work to get to the second body may be 5 more volts. The potential difference between the two bodies is 5 V. When the charge is positive and the charge transferred is also positive, positive work is done in moving this charge. The work is actually stored in the electric field.

1.8 ELECTRIC CURRENTS

Electric current is the charge passing through a given cross section of the conductor (wire) per unit time. Electrons are carriers of the charge. If charge dq passes through a hypothetical plane in time dt, then the current through that plane is

$$i = \frac{dq}{dt} \tag{1.8}$$

The SI unit for current is the ampere (A):

$$1 \text{ ampere } (1 \text{ A}) = 1 \text{ coulomb per second } (1 \text{ C}^{-1} \text{ s})$$

Current, as defined by Eq. (1.8) is a scalar, because both the charge and time are scalars. If a current splits into branches, the magnitude of the partial currents must add to yield the magnitude of the current in the original conductor (i_0), because the charge is conserved. Thus, for example, for two branches, this can be expressed as:

$$i_0 = i_1 + i_2 \tag{1.9}$$

where i_1 and i_2 are currents in the two branches, respectively (see Fig. 1.10)

In an electric loop, an addition of a battery imposes an electric potential between the ends of the loop that are connected to the terminals of the battery. The battery thus produces an electric field within the loop, from terminal to terminal and the field causes charges to move around the loop. This movement of charges is a current i. The direction of flow of the current is shown from the positive to the negative terminal. The positive

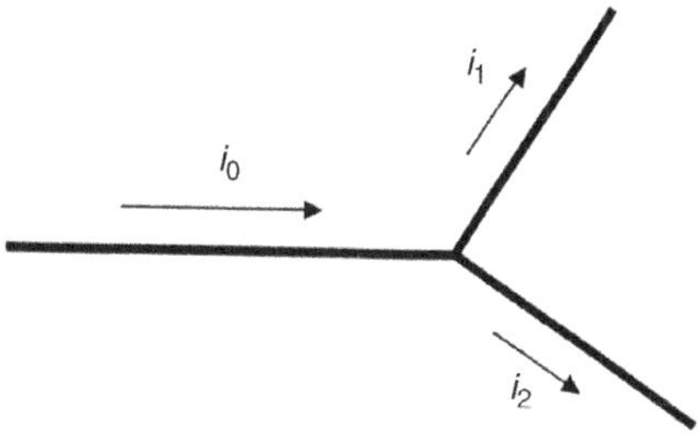

FIGURE 1.10 Current splitting at a junction into two branches.

charge carriers as they are often called will move away from the positive carrier and toward the negative terminal (see Section 1.7). Still, for historical reasons we use the following convention: *A current arrow is drawn in the direction in which positive charge carriers would move, even if the actual charge carriers are negative and move in the opposite direction* [5].

We can use this convention because in *most* situations, the assumed motion of positive charge carriers in one direction has the same effect as the actual motion of negative charge carriers in the opposite direction. When the effect is not the same, the convention is dropped and the actual motion is described.

1.9 RESISTANCE AND RESISTIVITY

If we apply the same potential difference between the ends of geometrically similar rods of copper and glass, very different currents result. The characteristic of the conductor that enters here is its resistance. We determine the resistance between any two points of a conductor by applying a potential difference V between those points and measuring the current I that results. The resistance R is then

$$R = \frac{V}{i} \tag{1.10}$$

The SI unit for resistance is the ohm (symbol Ω). That is,

$$1\ \Omega = 1 \text{ volt per ampere} = 1\ \text{VA}^{-1}$$

If we focus on the *electric field E* at a point in a resistive material instead on the *potential difference V* across a particular resistor, we deal with the *current density J* at the point in question instead of the current i through the resistor. Then we are dealing with the resistivity of the *material* instead of the resistance R of the object. The resistivity ρ of the material is defined as:

$$\rho = \frac{E}{J} \tag{1.11}$$

The unit of resistivity is

$$\frac{\text{unit }(E)}{\text{unit }(J)} = \frac{\text{V m}^{-1}}{\text{A m}^{-2}} = \left(\frac{\text{V}}{\text{A}}\right)\text{m} = \Omega\,\text{m (ohm-meter)}$$

The reciprocal of the resistivity of a material is its conductivity (σ) defined as:

$$\sigma = \frac{1}{\rho} \tag{1.12}$$

The SI unit of conductivity is siemens, S, where $1\ \text{S} = 1\ \Omega^{-1}$.

Values for the resistivity of some materials are listed in Table 1.7.

TABLE 1.7 Values of Resistivities of Some Materials at Room Temperature

Material	Volume Resistivity ($\Omega \cdot$ m)
Typical Metals	
Silver	1.62×10^{-8}
Copper	1.69×10^{-8}
Aluminum	2.75×10^{-8}
Tungsten	5.25×10^{-8}
Iron	9.68×10^{-8}
Platinum	10.6×10^{-8}
Manganin[a]	48.2×10^{-8}
Typical Semiconductors	
Silicon, pure	2.5×10^{3}
Silicon, n-type[b]	8.7×10^{-4}
Silicon, p-type[c]	2.8×10^{-3}
Typical Insulators	
Glass	10^{10}–10^{14}
Fused quartz	$\sim 10^{16}$
Polyethylene	$\sim 10^{19}$
Polytetrafluoroethylene	$>10^{20}$
Polyisobutylene	$>10^{17}$

[a] An alloy specifically designed to have a small value of α.
[b] Pure silicon doped with phosphorus impurities.
[c] Pure silicon doped with aluminum impurities.

1.10 OHM'S LAW

Ohm's law is valid for circuits, expresses the relationship between current, voltage, and resistance:

$$E = IR \tag{1.13}$$

where E = potential (in V); I = current (in A); and R = resistance (in Ω).

Ohm's law is an assertion that the current through a device is *always* directly proportional to the potential applied to the device. A conducting device obeys Ohm's law when the resistance of the device is independent of the magnitude and polarity of the applied potential difference. There are exceptions to Ohm's law, such as applications in modern microelectronics [6].

1.11 SEMICONDUCTORS

A *semiconductor* is a material that has an electrical conductivity due to flowing electrons (as opposed to ionic conductivity), which is intermediate in magnitude between that of a conductor and an insulator; this means roughly in the range 10^{1}–10^{-10} S m^{-1}) siemens per meter. Devices made from semiconductor materials are the foundation of modern electronics, including the radio, computers, telephones, and many other devices.

In a metallic conductor, current is carried by the flow of electrons. In semiconductors, current is often schematized as being carried either by the flow of electrons or by the flow of positively charged "holes" in the electron structure of the material. Actually, however, in both cases only electron movements are involved. In other words, semiconductors can be defined as materials with few conduction electrons, but with available conduction-level states that are close, in energy, to their valence bands.

Common semiconducting materials are crystalline solids, but amorphous and liquid semiconductors are also known. These include hydrogenated amorphous silicon and mixtures of arsenic, selenium, and tellurium in a variety of proportions. Such compounds share with better known semiconductors intermediate conductivity and a rapid variation of conductivity with temperature, as well as occasional negative resistance. Such disordered materials lack the rigid crystalline structure of conventional semiconductors, such as silicon, and are generally used in thin-film structures, which are less demanding when it concerns the electronic quality of the material and thus are relatively insensitive to impurities and radiation damage. Organic semiconductors, that is, organic materials with properties resembling conventional semiconductors, are also known.

Silicon is used to create most semiconductors commercially. Dozens of other materials are used, including germanium, gallium arsenide, and silicon carbide. A pure semiconductor is often called an "intrinsic" semiconductor. The electronic properties and the conductivity of a semiconductor can be changed in a controlled manner by adding very small quantities of other elements, called "dopants," to the intrinsic material. In crystalline silicon, typically this is achieved by adding impurities of boron or phosphorus to the melt and then allowing the melt to solidify into the crystal. This process is called "doping." The doping atoms contribute electrons to the conduction band [7].

1.12 SUPERCONDUCTIVITY AND SUPERCONDUCTORS

Superconductivity was discovered in 1911. The observation was that the resistivity of mercury absolutely disappears at temperatures below ~4 K (−452.5°F). Currents created in a superconducting ring, for example, have persisted for several years without diminution; the electrons making up the current require a force and source of energy at start-up time, but not thereafter [8].

The reason superconductivity was not applied often enough was that the cost of producing the extremely low temperatures required to achieve this effect was prohibitively high. In 1986, however, new ceramic materials were discovered that became superconductive at considerably higher (hence, cheaper to produce) temperatures.

Superconductivity is much different from conductivity. In fact, the best of the normal conductors (e.g., silver and copper) cannot become superconducting at any temperature, and the new ceramic superconductors are actually insulators when they are not at low enough temperatures to be in the superconducting state [8].

One explanation of superconductivity is that the electrons that make up the current move in coordinated pairs. One of the electrons in a pair electrically distort the molecular structure of the superconducting material as it moves through, creating a nearby short-lived concentration of positive charge. The other electron in the pair may then be attracted toward this positive charge. According to theory, such coordination between electrons would prevent them from colliding with the molecules and thus would eliminate electrical resistance [9].

1.13 PIEZOELECTRICITY

Piezoelectricity is the charge that accumulates in certain solid materials (notably crystals, certain ceramics, and biological matter (e.g., bone, deoxyribonucleic acid (DNA) and various proteins) in response to applied mechanical strain [10, 11]. The word piezoelectricity means electricity resulting from pressure. It is derived from the Greek *piezo* or *piezein*, which means to squeeze or press, and *electric* or *electron*, which stands for amber, an ancient source of electric charge. Piezoelectricity is the direct result of the *piezoelectric effect*.

The piezoelectric effect is understood as the linear electromechanical interaction between the mechanical and the electrical state in crystalline materials with no inversion symmetry [11, 12]. The piezoelectric effect is a reversible process in that materials exhibiting the direct piezoelectric effect (the internal generation of electrical charge resulting from an applied mechanical force) also exhibit the reverse piezoelectric effect (the internal generation of a mechanical force resulting from an applied electrical field).

Piezoelectricity is found in useful applications, such as the production and detection of sound, generation of high voltages, electronic frequency generation, microbalances, and ultrafine focusing of optical assemblies. It is also the basis for a number of scientific instrumental techniques with atomic resolution, the scanning probe microscopy techniques, sensors, transducers, microphones, loudspeakers, inkjet printers, fuel injectors, piezoelectric motors, and everyday uses (e.g., acting as the ignition source for cigarette lighters and push–start propane barbecue grills).

1.14 PYROELECTRICITY

Pyroelectricity (from the Greek *pyr*, fire, and electricity) is the ability of certain materials to generate a temporary voltage when they are heated or cooled [12]. The change in temperature slightly modifies the positions of the atoms within the crystal structure, such that the polarization of the material changes. This polarization change gives rise to a voltage across the crystal. If the temperature stays constant at its new value, the pyroelectric voltage gradually disappears due to leakage current (the leakage can be due to electrons moving through the crystal, ions moving through the air, current leaking through a voltmeter attached across the crystal, etc.) [13, 14]. Pyroelectricity should not be confused with thermoelectricity: In a typical demonstration of pyroelectricity, the whole crystal is changed from one temperature to another, and the result is a temporary voltage across the crystal. In a typical demonstration of thermoelectricity, one side of the material is kept at one temperature. The other side at a different temperature. The result is a *permanent* voltage across the crystal. All pyroelectric materials are also piezoelectric, the two properties being closely related. However, some piezoelectric materials have a crystal symmetry that does not allow pyroelectricity.

1.15 FERROELECTRICITY

Ferroelectricity is a spontaneous electric polarization of a material. Ferroelectric materials may exhibit a spontaneous dipole moment, the direction of which can be switched between equivalent states by the application of an external electric field. The term ferroelectricity is related to the concept of ferromagnetism, which has to do with the creation of a magnetic moment [15, 16]. Ferroelectrics are a subgroup of pyroelectrics [15].

Placing a ferroelectric material between two conductive plates creates a ferroelectric capacitor. Ferroelectric capacitors exhibit nonlinear properties and usually have very high dielectric constants. The fact that the internal electric dipoles can be forced to change their direction by the application of an external voltage gives rise to *hysteresis* in the "polarization versus voltage" property of the capacitor. In this case, polarization is defined as the total charge stored on the plates of the capacitor divided by the area of the plates. Hysteresis means memory and ferroelectric capacitors are used to make ferroelectric random access memory (RAM) for computers [17] and radio frequency identification (RFID) cards [18].

1.16 METHODS FOR GENERATING ELECTRIC ENERGY

1.16.1 Batteries

The simplest sources of electrical energy are *batteries*. These are devices in which energy is generated by chemical reactions once the active components are allowed to circulate their electrons in an external circuit. The energy that is stored chemically is potential energy available to do electrical work. In rechargeable batteries, the chemistry is available and energy can be put back into the battery. The most common batteries are

- *Zinc–carbon batteries* are the most common "flashlight" batteries, available in regular and heavy-duty types. Their electrodes are zinc and carbon with an acidic paste between them that serves as the electrolyte. They are not recommended for electronic application.

- *Alkaline batteries* last 300–800% longer than zinc batteries, depending on their application. The electrodes of alkaline batteries are zinc and manganese oxide, with an alkaline electrolyte.
- *Nickel–cadmium batteries* are very popular in the majority of commercially available rechargeable products. The electrodes are nickel hydroxide and cadmium, with potassium hydroxide as the electrolyte. Their shortcoming is their tendency to develop "memory." Therefore it is recommended to discharge them fully before recharging.
- *Lithium iodide batteries* are designed to supply a small amount of power for a long period of time. They are widely used in power back-up systems in computers and wrist watches, pacemakers and hearing aids.
- *Lithium ion batteries* have a favorable power/weight ratio and are used widely in high-end laptop computers and cell telephones. They are rechargeable. More details on this subject are in Section 6.5.
- *Nickel–metal hydride batteries* are rapidly replacing the nickel–cadmium types because they do not suffer from memory effect.
- *Silver–zinc batteries* are used in aeronautical applications and in some miniature consumer electronic devices.
- *Zinc–air batteries* are light weight and rechargeable.
- *Zinc–mercury oxide* batteries are light, supply a small amount of power over a long period of time, and are used in hearing aids.
- *Lead–acid batteries* contain one plate made of lead and another plate made of lead dioxide, with a strong sulfuric acid electrolyte in which the plates are immersed. They are used mainly in automobiles.
- *Gelled electrolyte batteries* fall into the same category as lead–acid types and are most widely used in the uninterruptible power supplies (UPS) for computer systems, burglar alarms, and emergency lights.

1.16.2 Fuel Cells

Other sources of electrical energy are *fuel cells*. Fuel cells are different from conventional electrochemical cell batteries in that they consume reactant from an external source, which must be replenished: a thermodynamically open system. By contrast, batteries store electrical energy chemically, and hence represent a thermodynamically closed system. A fuel cell is an electrochemical cell that converts a source fuel into an electric current. It generates electricity inside a cell through reactions between a fuel and an oxidant, triggered in the presence of an electrolyte. The reactants flow into the cell, and the reaction products flow out of it, while the electrolyte remains within it. Fuel cells can operate continuously as long as the necessary reactant and oxidant flows are maintained. Currently, the most commonly used fuel is hydrogen. More on this subject is found in Section 6.4.2.

1.16.3 Solar Cells

Solar cells (photovoltaic devices) generate power by absorbing sunlight. Currently, commercial solar cells are made from inorganic substances, such as single-crystal silicon, gallium, and others. Organic solar cells, including polymeric types are being developed. For more details, see Section 6.1.2.

1.16.4 Commercial Facilities

Electricity for industrial and residential use is most often generated at a power station by electromechanical generators, primarily driven by heat engines fueled by chemical combustion or nuclear fission, but also by other means, such as the kinetic energy of flowing water and wind. There are many other technologies that can be and are used to generate electricity, such as solar photovoltaics (see Section 1.16.3), and geothermal power.

Depending on the way the electric energy is generated, it can be in the form of direct current (dc) and alternating current (ac), the latter being defined by its frequency. For commercial electric utilities the most common frequencies are 60 and 50 Hz depending on the country. Currently in North America the frequency is 60 Hz and in most European countries 50 Hz. Furthermore, it is possible to generate alternating electric fields by specialized devices with frequencies much lower and higher than the above.

FURTHER READINGS

Halliday, D., Resnick, R., Walker, J., *Fundamental of Physics Extended*, 5th ed., John Wiley & Sons, Inc., New York, 1997.

Thomson, L.M., *Basic Electricity and Electronics for Control: Fundamentals and Applications*, 3rd ed., ISA, Research Triangle Park, NC, 2006.

Maloney, T.J., *Electricity: Fundamentals, Concepts & Applications*, Delmar Cengage Learning, Clifton Park, NY, 1992.

Morrison, R., *The Fields of Electronics: Understanding Electronics Using Basic Physics*, John Wiley & Sons, Inc., New York, 2002.

Van Valkenburgh, Nooger & Neville, Inc., *Basic Electricity*, Revised ed., Prompt Publications, Indianapolis, IL, 1992.

Nussbaum, A., *Electronic and Magnetic Behavior of Materials*, Prentice-Hall, Englewood Cliffs, NJ, 1967.

Generating Electricity in a Carbon-Constrained World (Sioshansi, F.P., Ed.), Academic Press, London, 2009.

REFERENCES

1. Morrison, R., *The Fields of Electronics: Understanding Electronics Using Basic Physics*, John Wiley & Sons, Inc., New York, 2002, p. 22.
2. Ku, C.C. and Liepins, R., *Electrical Properties of Polymers: Chemical Principles*, Hanser Publishers, Munich, 1987, p. 24.
3. Blythe, T. and Bloor, D., *Electrical Properties of Polymers*, 2nd ed., Cambridge University Press, Cambridge, UK, 2005, p. 40.
4. Blythe, T. and Bloor, D., *Electrical Properties of Polymers*, 2nd ed., Cambridge University Press, Cambridge, UK, 2005, p. 37.
5. Halliday, D., Resnick, R., and Walker, J., *Fundamental of Physics Extended*, 5th ed., John Wiley & Sons, Inc., New York, 1997, p. 653.
6. Halliday, D., Resnick, R., and Walker, J., *Fundamental of Physics Extended*, 5th ed., John Wiley & Sons, Inc., New York, 1997, p. 661.
7. Halliday, D., Resnick, R., and Walker, J., *Fundamental of Physics Extended*, 5th ed., John Wiley & Sons, Inc., New York, 1997, p. 667.
8. Halliday, D., Resnick, R., and Walker, J., *Fundamental of Physics Extended*, 5th ed., John Wiley & Sons, Inc., New York, 1997, p. 665.
9. Halliday, D., Resnick, R., and Walker, J., *Fundamental of Physics Extended*, 5th ed., John Wiley & Sons, Inc., New York, 1997, p. 666.
10. Holler, F.J., Skoog, D.A., and Crouch, S.A., *Principles of Instrumental Analysis*, 6th ed., Cengage Learning, Florence, KY, 2007, p. 9.
11. Tichý, J., Erhart, J., Kittinger, E., and Privratská, J., *Fundamentals of Piezoelectronic Sensorics*, Springer Verlag, Berlin Heidelberg, 2010, p. 1.
12. Tichý, J., Erhart, J., Kittinger, E., and Privratská, J., *Fundamentals of Piezoelectronic Sensorics*, Springer, Berlin, Heidelberg, 2010, p. 6.
13. Lang, S.B., *Phys. Today*, **58**(8), 31 (2005).
14. Gautschi, G., *Piezoelectric Sensorics*, Springer, Berlin, Heidelberg, 2002.

15. *Solid States Physics: Advances in Research and Applications*, Vol. 4 (Seitz, F., Das, T.P., Turnbull, D., and Hahn, E.L., Eds.), Academic Press, New York, 1957, p. 5.
16. Lines, M. and Glass, A., *Principles and Applications of Ferroelectric and Related Materials*, Clarendon Press, Oxford, 1979.
17. Scott, J.F., *Ferroelectric Memories*, Springer Verlag, Heidelberg, 2000.
18. Caraullo, M. et al., U.S. Patent 3,713,148 (January 1973) to Communications Services Corporation, Inc.

CHAPTER 2

POLYMERIC MATERIALS

2.1 INTRODUCTION

The word *polymer* originates in the Greek language. It is composed of two words, namely, *poly* = many and *meros* = part. Each individual part of this multitude is referred to as a monomer. A monomer is a compound composed of atoms or molecules capable of forming one or more characteristic and for each compound an unchangeable unit. Linking the monomers chemically produces a uniformly built polymer, which is also called a macromolecule. These compounds can be either manufactured synthetically or obtained or transformed from natural products.

Essentially, polymers are organic materials, prepared by processes in which simple organic molecules (monomers) are used as building blocks to produce large molecules (polymers or macromolecules). The number of monomer repeat units in a macromolecule is so large that the characteristics of a polymer are not significantly modified when removing or adding a single monomer repeat unit. Commercially available polymers are comprised of >1000 monomer repeat units that result in molecular weights ranging from 10^4 to 10^7. The number of monomeric units determines the *degree of polymerization* usually denoted by the symbol N. The structure of most macromolecules consists most frequently of one

Polymers for Electricity and Electronics: Materials, Properties, and Applications, First Edition. Jiri George Drobny.

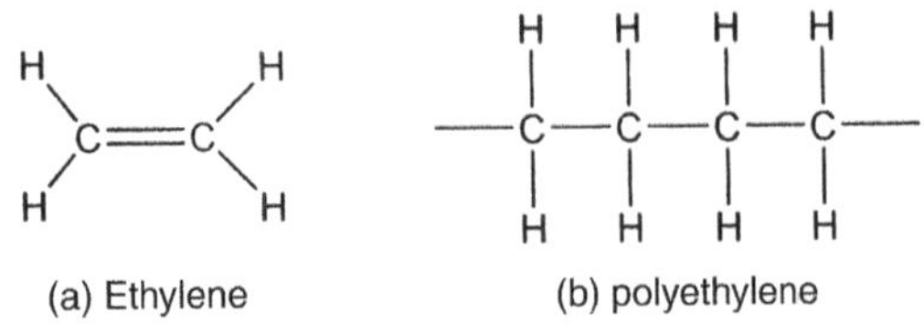

FIGURE 2.1 (a) Ethylene (monomer) and (b) polyethylene (polymer); simplified formulas.

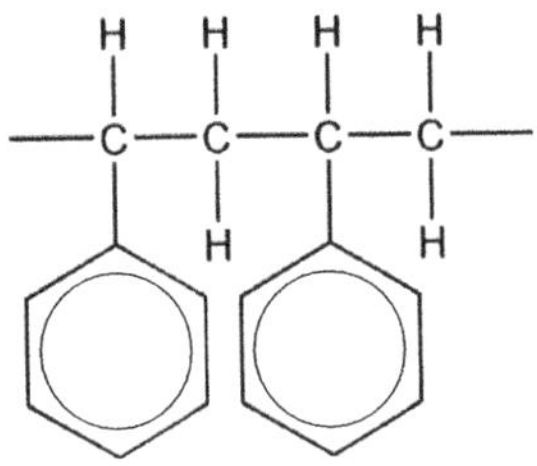

FIGURE 2.2 Polystyrene.

or two types of monomer repeat units, defined as either homopolymer or copolymer, respectively. A macromolecule composed of three different types of monomers is referred to as a terpolymer. The term macromolecule generally indicates a structure of covalently bound monomers, whereas the term polymer is reserved for defining synthetic materials. Figure 2.1 shows an example of a monomer (ethylene) and polymer (polyethylene).

Represented in their simplest form, monomers are ordered into a thread or chainlike structure. The concept of the chain is based on the macromolecular structure mentioned above. The structure is such that the movement of an individual element within the chain is transferred to neighboring elements. A linear macromolecule attempts to form an irregularly coiled state. However, the coiled state often becomes hindered by other macromolecules.

Substitution of individual atoms by different substituents can occur on the backbone of a linear macromolecule, which has a more or less regular structure. Such substituents are commonly composed of atoms or atomic groups, which are chemically different from the monomer. An example, shown in Fig. 2.2, is polystyrene, a polymer in which phenyl groups are attached as side groups to the –C–C– backbone. Thus the substituents (or side groups) produce a structure different from that of the base backbone. If the structure of the substituents is identical to that of the backbone, they are called *branches*. Substituents can make the main chain both stronger and stiffer or both looser and more flexible.

Macromolecules can be linked together by either physical or chemical bonds. Chemically linked macromolecules, are known as cross-linked macromolecules. Cross-linked macromolecules form a *network*. Physical bonds can develop between the individual macromolecular segments in addition to the existing chemical bonds to strengthen the entire network. The individual macromolecules prefer the coiled state and penetrate each other, thus forming an entangled chain structure. Entanglements will hold the macromolecules together and as a result the physical bonds become mobile.

Compared to the existence of physical bonds, chemical bonds are up to 10^3 times stronger [1]. One significant difference between physical and chemical bonds is that physical bonds are reversible when subjected to heat, solvents, or mechanical forces, whereas chemical bonds are irreversible. The chemical and physical structures that illustrate normal linear, branched, cross-linked, and entangled macromolecules are shown schematically in Fig. 2.3.

A selection of conventional polymers is shown in Table 2.1. These polymers are electrically neutral. If the polymeric chains are built from monomers that contain an ionizable group, that is, a group that can dissociate into a chain-fixed cation and anion and a mobile counterion bearing the opposite charge, a *polyelectrolyte* is obtained. Several examples of this kind of polymers are shown in Table 2.2. The first three are synthetic polymers; the other two are biopolymers, cellulose, and starch in the form of derivatives, which include ionizable substitutes.

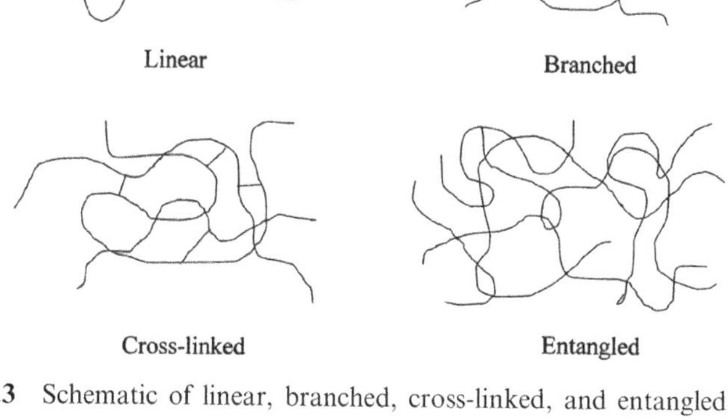

FIGURE 2.3 Schematic of linear, branched, cross-linked, and entangled macromolecular chains.

TABLE 2.1 Conventional Polymers, Part A

Structure	Name (abbrevation)
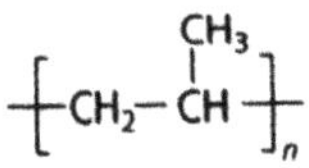	Polypropylene (PP)
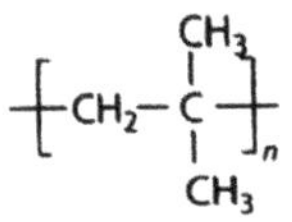	Polyisobutylene (PIB)
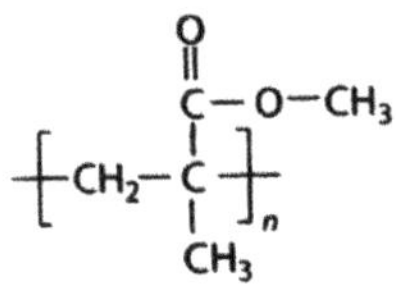	Poly(methylmethacrylate) PMMA
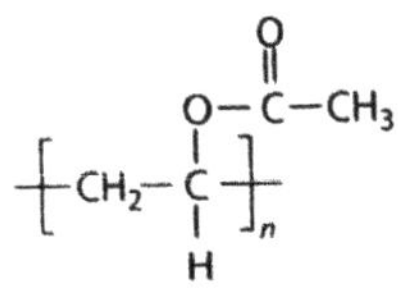	Poly(vinylacetate) (PVAc)
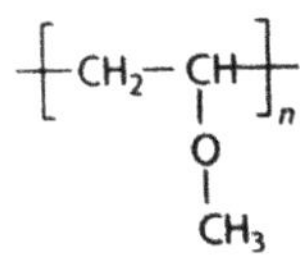	Poly(vinylmethylether) (PVME)
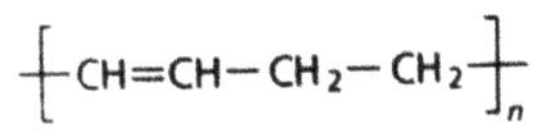	Polybutadiene (PB)
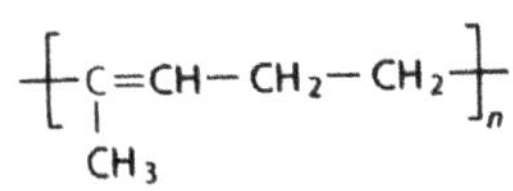	Polyisoprene (PI)
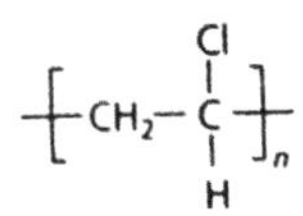	Poly(vinyl chloride) (PVC)
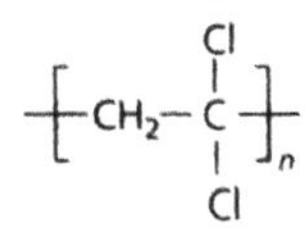	Poly(vinylidene chloride) (PVDC)
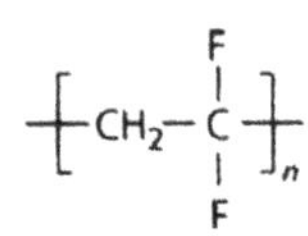	Poly(vinylidene fluoride) (PVDF)
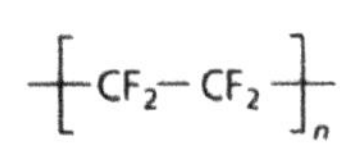	Poly(tetrafluoroethylene) (PTFE)
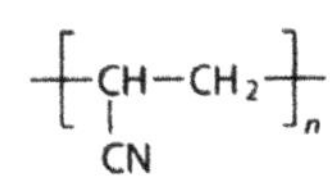	Poly(acrylonitrile) (PAN)

TABLE 2.1 Conventional Polymers, Part B

Structure	Name (abbrevation)
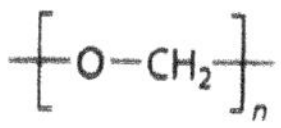	Poly(oxymethylene) (POM)
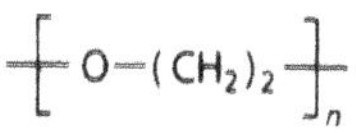	Poly(ethyleneoxide) (PEO)
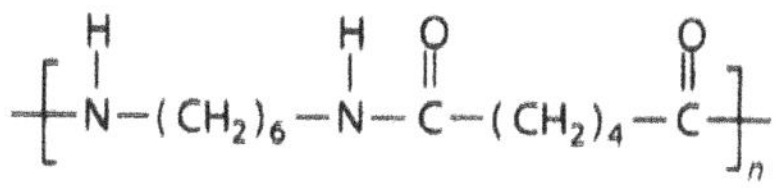	Poly(hexamethylene adipamide) (nylon 6,6)
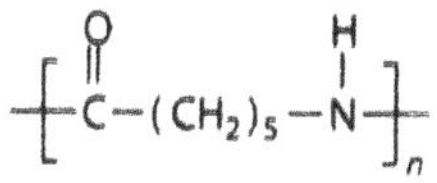	Poly(ε-caprolactam)(nylon 6)
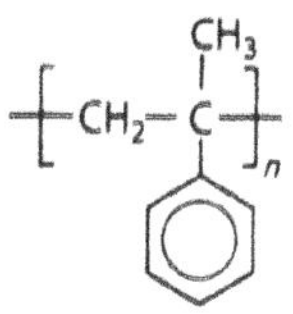	Poly(α-methylstyrene)
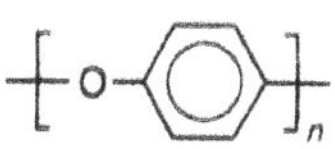	Poly(α-phenylene oxide) (PPO)
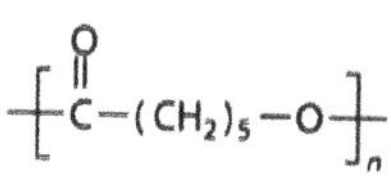	Poly(ε-caprolactone) (PεCL)
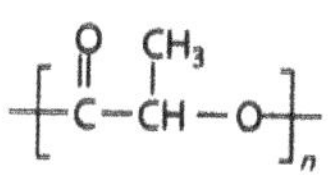	Poly(L-lactic acid) (PLLA)
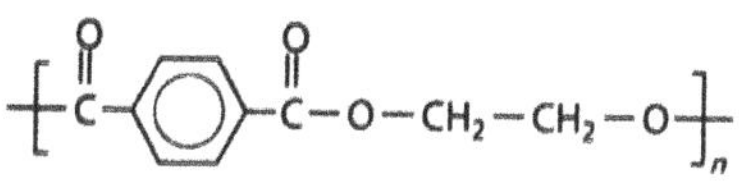	Poly(ethylene terephthalate) (PET)
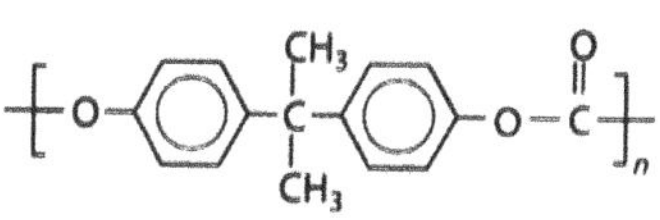	Polycarbonate (PC)
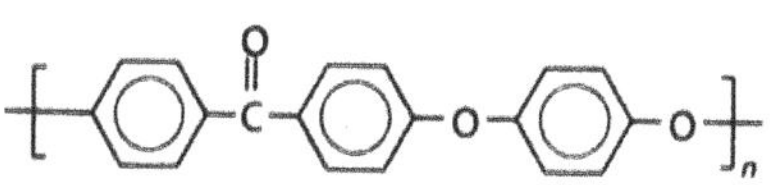	Poly(ether ether ketone) (PEEK)
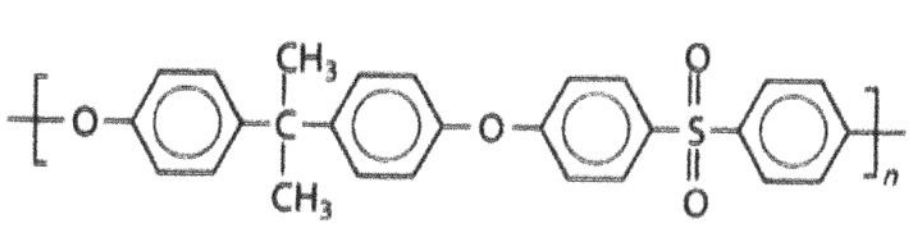	Polysulfone (PSF)
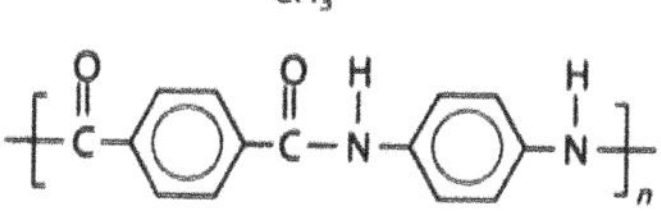	Poly(*p*-phenylene-terephthalamide) (Kevlar)

(Continued)

TABLE 2.1 Conventional Polymers, Part C

Structure	Name (abbrevation)
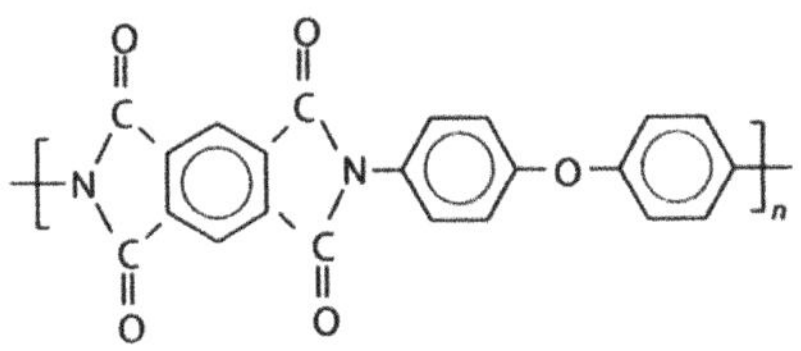	Polyimide
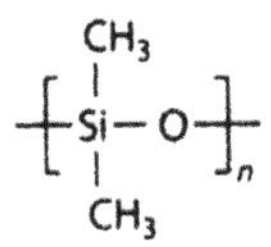	Poly(dimethylsiloxane) (PDMS)
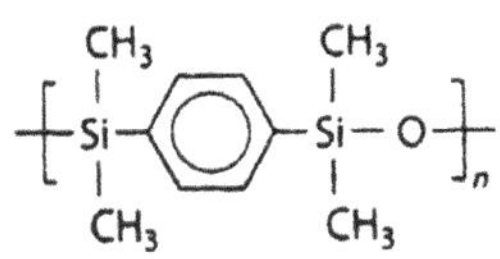	Poly(tetramethyl-*p*-silphenylene siloxane) (TMPS)

Strobl, G., *The Physics of Polymers: Concepts for Understanding Their Structures and Behavior*, Springer-Verlag Berlin Heidelberg, 2007, pp. 3–6.

TABLE 2.2 Polyelectrolytes (Strobl, G., *The Physics of Polymers: Concepts for Understanding Their Structures and Behavior*, Springer-Verlag Berlin Heidelberg, 2007, p. 7)

Structure	Name (abbrevation)
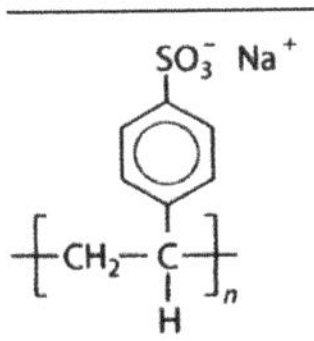	Sodium poly(styrene sulfonate) (NaPSS)
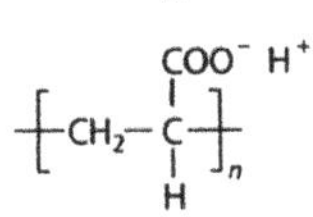	Poly(acrylic acid) (PAA)
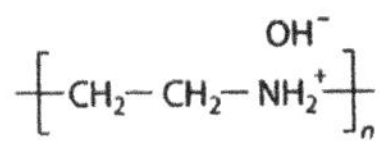	Poly(ethylene imine)
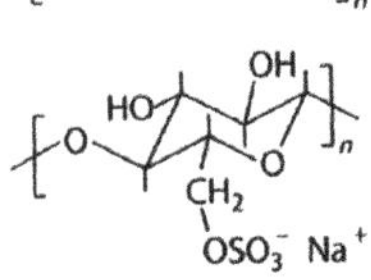	Sodium cellulose sulfate
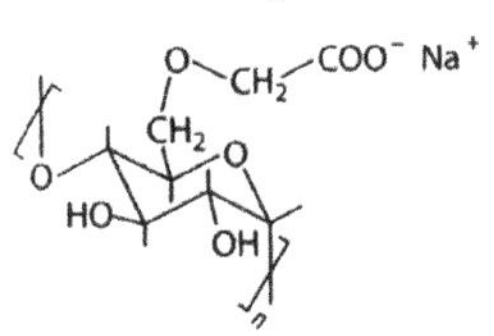	Sodium carboxymethyl amylose

Charges on a chain also can be created by *doping processes*. This process is particularly easy to do for *conjugated polymers*, that is, with chains with conjugated carbon-to-carbon double bonds. Even more importantly, the produced charges are mobile and thus provide electrical conductivity [2]. This subject is discussed in some detail in Section 4.1.2.

2.2 POLYMERIZATION

As pointed out earlier, macromolecules (polymers) are formed by the reaction of one or more types of monomers. The result can be linear, branched, or three-dimensional (3D) cross-linked chain molecules with molecular weights typically $>10^4$. The monomers represent building blocks that are linked together in large numbers by chemical bonds. The number of monomer units in a macromolecule is referred to as the *degree of polymerization*. Typical degrees of polymerization are

Polycarbonate (PC): 80–160
Polyamide (PA) 6: 100–300
PVC: 800
Polystyrene (PS): 1500–4000
PP: 3500–35000
Ultrahigh Molecular Weight Polyethylene (PE–UHMW): 200,000

If only one type of monomer is involved in the polymerization reaction, the resulting product is a homopolymer:

$$\left[\text{A–A–A} \cdots \text{A–A–A} \right]_n$$

If two or more monomers are involved, the resulting product is a copolymer. For the coupling of the two monomeric units in the chain, two limited cases exist. In the first, the coupling is statistical and determined by the probabilities of attachment of the two monomers on the growing chain. Chains of this kind are called *statistical copolymers*. They exhibit a short-range order with preferred sequences, and thus are different from a random mixing, but possess no order in the chemical composition over the long range. Block copolymers are the second limiting case, which are formed by coupling long macromolecular sequences of uniform composition, and depending on the number of sequences, di-, tri-, and multiblock copolymers may be prepared. The remaining structures, which are most frequently represented, are *random* and *alternating copolymers*. The typical copolymer structures are shown below:

$\left[\text{A–B–B–A–A–B–A–B–A–A}\right]_n$	Random copolymer
$\left[\text{A–B–A–B–B} \cdots \text{A–B–A–A–B}\right]_n$	Statistical copolymer
$\left[(\text{A–A–A–A})_n\text{ –}(\text{B–B–B–B})_m\right]_n$	Block copolymer
$\left[\text{A–B–A–B–A–B–A–B–A–B}\right]_n$	Alternating copolymer
$\left[\text{A–}\underset{[\text{B}]_m}{\underset{\mid}{\text{A}}}\text{–A–A–}\underset{[\text{B}]_m}{\underset{\mid}{\text{A}}}\text{–A–A–}\underset{[\text{B}]_m}{\underset{\mid}{\text{A}}}\text{–A–A}\right]_n$	Graft copolymer

Graft copolymers also consist of blocks of two different monomers, except that the main chain is composed of one type of monomer units and the branches of the other. The second monomer is polymerized onto the trunk of the chain or attached through reactive groups similar to grafting a branch of one tree onto the trunk of another.

The two main types of polymer synthesis are (1) chain polymerization (2) step reaction polymerization.

2.2.1 Chain Polymerization

The most important types of chain polymerization are radical polymerization, ionic (anionic and cationic) polymerization, and coordination polymerization.

2.2.2 Step Reaction Polymerization

Polycondensation and polyaddition are both step reaction polymerizations. The details regarding the polymerization methods can be found in many publications, for example, Flory (1953), and/or Cowie and Arrighi (2008).

2.3 MOLECULAR WEIGHTS AND MOLECULAR WEIGHT DISTRIBUTIONS

One of the most important features that distinguish a polymer from a simple molecule is the inability to assign an exact molecular weight to a polymer. This occurs because in a polymerization reaction, the length of the chain formed is determined entirely by random events. In a condensation reaction, it depends on the availability of a suitable reactive

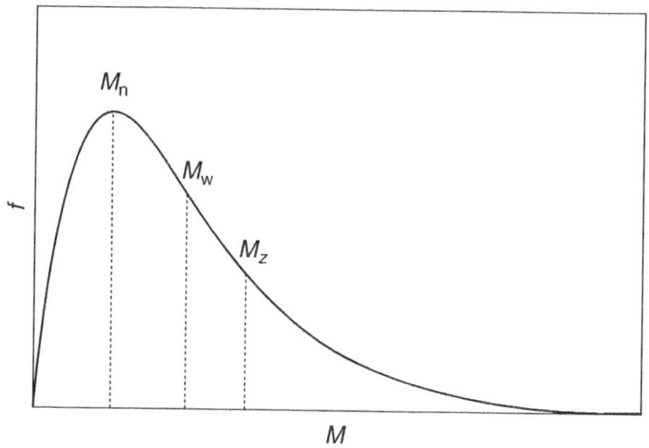

FIGURE 2.4 Typical distribution of molecular weight for a synthetic polymer sample, where *f* is the molar fraction of polymer in each interval of *M* considered. (Cowie, J.M.G. and Arighi, V., *Chemistry and Physics of Modern Materials, 3rd Edition*, 2008, p. 8)

group. In an addition reaction it depends on the lifetime of the chain carrier. Inevitably, because of the random nature of the growth process, the product is a mixture of chains of different length: a *distribution* of chain lengths, which in many cases can be calculated statistically.

Thus the polymer is characterized best by a molar weight distribution and the associated molar weight distribution, rather than a single molecular weight. The typical distributions shown in Fig. 2.4 can be described by a variety of averages. As the methods used for estimating the molecular weight of polymers employ different averaging procedures, it is safer to use more than one technique to obtain two or more averages, thereby characterizing the sample more fully.

A colligative method, such as osmotic pressure, effectively counts the number of molecules present and provides a *number average* molecular weight, M_n, defined by

$$M_\mathrm{n} = \frac{\sum N_i M_i}{\sum N_i} = \frac{\sum w_i}{\sum (w_i/M_i)} \tag{2.1}$$

where N_i is the number of molecules of species i of molecular weight of M_i. The alternative expression is in terms of weight $w_i = N_i M_i / N_\mathrm{A}$, where N_A is Avogadro's constant.

Light scattering measurements, a method depending on the size, rather than the number of molecules, provides the value of the weight average molecular weight M_w. This is defined as:

$$M_\mathrm{w} = \frac{\sum N_i M_i^2}{\sum N_i M_i} = \frac{\sum w_i M_i}{\sum w_i} \tag{2.2}$$

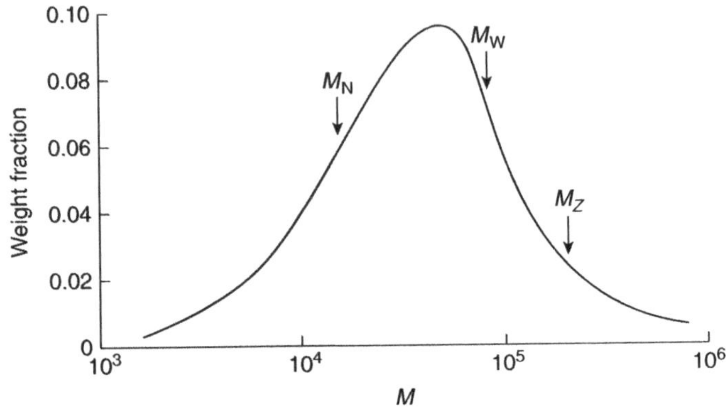

FIGURE 2.5 The molecular weight distribution determined by size-exclusion chromatography.

Statistically, M_n is simply the first moment and M_w is the ratio of the second to the first moment of the number distribution. A higher average, the M_z, can be measured in the ultracentrifuge, as well as M_{z+1} used mainly for describing mechanical properties.

Another method that mainly uses the molecular weight distribution is size-exclusion chromatography. It is a liquid column chromatographic technique, which sorts molecules according to their size. The sample solution is introduced onto the column, which is filled with a rigid-structure porous particle column packing and is carried by solvent (mobile phase) through the column. The size sorting takes place by repeating exchange of the solute molecules between the bulk solvent of the mobile phase and the stagnant liquid phase within the pores of the packing. The result is a distribution curve (Fig. 2.5) somewhat similar to that in Fig. 2.4.

The width of the distribution can often be gauged by establishing the *heterogeneity index* M_w/M_n. For many polymerizations, the most probable value is ~2.0, but both larger and smaller values are obtained, and it is at best only a rough guide [3].

2.4 CONFIGURATION

2.4.1 Isomerism

For polymers, two types of configurational isomers are of importance: (1) geometric isomerism (e.g., cis and trans) and (2) stereoisomerism. The polymerization of monomers with two double bonds (e.g., butadiene or isoprene), leads to polymer chains with a residual double bond per monomer unit: cis–trans isomerism is possible. Figure 2.6 shows the

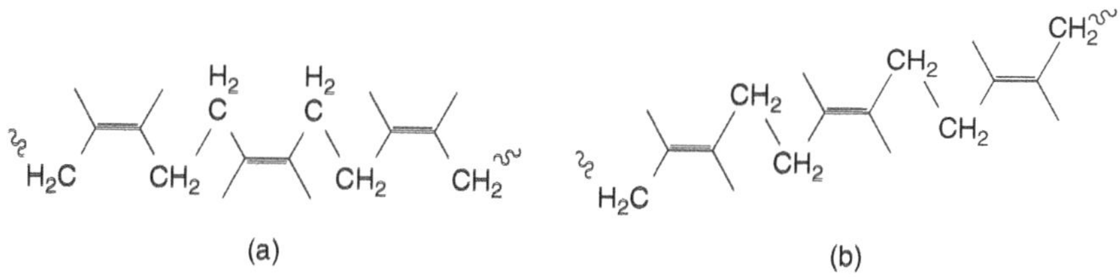

FIGURE 2.6 Geometric isomers of 1,4-polybutadiene (a) cis-, (b) trans-configuration. (Cowie, J.M.G. and Arighi, V., *Chemistry and Physics of Modern Materials, 3rd Edition*, 2008, p. 13)

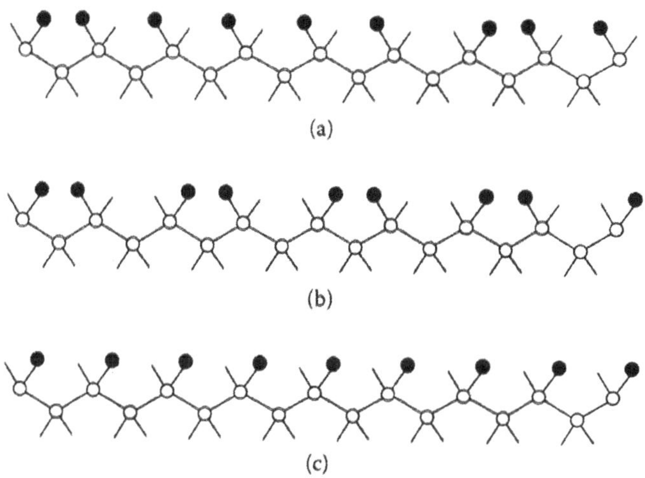

FIGURE 2.7 Section of polymer chains having different tacticity (a) atactic, (b) syndiotactic, and (c) isotactic. (Cowie, J.M.G. and Arighi, V., *Chemistry and Physics of Modern Materials, 3rd Edition*, 2008, p. 14)

difference. The regularity of a trans-configuration makes this type of isomer more crystalline, with a higher melting point compared to the cis-configuration.

If the polymer chain contains carbon atoms with two different substituents, then the C atom is asymmetrical. The situation is encountered for vinyl monomers (e.g., polypropylene). For a chain containing asymmetric centers, three different stereoisomers are possible (see Fig. 2.7):

1. The *isotactic* configuration, in which each substituent is placed on the same side of the chain.
2. The *syndiotactic* structure, in which substituents lie on alternate sides of the backbone.
3. A disordered, *atactic* configuration in which substituent groups are placed randomly on either side of the chain.

Of the three stereoisomers, atactic polypropylene was the first to be synthesized, because it is the atactic structure that is obtained by the conventional polymerization of the monomers when no optically active catalyst is used. Only when the Ziegler–Natta catalysts were introduced in the 1950s did the production of stereoregular polymers became possible. Isotactic polypropylene is one of the most important commercial polymers.

2.4.2 Polarity

The polarity of polymers is essentially determined by the structure of the polymer chains and the nature of the substituents on them. If groups of the same charge sign are located along the polymer chain, they repel one another and the molecule is stretched and becomes rigid. As a result of this interaction between molecules, polarity limits the mobility of the polymer chains, which determines their rigidity and cohesion, which in turn affect important properties (e.g., glass transition temperature, the tendency to crystallize, solubility, hardness, scratch resistance, elasticity, and adhesion) [4]. Examples of polar polymers are PVC, PVDF, PA, acrylates, polychloroprene rubber, nitrile–butadiene rubber, and fluoroelastomers).

2.4.3 Effects of Large Substituents

Substituents that occupy a relatively large volume, such as phenyl groups on the polystyrene chains, also reduce mobility for steric reasons and have the same effect: Polystyrene is harder and more rigid than normal polyolefins. However, the space-occupying substituents on a polymer chain increase the free volume of the polymer molecule and this causes a reduction of the glass transition temperature, which counters its increase due to reduced molecular mobility [4].

2.5 GLASS TRANSITION TEMPERATURE AND MELTING TEMPERATURE

2.5.1 Amorphous and Crystalline Polymers

At sufficiently low temperatures, all polymers are hard rigid solids. As temperature rises, each polymer eventually obtains sufficient thermal energy to enable its chains to move freely enough for it to behave like a viscous liquid (assuming no degradation has occurred).

There are two ways in which a polymer can pass from the solid to the liquid state, depending on the internal organization of the chains in the material. The different thermal responses are based on the change in specific volume (Fig. 2.8).

A polymer may be completely *amorphous* in the solid state, in which case the chains are arranged in a totally random fashion.

In a perfectly crystalline polymer, all chains would be incorporated in regions of 3D order, called *crystallites*. However, perfectly crystalline polymers are not common in practice, but there is always a certain proportion of amorphous regions. Commonly, such partially crystalline polymers are referred to as *semicrystalline* polymers.

In order for a polymer to crystallize, its chains must be able to come close together and fold to make a tight package. For that, the groups on adjacent chains must interact and are held together by molecular interaction. The crystalline regions have defined melting characteristics within a narrow temperature range (for details, see below).

The specific volume of a completely amorphous polymer follows the line ABCD. In the C–D region, the polymer is a glass. As it is heated further, it goes through the *glass transition temperature* T_g at which point it softens and becomes rubbery. This is an important temperature because it represents the point where important property changes take place; the material may be more easily deformed or become ductile above T_g. A continuing increase in temperature along the C–B–A

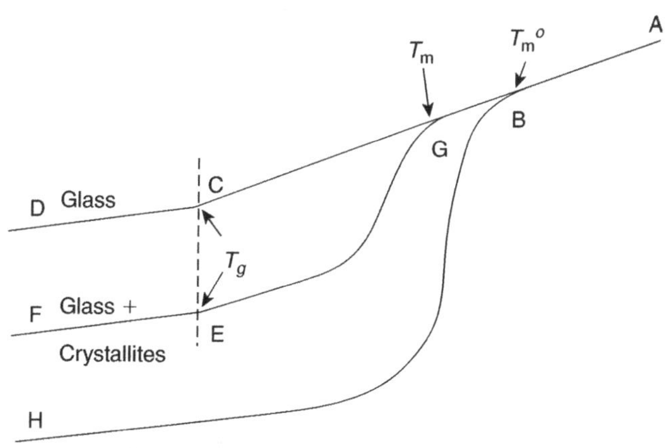

FIGURE 2.8 Change of a specific volume of different volume of different polymers with temperature. (Cowie, J.M.G. and Arighi, V., *Chemistry and Physics of Modern Materials, 3rd Edition*, 2008, p. 15)

line shows how the material changes from a rubbery state to a viscous liquid [5].

The T_g is a second-order transition, associated with the above changes. At this temperature, the main chains become mobile. Consequently, a number of material properties change including moduli, specific volume, refractive index, density, dielectric constant, loss factor, and thermal conductivity. The T_g can be influenced by the polarity of the monomeric units, the size of substituents on the main chain, the nature of copolymerization, tacticity, rigidity of the polymer chain, molecular weight, degree of branching, and degree of cross-linking.

When heated the perfectly crystalline polymer will follow the curve H–B–A; at the temperature T_m^o, it would melt and would turn into a viscous liquid.

The semicrystalline polymers usually exhibit both T_g and T_m, corresponding to the ordered and disordered portions and follow curves similar to F–E–G–A. The melting temperature T_m is lower than T_m^o, and often represents a melting range, because the semicrystalline polymer contains a spectrum of chain lengths and crystallites of various sizes with many defects [5].

The occurrence of significant crystallinity in a polymer sample is of considerable consequence. The properties of the material (the density, optical clarity, stiffness (modulus), and general mechanical response) all change dramatically when crystallites are present and the polymer is no longer subject to the rules of viscoelasticity, which apply to amorphous polymers [6]. These differences depend on the degree of crystallinity. The difference between a completely amorphous and a semicrystalline polymer is shown in Fig. 2.9.

2.5.2 Crystallization and Melting

Crystallization of polymers begins with considerable supercooling, at first slowly and then accelerating on decreasing the temperature further. Under practical circumstances, the solidification thus extends over a large temperature range. The formation of a 3D ordered phase from a disordered state is a two-stage process. The first step in crystallite formation is the creation of a stable nucleus by the ordering of chains in a parallel array. The second stage is the growth of the crystalline region, the size of which is governed by the rate of addition of other chains to the nucleus.

Measurable rates of crystallization occur between ($T_m - 10$ K) and ($T_g + 30$ K), a range at which the thermal motion of the polymer chains is conducive to the stable ordered regions [7]. The growth rate of crystalline

(a)

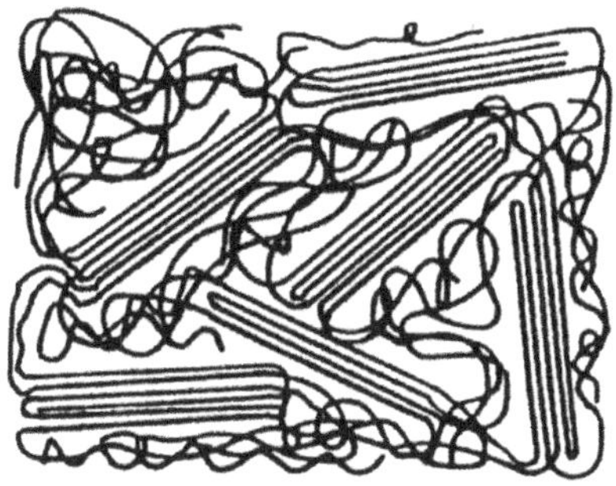

(b)

FIGURE 2.9 Amorphous polymer (a) and semi-crystalline polymer (b).

areas usually passes through a maximum in this range, as illustrated by Fig. 2.10. The melt usually has to be supercooled by ~5–20 K before a significant number of nuclei appear that possess the critical dimensions required for stability and further growth. If a *nucleating agent* is added to the system, crystallization can be induced at higher temperatures. This is known as *heterogeneous nucleation* and only affects the crystallization rate, not the spherulitic growth rate, at a given temperature.

Since polymers are never perfectly crystalline, they never show a marked volume change and a well-defined melting temperature, characteristic for a perfectly crystalline material. They contain disordered regions and crystallites of varying size. The process is normally incomplete because crystallization takes place when the polymer is a viscous liquid. In this state, chains are highly entangled and sufficient time must be allowed for the chains to diffuse into the 3D-order required for crystalline formation. Thus rapid cooling from the melt usually prevents the development of significant crystallinity. Another factor affecting the melting behavior of semicrystalline polymers is the size of crystallites. Small crystals usually melt at a lower temperature than large crystals;

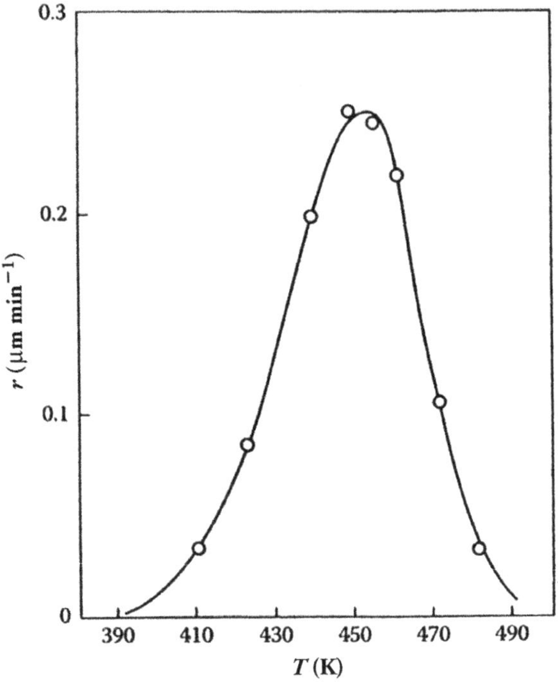

FIGURE 2.10 Radial growth rate *r* of spherulites of isotactic polybutadiene as a function of the crystallization temperature. (Cowie, J.M.G. and Arighi, V., *Chemistry and Physics of Modern Materials, 3rd Edition*, 2008, p. 281)

this difference may be up to 30 K (−405.7°F) [8]. The way to achieve a high degree of crystallinity is a careful annealing at the appropriate temperature.

2.6 SOLUTIONS OF POLYMERS

In some applications (e.g., certain adhesives, sealants, coatings, and potting compounds) polymers are dissolved in volatile organic solvents. The volatile components of adhesives and coatings are used to reduce viscosity of the solution to be low enough for application (usually in the range 0.05–1 Pa.s) and to control viscosity changes during the application and film-formation process.

2.6.1 Solubility of Polymers

Because of the large number of polymers, the selection of the solvents has become challenging. Currently, the selection of solvents or solvent systems is based on *solubility parameters*. The concept of solubility parameters is derived from the theory of mixing based on thermodynamics involving the *Gibbs free energy equation*:

$$\Delta G_{\mathrm{m}} = \Delta H_{\mathrm{m}} - T\Delta S_{\mathrm{m}} \tag{2.3}$$

where ΔG_{m} is the change of the free energy of mixing, ΔH_{m} is the change of the enthalpy of mixing, ΔS_{m} is the change of entropy of mixing, and T is the absolute temperature in Kelvin (K). Mixing is favored when the change of the free energy of mixing is negative.

Solubility parameter δ is defined as the square root of cohesive energy density (CED) of a solvent:

$$\delta = \sqrt{\mathrm{CED}} = \frac{\sqrt{(\Delta H_v - RT)}}{V} \tag{2.4}$$

where ΔH_v is the heat of vaporization, V is the molar volume, R is the universal gas constant, and T is the absolute temperature in Kelvin.

A pair of solvents is miscible if the square of the difference of their solubility parameters approaches zero. The Systéme International (SI) unit for the solubility parameter has the dimension $(\mathrm{MPa})^{1/2}$. An older unit has the dimension $(\mathrm{cal\ cm}^{-3})^{1/2}$, and is sometimes designated as the hildebrand, h. It is still used more frequently than the SI unit.

Solubility parameters have been derived from thermodynamic relations (e.g., Eq. 2.4) under the assumption that polar forces and hydrogen bonds play a minor role in comparison with dispersion forces. In real systems, this is case is an exceptional; the resins and solvents used in coatings frequently carry polar groups that are capable of hydrogen bonding [9]. Therefore, the values of solubility parameters for such systems include additional variables (e.g., dipole moment, hydrogen-bonding indices, polarity variables, and polarizability) to obtain more accurate statements that correspond better to the real conditions [10–12]. Values of δ for solvents and resins are listed in the literature, for example in [13].

Resins having solubility parameters that are comparable with those of the solvents or solvent mixtures dissolve well in these solvents. If the solubility parameter of the resin deviates greatly from that of the solvent, gelation or insolubility must be expected [14].

Examples of values for solubility parameters of polymers and solvents are shown in Tables 2.3 and 2.4, respectively.

TABLE 2.3 Average Solubility Parameters for Some Common Polymers

Polymer	Solubility Parameter, δ (J m^{-3})$^{1/2}$
Polytetrafluoroethylene	12.7
Polyisobutylene	16.3
Polyethylene	16.4
Polyisoprene	16.7
Polybutadiene	17.1
Polypropylene	17.8
Polystyrene	18.7
Poly(methyl methacrylate)	19.0
Poly(vinyl acetate)	19.2
Poly(vinyl chloride)	20.7
Polyamide 66	27.8
Poly(acrylonitrile)	28.7

TABLE 2.4 Solubility Parameters of Some Common Solvents

Solvent	Solubility Parameter, δ (J m^{-3})$^{1/2}$
Perfluoroheptane	11.9
Decane (normal)	13.5
Heptane (normal)	14.9
Diethyl ether	15.1
Octane (normal)	15.6
Dodecane	16.2
Cyclohexane	16.8
Carbon tetrachloride	17.6
Dimethyl ether	18.0
Xylene	18.0
Toluene	18.2
Ethyl acetate	18.6
Tetrahydrofurane	18.6
Benzene	18.8
Chloroform	19.0
Methyl ethyl ketone	19.0
Chlorobenzene	19.4
Acetone	20.3
Propyl alcohol	23.5
Ethyl alcohol	26.0
Methyl alcohol	29.7
Water	47.9

2.6.2 Viscosity of Polymer Solutions

Liquids flow in various ways. The most common type in coatings is *shear flow*. Shear flow can be visualized as liquid between two plates; one fixed and the other movable. If the movable plate is pulled or pushed, the liquid is subjected to *shear stress* (i.e., a force per unit area), which is applied at some *shear rate*, velocity per unit thickness. *Viscosity* (dynamic viscosity in this case) is a measure of resistance to flow and is defined as sheer stress divided by shear rate. The SI unit of viscosity is pascal seconds (Pa.s); the older centimeter–gram–second (cgs) unit, still used occasionally is the poise (P); the relationship between these two units is 1 Pa.s = 10 P. When flow is driven by gravity, *kinematic viscosity* is measured. The SI unit of kinematic viscosity is square meter per second ($m^2 s^{-1}$), formerly stokes (St); the relationship is 1 $m^2 s^{-1} = 10^4$ St. Kinematic viscosity can be converted into dynamic viscosity by multiplying kinematic viscosity by the density of the liquid.

In the case of ideal liquids, the ratio of shear stress to shear rate is constant for any shear stress. Such liquids are said to exhibit *Newtonian flow*. Almost all low molecular weight liquids and resins in good solvents, and dispersions of unassociated rigid particles in Newtonian fluids, exhibit non-Newtonian flow. In most of these cases that are important in coatings, viscosity at a high shear rate is lower than it is at low shear rates [15]. Such liquids are called *shear thinning fluids*. In some shear thinning fluids, the extent of shear thinning at any given shear rate is dependent on shear history and duration of application of shear stress; these systems exhibit so-called *thixotropy*. Other fluids exhibit *shear thickening*. The Newtonian and non-Newtonian flow curves are shown in Fig. 2.11.

The molecular chains of a polymer are in general entangled [16–18]. To separate them, for example, by the action of solvents or by heating until the polymer flows, there are cohesive forces that must be overcome. The longer the individual molecular chain, the more entangled it is and the more difficult it is to separate it from other molecules. Thus the higher the molecular weight of the polymer the more difficult it is to dissolve it and the higher is the viscosity of the solution.

Another factor affecting the viscosity of a polymer solution is the state of the chain in the solution. It can be stretched out or in a bundle or may adopt any conformation between these two extremes. Solutions containing stretched chains exhibit a higher viscosity than the entangled bundles. The conformation of the chain in the solution depends on the individual solvent, its molecular structure, polarity, and compatibility with the given polymer.

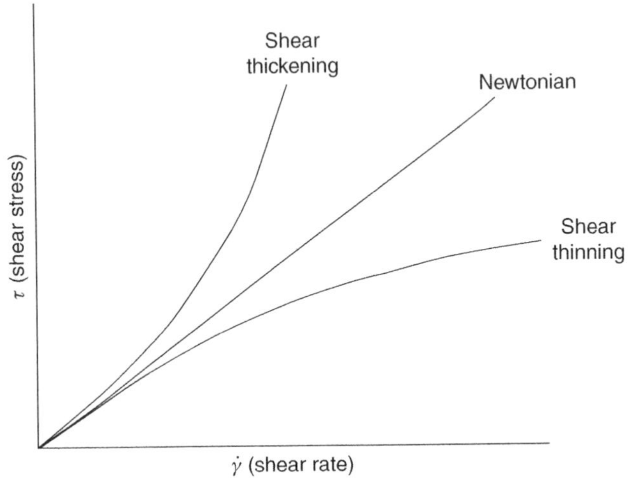

FIGURE 2.11 Different types of flow.

The viscosity of solutions of polymers in good solvents increases with increasing concentration. The range from 0.01 to 10 Pa.s is of greatest importance for liquid adhesives and coatings. The log of the viscosity increases approximately in proportion with the concentration [19]. The effect of temperature on viscosity is important: Viscosity of solutions almost always decreases as temperature increases.

The relative viscosity depends on the volume proportion of the dissolved or dispersed phase ρ, as shown by the Einstein's viscosity law (Eg. 2.5) [20]:

$$\eta_{rel} = 2.5\rho + 1 \tag{2.5}$$

The relative viscosity is the ratio of inherent viscosity of the solution η_S and the inherent viscosity of the solvent η_0:

$$\eta_{rel} = \frac{\eta_S}{\eta_0} \tag{2.6}$$

The specific viscosity is

$$\eta_{sp} = \eta_{rel} - 1 = \frac{\eta_S}{\eta_0} - 1 = \frac{\eta_S - \eta_0}{\eta_0} \tag{2.7}$$

The intrinsic viscosity $[\eta]$ is the specific viscosity for the concentration of the polymer reduced to zero:

$$[\eta] = \frac{\eta_{sp}}{c} \quad \text{for} \quad c \to 0 \tag{2.8}$$

The intrinsic viscosity is related to the molecular weight, as shown by the *Mark–Houwink equation* [21, 22]:

$$[\eta] = kM_{\mathrm{v}}^{\mathrm{a}} \tag{2.9}$$

where M_{v} is the viscosity average molecular weight, k and a are temperature-dependent constants.

2.7 POLYMER COMPATIBILITY

In a practical industrial situation, different polymers are blended to achieve required processing behavior and/or mechanical or physical properties. The simplest case is blending two grades of the same type polymer with different melt viscosities. In other cases, two or more polymers are blended for the same reasons and often for cost reduction. For given polymers, their compatibility with one another is dependent on the polymer amount in the mixture and its molecular weight [9]. The reason for an incompatibility is a decrease in the entropy of mixing as the molecular weight increases, so that in accordance with Eq. 2.3 the amount of ΔH_{m} is not compensated by the amount of $T\Delta S_{\mathrm{m}}$, and the value of ΔG_{m} becomes positive.

The enthalpy of mixing depends on interactive forces between the components and can be described by the parameter δ [9, 23]. The measure of compatibility is the *interaction parameter* χ [24–26] defined as:

$$\chi_{\mathrm{AB}} = \frac{V_{\mathrm{s}}}{RT}(\delta_{\mathrm{A}} - \delta_{\mathrm{B}})^2 = \frac{(\delta_{\mathrm{A}} - \delta_{\mathrm{B}})^2}{6} \tag{2.10}$$

V_{s} is the volume of the polymer segment taken equal to the repeating unit of the polymer chain (the same for both polymers) [27].

According to Eq. 2.10, polymers are only compatible with one another if the corresponding interaction parameter does not exceed a defined critical value that, depending on molecular weight, is between 0.02 and 0.002 [9].

2.8 DEFORMATION BEHAVIOR

2.8.1 Introduction

The mechanical behavior of a material is characterized by the deformation that results from the application of an external load. With polymers, three distinct, although fundamentally related types of deformation, are differentiated: (1) elastic deformation (instantaneous and completely reversible), (2) viscoelastic deformation (time dependent and reversible), and (3) viscous deformation (time dependent and irreversible).

When subjected to extension by an external force, the amount of elongation depends on the magnitude of the applied force. It is common to use the tensile stress, σ (force/original area of a sample in lb in.^{2} of MPa) and the elongation (or strain), ε, the amount of extension divided by original length, expressed as a percentage:

$$\sigma = \frac{F}{A_0} \tag{2.11}$$

$$\varepsilon = \frac{l - l_0}{l_0} \tag{2.12}$$

where F is the force in pounds or newtons, A_0 the original area is in square inches or square meters; l is the extended length, and l_0 is the original length of the sample.

The relationship between stress and strain of elastic materials is expressed mathematically by *Hooke's law* stating that the deformation (elongation or compression) of an elastic material is proportional to the applied stress. For simple extension, the strain is directly proportional to the stress [28], thus:

$$\varepsilon = \frac{\sigma}{E} \tag{2.13}$$

where E is the *modulus of elasticity* (*Young's modulus*.).

The test result can be illustrated by a *stress–strain curve*. The stress–strain curves are different for different materials. Typical elastomers can be extended much more than typical plastics, because they are softer and their structure allows that.

Elastic materials can exhibit elongations anywhere between 100 and 1000% before they rupture; the stresses at specific elongations (typically 100 and 300%) are called *moduli* and characterize the stiffness of the given material (see Fig. 2.12). Plastic materials typically respond to an external force slower and exhibit a yield point before they rupture (see Fig. 2.13). A brittle plastic extends only very little, exhibits a distinct yield point, and ruptures almost immediately after reaching the yield point (see Fig. 2.14). In *compression*, elastomers respond by decreasing their original height and increase their original area. Plastics are compressed much less and considerable loads are required to either deform or crush them.

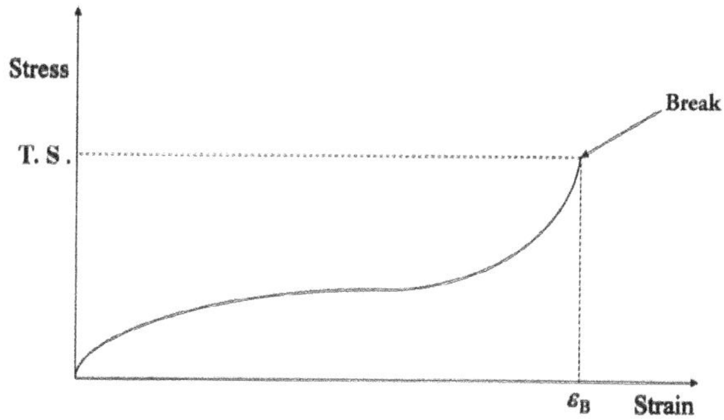

FIGURE 2.12 Typical stress–strain plot for an elastomer.

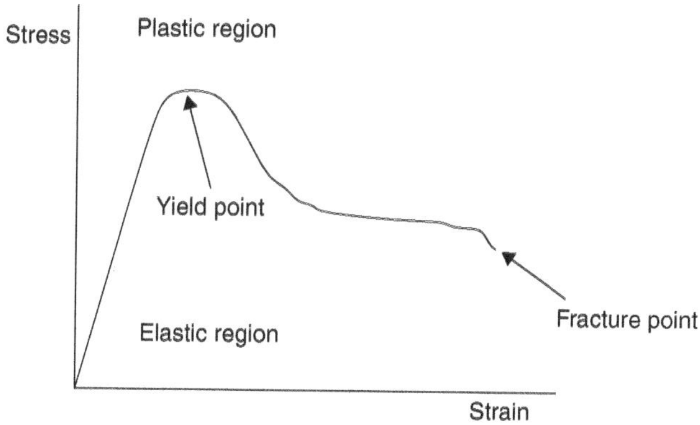

FIGURE 2.13 Typical stress–strain plot for a flexible plastic.

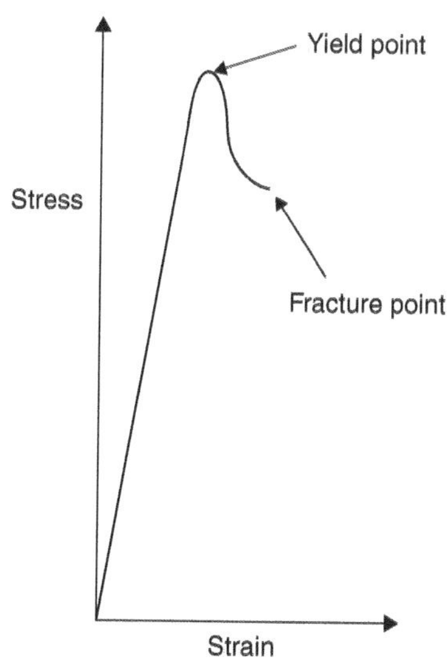

FIGURE 2.14 Typical stress–strain plot for a brittle plastic.

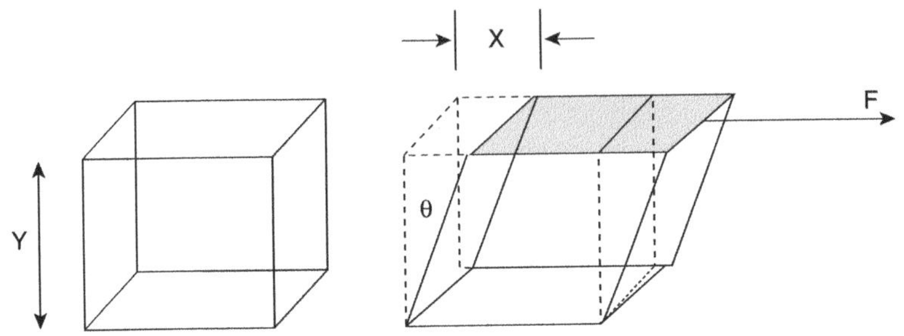

FIGURE 2.15 Shear of an elastic material. (Cowie, J.M.G. and Arighi, V., *Chemistry and Physics of Modern Materials, 3rd Edition*, 2008, p. 398)

In *simple shear*, an elastic body is subjected to a force applied to its surface. The bottom of the sample is stationary (see Fig. 2.15). The resistance to deformation is defined in terms of *shear stress*, τ:

$$\tau = \frac{F}{A} \tag{2.14}$$

given by the ratio between the force, F, and the cross-sectional area A. The *shear strain* γ for small deformations is equal to

$$\gamma = \tan\theta = \frac{X}{Y} \tag{2.15}$$

A similar situation is applicable to shear in fluids and is the basis for the development of the concept of their viscosity.

2.8.2 Viscoelastic Response

Bulk polymers combine elastic and viscous properties in both the fluid and solid state. Therefore they are generally addressed as *visco-elastic* and, in fact, polymers are the main representatives of this special class of materials.

Viscoelastic behavior does not mean just a superposition of independent viscous and elastic forces. In addition, it includes a new phenomenon known as *anelasticity*, where both become coupled. Thus anelasticity is the property of a solid in which deformation depends on the time rate of change of stress, as well as on the stress itself [29].

The contribution of perfect elasticity, anelasticity, and viscous flow to the total mechanical response of a sample is different for different polymers and depends largely on temperature. As a consequence of these changes, the temperature range for a certain polymer is limited. The most important limitation results from the T_g, where the elasticity and strength shown by a glassy solid are lost and the polymer becomes meltlike or turns into a rubber.

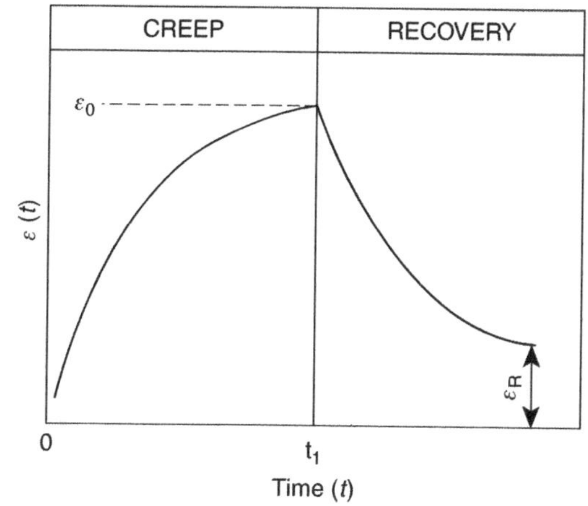

FIGURE 2.16 Creep.

2.8.2.1 Creep A direct simple method of determining the properties of viscoelastic properties of a material is the creep experiment. This is carried out by instantaneously applying a constant force, which is then followed by a measurement of the resulting deformation as a function of time (see Fig. 2.16). The resulting *creep curve* represents three contributions: (1) a perfectly elastic (i.e., instantaneous response), (2) a retarded elastic deformation (i.e., an anelastic deformation) and (3) viscous flow.

The first two contributions are reversible and the third is irreversible.

If the applied force is sufficiently small, then the creep curve becomes proportional to the force. In this *linear viscoelastic range*, the time-dependent elongation and the force for the description of the response can be used, For a sample under tension, the *tensile creep compliance* $J(t)$ is introduced as:

$$J(t) = \frac{\varepsilon(t)}{\sigma} \tag{2.16}$$

The *creep modulus* is

$$E(t) = \frac{\sigma}{\varepsilon(t)} \tag{2.17}$$

2.8.2.2 Stress–Relaxation In the *stress–relaxation experiment* a certain constant strain is instantaneously imposed on a sample and the stress induced by this procedure is measured as a function of time (see Fig. 2.17). The tensile stress has its maximum directly after the stretch

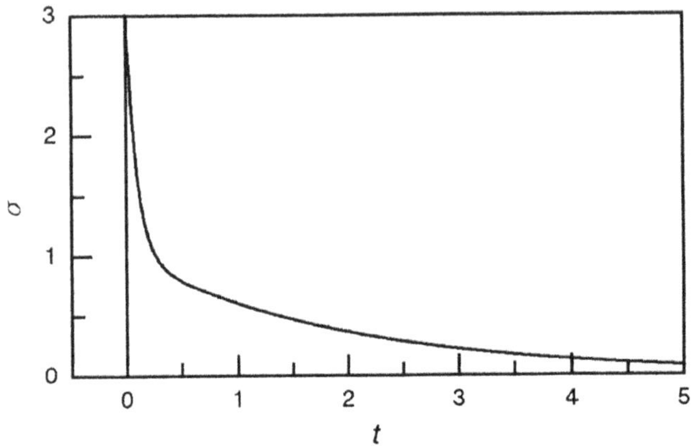

FIGURE 2.17 Stress–relaxation curve.

and then decays. Anelastic components first produce a downward step. If the sample can flow, the stress will further decrease and eventually vanish completely. The result from such an experiment can be described by the *time-dependent tensile modulus* $E(t)$, defined as:

$$E(t) = \frac{\sigma(t)}{\varepsilon_0} \qquad (2.18)$$

where ε_0 is the imposed longitudinal strain.

2.8.2.3 Dynamic-Mechanical Response In the *dynamic-mechanical experiments*, samples are exposed to a periodically varying stress field, for example, to a tensile strain for an elastic body (Fig. 2.18a) [30]:

$$\sigma = \sigma_0 \sin(\omega t) \qquad (2.19)$$

and for a viscoelastic body (Fig. 2.18 b):

$$\sigma(t) = \sigma_0 \sin(\omega t + \delta) \qquad (2.20)$$

Corresponding strain in an elastic body is

$$\varepsilon = \varepsilon_0 \sin(\omega t) \qquad (2.21)$$

and in a viscoelastic body

$$\varepsilon = \varepsilon_0 \sin(\omega t + \delta) \qquad (2.22)$$

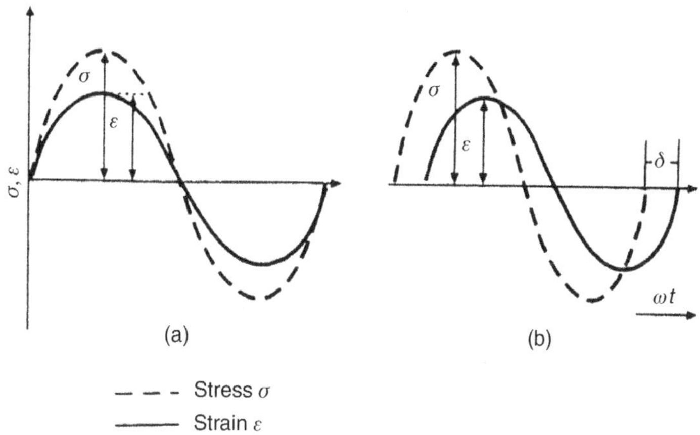

FIGURE 2.18 Dynamic stress–strain for an elastic (a) and a viscoelastic (b) polymeric material.

where ω is the angular frequency, and σ_0 and ε_0 are the stress and strain amplitudes.

With trigonometric expansion of Eq. 2.20, the stress may be resolved into two components [31]:

$$\sigma(t) = \sigma_0[\sin(\omega t)\cos\delta + \sin\delta\cos(\omega t)]$$

that is,

$$\sigma(t) = (\sigma_0 \cos\delta)\sin(\omega t) + \sigma_0 \sin\delta\ \cos(\omega t)$$

Therefore

$$\sigma(t) = (E_1\varepsilon_0)\sin(\omega t) + (E_2\varepsilon_0)\cos(\omega t) \tag{2.23}$$

Since the elastic moduli, which are assumed to be tensile in this case, are given by

$$E_1 = \sigma_0/\varepsilon_0 \cos\delta \qquad E_2 = \sigma_0/\varepsilon_0 \sin\delta$$

The elastic modulus can be expressed by complex notation to relate stress and strain. Thus E_1 describes stress–strain relationships, which are in-phase, while the out-of-phase component is characterized by E_2 and

$$\tan\delta = \frac{E_2}{E_1} \tag{2.24}$$

Using complex notation, we can write

$$\varepsilon(t) = \varepsilon_0 \exp(i\omega t) \qquad \sigma(t) = \sigma_0 \exp[i(\omega t + \delta)] \quad \text{and}$$

the *complex dynamic modulus* E^*, consisting of a real and imaginary component, is expressed as follows:

$$E^* = \frac{\sigma(t)}{\varepsilon(t)} = \frac{\sigma_0}{\varepsilon_0} \exp(i\delta) = \frac{\sigma_0}{\varepsilon_0} (\cos\delta + i \sin\delta)$$

Therefore

$$E^* = E_1 + iE_2 \tag{2.25}$$

Under high-frequency dynamic loading conditions, polymeric materials are characterized by two modulus components, and by the phase angle that separates the sinusoidal stress and strain cycles. The real part of the complex modulus, E_1, is often referred to as the *storage modulus*, since it represents the stored elastic energy in the sample. The imaginary component, E_2, arises due to viscous dissipation and is called the *loss modulus* (see Fig. 2.19).

Tan δ represents the viscous dissipation in the material (at a given test temperature and frequency) and is often the parameter chosen to relate dynamic data to molecular or structural motion in polymeric materials.

The *cyclic energy dissipation* per unit volume can be estimated from the following relation [32]:

$$U = \pi E_2(\varepsilon_0)^2 \tag{2.26}$$

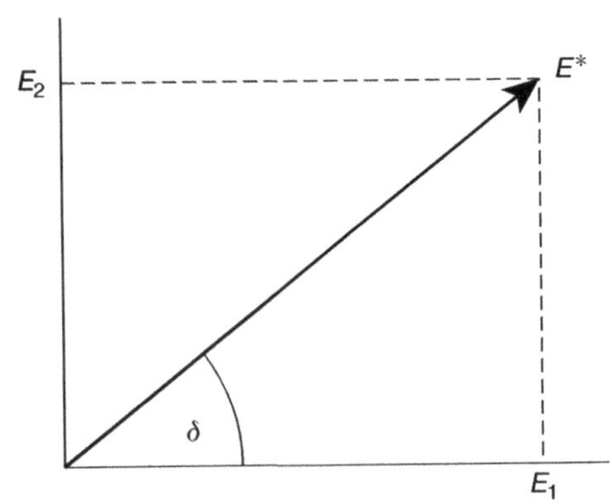

FIGURE 2.19 Complex modulus and its components.

2.9 POLYMERIC FOAMS

Polymeric foams generally consist of a minimum of two phases: a solid-polymer matrix and a gaseous phase derived from the blowing agent or air entrapment [33]. They comprise a wide variety of materials with densities ranging typically from as low as 1.6 up to 960 kg m^{-3}(0.1–60 lb $ft.^{-2}$) [34]. Their main applications include furniture, transportation, bedding, carpet underlay, packaging, textiles, toys and novelties, automotive, appliances, business machine housings, food and drink containers, and so on. There are two major classes of polymeric foams: thermoplastic and thermoset. While thermoplastic foams can be recycled easily and reprocessed, the thermoset are intractable since they are heavily cross-linked. Within these two classes, the polymeric foams are further classified as rigid, flexible, and semirigid (semiflexible), and can be fabricated to any desired degree of hardness, depending on their compositions, cellular morphology, and other physical and thermal characteristics, such as T_g, degree of crystallinity, extent of cross-linking, and so on [33, 35].

Solid polymeric foams can either consist of *closed* or *open cells.* Closed-cell foams have a cellular structure in which continuous air bubbles are entrapped within a continuous macromolecular phase. For example, a polystyrene coffee cup consists entirely of closed foam cells. Closed-cell foams have a cellular network, in which continuous channels are available throughout the solid macromolecular phase for air to flow through at will. An example of an open cell is a polyurethane seat cushion [36].

Structural foams have a somewhat different structure than the other solid foams: They are most frequently molded or extruded parts that have a cellular foam core and an integral solid skin; the transition from skin to core is gradual. The solid skin gives the shaped part its form and toughness, while the cellular core contributes to the high strength/weight ratio. Structural foams can be made from thermoset materials (polyurethanes formed from polyaddition of reactive component, e.g., polyols and isocyanates or from thermoplastic resins using physical or chemical blowing agents). The latter type is produced in larger volumes than the former [37].

Syntactic foams are materials consisting of hollows sphere fillers in a resin matrix [38]. The name is derived from the Greek word 'syntactikos' meaning "to arrange together" [39]. The term "foam" is used because of the cellular nature of the material. Syntactic foams are also known as "foam composites" since hollow microspheres can be regarded as reinforcements in a matrix [40]. The microspheres can be made mainly from glass, various polymers (e. g., polystyrene and phenolics), carbon, ceramics, and metals [41]. Matrix materials include epoxy resins, phenolic resins, cyanate ester resins, polyimide, polyurethanes, polyesters, siolicones, polyamides, polypropylene, and polybutadienes. Particularly interesting are carbon–carbon syntactic foams; here the precursor matrix is a phenolic resin, which is carbonized at temperatures close to 1000°C (1832°F) [40]. In addition to low densities, high strength, good impact behavior [43], and syntactic foams have very low dielectric constants and very low loss tangents. Such foams are used in radom applications where minimum absorption of electromagnetic energy is required [44, 45].

Most polymeric foams are produced by one of the several known manufacturing techniques, including continuous slabstock, pour-in-place, compression, transfer and injection molding, reaction injection molding, extrusion, spraying, and lamination.

The gaseous phase in any polymeric foam material derives from the use of blowing agents in the foam manufacturing process. There are two types of blowing agents: chemical and physical. The first type includes chemical compounds that give off gases under the process conditions either due to decomposition or chemical reaction. The second type is essentially simple inert gases (nitrogen, carbon dioxide, volatile hydrocarbons, low-boiling chlorofluorocarbons, hydrofluorocarbons, and hydrochlorofluorocarbons). Latex foam is produced mainly by frothing of the latex, gelling the frothed latex, and then vulcanizing it in the expanded state.

A relatively large number of polymers are used for foams, including polyolefins, acrylonitrile–butadiene–styrene (ABS), acrylics, urea–formaldehyde, polyimides, polyphenylene oxide, polychloroprene, nitrile rubber, natural rubber, thermoplastic elastomers, epoxy resins, silicones, and polyurethanes. Foams based on epoxy resins, silicones, and flexible polyurethane foams are used widely in electrical and electronic applications.

2.10 LIQUID CRYSTALS

2.10.1 Introduction

Liquid crystals (LCs) have properties between those of a conventional liquid and those of a solid crystal. For example, an LC may flow like a liquid, but its molecules may be oriented in a crystal-like way. There are many different types of LC phase, which can be distinguished by their different optical properties (e.g., birefringence). When viewed under a microscope using a polarized light source, different LC phases will appear to have distinct textures. The contrasting areas in the textures correspond to domains where the LC molecules are oriented in different directions. Within a domain, however, the molecules are well ordered.

Liquid crystals can be divided into *thermotropic*, *lyotropic*, and *metallotropic* phases. Thermotropic and lyotropic LCs consist of organic molecules. Thermotropic LCs exhibit a phase transition into the LC phase as temperature is changed. Lyotropic LCs exhibit phase transitions as a function of both temperature and concentration of the LC molecules in a solvent. Metallotropic LCs are composed of both organic and inorganic molecules; their LC transition depends not only on temperature and concentration, but also on the inorganic/organic composition ratio. Examples of LCs can be found both in the natural world and in technological applications. For example, most modern electronic displays are LC based.

Most LC compounds exhibit *polymorphism*, or a condition where more than one phase is observed in the liquid-crystalline state. The term *mesophase* is used to describe the "subphases" of LC materials. Mesophases are formed by changing the amount of order in the sample, either by imposing order in only one or two dimensions, or by allowing the molecules to have a degree of translational motion. The following section describes the mesophases of LCs in greater detail as applicable to macromolecules.

2.10.2 Liquid-Crystalline Polymers

Polymers with highly ordered, stiff, crystalline regions in the liquid phase are known as liquid-crystalline polymers (LCPs). Thus a LCP is very different from the usual polymer whose macromolecules are disordered in the melt. The LCPs form a liquid-crystalline intermediate state, called *mesomorphous*, between the liquid and the solid state. The stiff crystalline regions of a chain molecule are known as *mesogens*. An LCP that consists only of mesogens in the main chain cannot be processed like a thermoplastic because it melts at 400–600°C (750–1110°F) only (i.e., above its decomposition temperature). By selectively creating points of disturbance between the mesogenic regions of the macromolecule, thermotropic LCP chains, which melt at 250–400°C (480–750°F), can be obtained [46].

In polymers containing branches, the mesogenic regions are in the branches that are grafted onto the main chains through flexible spacer molecules. Mesogens in a crystalline melt can be aligned under the influence of an electric field and will exhibit birefringence. Therefore they are suitable for storing electrooptical information that can be frozen in and later erased by melting. Figure 2.20 schematically shows the essential

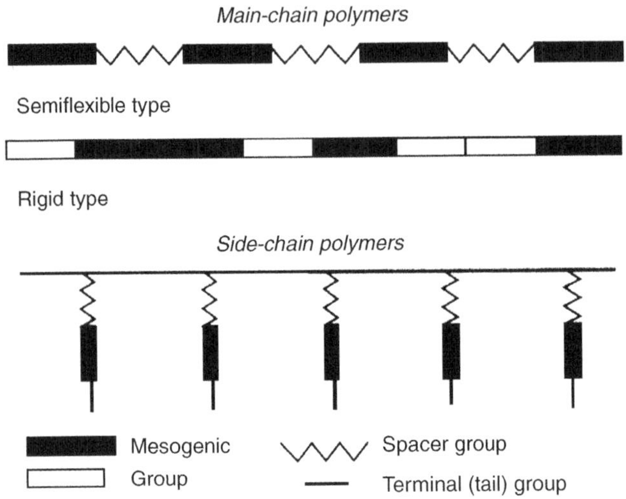

FIGURE 2.20 Main- and side-chain LCPs. (Gedde, U.W., *Polymer Physics*, Chapman & Hall, 1995, p. 111)

features of the structure of main- and side-chain polymers. The mesogenic units give the polymers their anisotropic rod shape [49].

The liquid-crystalline state involves on many structures. Different organizations between the isotropic, amorphous structure and the perfectly organized 3D crystal may exist [47]. The most probable LCP to be formed directly from a isotropic melt is the *nematic*, which is characterized by long-range orientational order and positional disorder (see Fig. 2.21). For a nematic LC domain, when selecting the average direction of the mesogens as the reference direction (director); this order parameter takes values between 0.3 and 0.8. The order parameter decreases with increasing temperature until the isotropization temperature is reached and the order parameter falls to zero [47].

The phase and thermal transitions in a typical so-called *thermotropic* LCP such that the liquid-crystalline phase is stable within a certain temperature range, are [48]:

$$\text{Low temperature} \longleftrightarrow \text{High temperature}$$

$$\text{Crystal} \longleftrightarrow \text{Smectic} \longleftrightarrow \text{Nematic} \longleftrightarrow \text{Isotropic liquid}$$

When processing LCPs the molecules are oriented by shear or drawing resulting in highly anisotropic properties and self-reinforcement in the direction of orientation. Examples of LCPs are of partially crystalline aromatic polyesters based on *p*-hydroxybenzoic acid and related

FIGURE 2.21 Nematic structure with the director indicated by an arrow. (Gedde, U.W., *Polymer Physics*, Chapman & Hall, 1995, p. 112)

TABLE 2.5 Examples of Properties of a Typical Solid LCP

Property	Value
Specific gravity	1.38–1.95
Modulus of elasticity (Mpa)	8,530–17,200
Elongation at break (%)	0.26–6.2
Izod impact strength, notched (kJ m^{-2})	21.0–82.5

monomers, LCPs are capable of forming regions of highly ordered structure while in the liquid phase. However, the degree of order is somewhat less than that of a regular solid crystal. Typically, LCPs have a high mechanical strength at high temperatures, extreme chemical resistance, inherent flame retardance, and good weatherability. Examples of properties of typical solid LCPs are shown in Table 2.5. The subject of LCPs is discussed in some length in Section 4.3.

2.11 NANOCOMPOSITES

Nanocomposites contain reinforcing materials with dimensions on the order of 1–100 nm. They represent a new class of heterogeneous composites. Addition of nanoparticles, for example, clay nanoparticles to the polymer, results in materials with better mechanical properties, improved thermal and barrier properties without the loss of transparency, and toughness. When 3–5% by weight of nanoparticles are added to the matrix, the same stiffness corresponding to 15–20% by weight glass fiber reinforcement can be obtained. Particles ~1 nm thick are typically ~200 nm long [49]. Their platelet-like shape not only makes them suitable for reinforcement, but also explains their barrier properties because the gas molecules have to move around the particles and consequently their diffusion path becomes much longer. For example, an addition of 2% by weight of nanoparticle platelets decreases the permeability of polyamide resin by 50% [49]. One of the first important applications of nanocomposites are films with improved barrier properties.

A publication suitable for further reading on the subject is *Introduction to Nanotechnology* by Charles P. Poole, Jr. and Frank J. Owens, John Wiley & Sons, New York, 2003.

2.12 COMMERCIAL POLYMERS

Numerous commercial polymers cover a wide range of properties. The majority of them are synthetic, although some are still of natural origin. The natural polymers may be used as they are or may be modified. Synthetic polymers include plastics, elastomers, and fibers. There is no firm dividing line between the groups. However, some classification is useful from the technological point of view. One of the useful methods of defining a member of the given group is their stress–strain behavior.

Rigid plastics and fibers resist deformation, exhibit high moduli, and low elongations. Elastomers, on the other hand can be easily stretched and exhibit large reversible elongations under relatively low stress (i.e., they are more or less elastic). A schematic showing the approximate demarcation lines for dividing some common polymers is shown in Table 2.6.

TABLE 2.6 Demarcation Lines of Some Common Plastics, Elastomers, and Fibers

Elastomers	Plastics	Fibers
Polyisoprene	Polyethylene	
Polyisobutylene	Polytetrafluoroethylene	
Polybutadiene	Polystyrene	
	Poly(methlmethacrylate)	
	Phenol–formaldehyde	
	Urea–formaldehyde	
	Melamine–formaldehyde	
←——— Poly(vinyl chloride)	———→	
←——— Polyurethanes	———→	
←——— Polysiloxanes	———→	
	←——— Polyamide	———→
	←——— Polyester	———→
	←——— Polypropylene	———→

(Cowie, J.M.G. and Arighi, V., *Chemistry and Physics of Modern Materials, 3rd Edition*, 2008, p. 17).

It is clearly evident that the division of common commercial polymers into the three major groups is not clear-cut.

2.12.1 Plastics

Plastics are synthetic materials that are used mostly as replacement of conventional materials (e.g., metals, wood, stone, and ceramics), although there are some applications where only plastics can be used.

Total world demand for plastics is well above 255 million metric tons (560 billion lb) [50]. The industrially important plastics consist of 30–40 different types of polymers that are currently offered in 13,000 variations under ~25,000 different trade names [51].

Essentially, plastics can be classified into following three major groups:

1. Commodity or bulk resins, which include PVC, several grades of PE, PP, and PS.
2. Engineering plastics, which offer improved mechanical, electrical, and thermal characteristics when compared to commodity resins.
3. High-performance resins offering high-quality characteristics, partially in respect to high temperature.

Estimated percentages of world consumption of industrial plastics (excluding plastics used as additives) are according to [52]:

Polyethylene, low density (LDPE), linear low density (LLDPE)	17%
PP	19%
Polyethylene, high density (HDPE)	12%
PVC	13%
PS and expanded polystyrene (EPS)	7%
Polyesters	7%
Polyurethanes	6%
Others	19%

A comparison of various polyethylene grades is found in Table 2.7.

TABLE 2.7 Comparison of Various Polyethylene Grades

Property	LDPE	LLDPE	HDPE
Melting point (°C)	105–115	123–126	131–133
Density	0.918–0.925	0.918–0.94	0.94–0.97
Tensile strength [MPa (film)]	20–25	24–30	40–45
Branching degree	High	Very low	Low

2.12.1.1 Engineering Plastics Engineering plastics are a group of plastic materials that exhibit superior mechanical and thermal properties under a wide range of conditions over and above more commonly used commodity plastics. The term usually refers to thermoplastic materials rather than thermosetting ones. Engineering plastics are used for parts rather than containers and packaging. Depending on performance, we can divide them into two categories: (1) regular service engineering plastics and (2) high-performance engineering plastics.

2.12.1.1.1 Regular Service Engineering Plastics Examples of widely used engineering plastics include the following:

1. ABS
2. PCs
3. PAs
4. Polybutylene terephthalate (PBT)
5. Polyethylene terephthalate (PET)

1. Acrylonitrile–Butadiene–Styrene is a copolymer made by polymerizing styrene and acrylonitrile in the presence of polybutadiene. The proportions can vary from 15 to 35% acrylonitrile, 5 to 30% butadiene, and 40 to 60% styrene. For the majority of applications, ABS can be used between −25 and 60°C (−13° and 140°F) as its mechanical properties vary with temperature. The properties are created by rubber toughening, where fine particles of elastomer are distributed throughout the rigid matrix.

   The most important mechanical properties of ABS are impact resistance and toughness. A variety of modifications can be made to improve impact resistance, toughness, and heat resistance. The impact resistance can be amplified by increasing the proportions of polybutadiene in relation to styrene and acrylonitrile, although this causes changes in other properties. Impact resistance does not fall off rapidly at lower temperatures. Stability under load is excellent with limited loads. Thus, changing the proportions of its components ABS can be prepared in different grades. Two major categories could be ABS for extrusion and for injection molding, then high and medium impact resistance. Generally, ABS would have useful characteristics within a temperature range from 10 to 80°C (50 to 176°F).

   Even though ABS plastics are used largely for mechanical purposes, they also have electrical properties that are fairly constant over a wide range of frequencies. These properties are little affected by temperature and atmospheric humidity in the acceptable operating range of temperatures.

The ABS polymers are resistant to aqueous acids, alkalis, concentrated hydrochloric and phosphoric acids, alcohols, and animal, vegetable, and mineral oils, but they are swollen by glacial acetic acid, carbon tetrachloride (CCl_4) and aromatic hydrocarbons, and are attacked by concentrated sulfuric (H_2SO_4) and nitric (HNO_3) acids. They are soluble in esters, ketones, and ethylene dichloride.

2. Polycarbonate is produced by the reaction of bisphenol A and phosgene ($COCl_2$).

   Polycarbonate has a T_g of ~150°C (302°F), so it softens gradually above this point and flows above ~300°C (572°F). Tools must be held at high temperatures, generally >80°C (176°F) to make strain- and stress-free products. Low molecular mass grades are easier to mold than higher grades, but their strength is lower as a result. The toughest grades have the highest molecular mass, but are much more difficult to process.

   Unlike most thermoplastics, polycarbonate can undergo large plastic deformations without cracking or breaking. As a result, it can be processed and formed at room temperature using sheet metal techniques (e.g., forming bends on a brake).

3. Polyamides include a number of polymers identified by the number of carbon atoms in the monomeric units. Where two monomers are involved, the polymer will carry two numbers (e.g., PA 6/6). Several different types of polyamides (sometimes referred to as "nylons"-based on the original trademark of a product from DuPont) are on the market; the two most widely used are PA 6/6 produced by polycondensation of hexamethylene diamine and adipic acid and PA6 produced by polymerization of caprolactam.

   Generally, crystalline polyamides have a high tensile strength and flex modulus, impact strength, and abrasion resistance. They are resistant to nonpolar solvents, including aromatic hydrocarbons, esters, and essential oils [53], but are softened by and absorbed by polar liquids (e.g., alcohols, glycols, and water). Moisture pickup is the major limitation of PAs because it results in dimensional changes and reduced mechanical properties. Polyamide 6 is more prone to absorb moisture than PA 6/6.

   Special amorphous optically clear polyamides are commercially available. Other types of polyamides include 6/9, 6/10, 6/12, 11, 12, 4/6, and 12/12.

4. Polybutylene Terephthalate is a thermoplastic semicrystalline polymer resistant to solvents, with low shrinkage, good mechanical properties, and is heat-resistant up to 150°C (302°F).

   The PBT is closely related to other thermoplastic polyesters. Compared to PET, PBT has slightly lower strength and rigidity, slightly better impact resistance, and a slightly lower T_g. Both PBT and PET are sensitive to hot water >60°C (140°F). Here PBT and PET need ultraviolet (UV) protection if used outdoors, and most grades of these polyesters are flammable, although additives can be used to improve both UV and flammability properties.
5. Polyethylene terephthalate or poly(ethylene terephthalate) is a thermoplastic polymer resin of the polyester family. Depending on its processing and thermal history, PET may exist both as an amorphous (transparent) and as a semicrystalline polymer. The semicrystalline material might appear transparent (particle size <500 nm) or opaque and white (particle size up to a few microns) depending on its crystal structure and particle size. Its monomer (bis-β-hydroxyterephthalate) can be synthesized by the esterification reaction between terephthalic acid and ethylene glycol with water as a byproduct, or by a transesterification reaction between ethylene glycol and dimethyl terephthalate with methanol (CH_3OH) as a byproduct. Polymerization is through a polycondensation reaction of the monomers (done immediately after esterification–transesterification) with ethylene glycol as the byproduct (the ethylene glycol is directly recycled in production).

2.12.1.1.2 High-Performance Engineering Plastics These materials are suitable for service at temperatures typically >150°C (302°F) and/or exhibit exceptional mechanical properties, high chemical resistance, flame resistance, dimensional stability, and so on. The following commercial polymers are included in this group:

Poly(p-phenylene oxide) (PPO) or poly(p-phenylene ether) (PPE): This is a high-temperature thermoplastic. It is rarely used in its pure form due to difficulties in processing. It is mainly used as a blend with polystyrene, high-impact styrene–butadiene copolymer, or polyamide. While it was one of the cheapest high-temperature resistant plastics, processing was difficult and the impact and heat resistance decreased with time. Mixing it with polystyrene in any ratio could compensate for the disadvantages. The PPE is an amorphous high-performance plastic. The T_g is 215°C (419°F), but it can be varied by mixing with polystyrene.

Polysulfone (PSU): Polysulfone describes another family of thermoplastic polymers. These polymers are known for their toughness and stability at high temperatures. They contain the subunit aryl-SO_2-aryl, the defining feature of which is the sulfone group. Polysulfones are used in specialty applications and often are a superior replacement for polycarbonates.

A typical polysulfone is produced by the reaction of a diphenol and bis(4-chlorophenyl) sulfone, forming a polyether by elimination of sodium chloride:

Polysulfones are rigid, high-strength, and transparent polymers that retaining these properties between −100 and +150°C (−148 and 302°F). It has very high-dimensional stability; the size change when exposed to boiling water or 150°C (302°F) air or steam generally falls >0.1%. Its T_g is 185°C (365°F). Polysulfone is highly resistant to mineral acids, alkali, and electrolytes, in the pH rang from 2 to 13. It is resistant to oxidizing agents; therefore it can be cleaned by bleaches. Polysulfone is also resistant to surfactants and hydrocarbon oils, but is not resistant to low-polar organic solvents (e.g., ketones and chlorinated hydrocarbons), and aromatic hydrocarbons. It is also stable in aqueous acids and bases and many nonpolar solvents; however, it is soluble in dichloromethane and methylpyrrolidone.

Polyether ether ketone (PEEK): These polymers are obtained by step-growth polymerization by the dialkylation of bisphenolate salts. Typical is the reaction of 4,4′-difluorobenzophenone with the disodium salt of hydroquinone which is generated *in situ* by deprotonation with sodium carbonate. The reaction is conducted ~300°C in polar aprotic solvents (e.g., diphenyl sulfone).

Polyether ether ketone is a semicrystalline thermoplastic with excellent mechanical and chemical resistance properties that are retained at high temperatures. The Young's modulus is 3.6 GPa (56500 psi) and its tensile strength is 90–100 MPa (13,050–14,500 psi). The PEEK has a T_g ~143°C (289°F) and melts ~343°C (662°F). It is highly resistant to thermal degradation, as well as attack by both organic and aqueous environments. It is attacked by halogens and strong Brønsted and Lewis acids, as well as some halogenated compounds and aromatic hydrocarbons at high temperatures.

Polyetherketone (PEK): Also known as polyaryletherketone, it is an advanced thermoplastic resin having both ether and ketone linkages in its chains. The PEK materials are semicrystalline aromatic polyethers with excellent mechanical and dielectric properties. Polyetherketone is also highly fire resistant, has good chemical resistance, and can be used for high-temperature applications. In general, PEK polymers are similar to PEEK, but have slightly higher resistance to heat (~5°C or 10°F).

Polyimide (PI): This polymer is produced by polymerization of imide monomers. Typical monomers include pyromellitic dianhydride and 4,4′-oxydianiline. According to the composition of their main chain, polyimides can be aliphatic (linear polyimides) and semiaromatic. According to the type of interactions between the main chains, polyimides can be (1) thermoplastic: very often called *pseudothermoplastic* and (2) thermosetting: commercially available as uncured resins, polyimide solutions, stock shapes, thin sheets, laminates, and machined parts.

Several methods are possible to prepare polyimides. The following are among them: the reaction between a dianhydride and a diamine (the most used method) and the reaction between a dianhydride and a diisocyanate.

Normal operating temperatures for thermoplastic PI range from cryogenic to those exceeding 500°F (260°C). Polyimides are also inherently resistant to flame and do not usually need to be mixed with flame retardants. Typical polyimide parts are not affected by commonly used solvents and oils, including hydrocarbons, esters, ethers, alcohols, and freons. They also resist weak acids, but are not recommended for use in environments that contain alkalis or inorganic acids. Some polyimides are solvent soluble and exhibit high optical clarity. The solubility properties lend them toward spray and low-temperature cure applications.

Polyamide-imide (PAI): This polymer is one of the thermoplastics with highest high-temperature resistance. Polyamide-imide resins have a useful service temperature range from cryogenic to almost 260°C (500°F). In their heat resistance, they approach that of polyimides, but their mechanic properties are distinctly better than those of polyimides. Polyamide-imide is inherently flame retardant and possess an excellent chemical resistance; it is generally not affected by aliphatic and aromatic hydrocarbons, acids, bases, and halogenated solvents. However, at high temperatures it is attacked by strong acids and bases.

Polyphenylene sulfide (PPS): This organic polymer consists of aromatic rings linked with sulfides. It is formed by reaction of sodium sulfide with *p*-dichlorobenzene: This polymer is known to resist chemical and thermal attack. Polyphenylene sulfide can be molded, extruded, or machined to high tolerances. In its pure solid form, it may be opaque white to light tan in color. Maximum service temperature is 218°C (424°F). It has not been found to dissolve in any solvent at temperatures below ~200°C (392°F).

It is also one of the most important high-temperature polymers because it exhibits a number of desirable properties that include resistance to heat, acids and alkalis, and to mildew, bleaches, aging, sunlight, and abrasion. It absorbs only small amounts of solvents and resists dyeing.

Polyarylates: These are amorphous aromatic polyesters. They fall between polycarbonates and and polysulfones in terms of temperature resistance. Other features are toughness, UV resistance, and good electrical properties. Polyarylates are transparent, but tend toward yellowness, although water-white materials are available.

Polyetherimide (PEI): Polyetherimide is an amorphous, transparent amber polymer combining high-temperature resistance, rigidity, high-impact strength, and creep resistance. It has a T_g of 215°C (420°F). It is soluble in partially halogenated solvents, but is resistant to alcohols, acids, and hydrocarbons. It performs well under humid conditions and withstands UV and ionizing high-energy radiation.

Polyoxymethylene (POM): Also known as acetal, polyacetal, or polyformaldehyde, it is an engineering thermoplastic used in precision parts that require high stiffness, low friction, and excellent dimensional stability.

Advantages of acetal-based plastics are chemical resistance, and very low water absorption resistance to hydrolysis by bases. Disadvantages are low-impact strength, a low-melting temperature, a very high thermal expansion, and they are sensitive to acid hydrolysis and chlorine attack.

Polyoxymethylene is a semicrystalline polymer (75–85% crystalline) with a melting point of 175°C (347°F) as a homopolymer. A copolymer version has a slightly lower melting point of 163°C (325°F). It is a tough material with a very low coefficient of friction.

Fluoroplastics: These are a rather unique group of polymeric materials. Several of the monomers used in their synthesis are derived from compounds used in refrigeration. The first fluoropolymer was PTFE, which was discovered in 1938 in the laboratories of E.I. DuPont de Nemours & Co. and commercialized in the early 1950s.

Since that time, many different types of fluoropolymers were developed with a relatively large proportion being made in the last three decades. Some of them were developed from the original PTFE, but others contain other elements (e.g., hydrogen, chlorine, silicone, oxygen, and nitrogen). The two main factors determining the unique properties of fluoroplymers are the strong bond between fluorine and carbon (see Table 2.8) and shielding of the carbon backbone by relatively large fluorine atoms [54]. These factors are the strongest in *perfluorinated* (fully fluorinated) polymers, that is, having their molecular backbones consisting of carbon atoms with only fluorine atoms attached. Monomers used in commercial fluoropolymers are listed in Table 2.9. Most of the fluoropolymers are produced by emulsion polymerization. Final products are homopolymers, copolymers, and terpolymers. The main industrial fluoroplastics are listed in Table 2.10.

Fluoroplastics as a group possess high chemical inertness, low friction coefficient, and antistick properties; several of them have very high

TABLE 2.8 Carbon Bond Energies

Bond	Bond Energy (kJ mol^{-1})
C—F	485
C—H	403
C—O	342
C—C	338
C—Cl	312
C—Br	269
C—I	232

TABLE 2.9 Monomers Used for Current Commercial Fluoropolymers

Compound	Formula
Ethylene	$CH_2{=}CH_2$
Tetrafluoroethylene	$CF_2{=}CF_2$
Chlorotrifluoroethylene	$CF_2{=}CClF$
Vinylidene fluoride	$CH_2{=}CF_2$
Vinyl fluoride	$CFH{=}CH_2$
Propene	$CH_3CH{=}CH_2$
Hexafluoropropene	$CF_3CF{=}CF_2$
Perfluoromethylvinyl ether	$CF_3OCF{=}CF_2$
Perfluoropropylvinylether	$CF_3CF_2CF_2OCF{=}CF_2$

TABLE 2.10 Major Industrial Fluoroplastics

Abbreviation	Name
ETFE	Copolymer of ethylene and tetrafluoroethylene
ECTFE	Copolymer of ethylene and chlorotrifluoroethylene
EFEP	Terpolymer of ethylene, tetrafluoroethylene and hexafluoropropylene
FEP	Copolymer of tetrafluoroethylene and hexafluoropropylene
MFA	Copolymer of tetrafluoroethylene and perfluoromethylvinyl ether
PCTFE	Poly(chlorotrifluoroethylene)
PFA	Copolymer of tetrafluoroethylene and perfluoropropylvinyl ether
PTFE	Poly(tetrafluoroethylene)
PVDF	Poly(vinylidene fluoride)
PVF	Poly(vinyl fluoride)
THV	Terpolymer of tetrafluoroethylene, hexafluoropropylene and vinylidene fluoride

melting points, notably PTFE and perfluoroalkoxy (PFA) resins, some are excellent dielectrics and most of them exhibit a very high resistance to common solvents, UV radiation, and aggressive chemicals.

- Polytetrafluoroethylene is produced from gaseous tetrafluoroethylene (TFE) via a free radical polymerization. In principle, there are two distinct methods for polymerization of PTFE. One method, suspension polymerization, produces granular resins, the other one, the emulsion polymerization, yields fine powders and aqueous dispersions [55]. Each of these products is processed by a specific processing method: granular resins are compressed and sintered in a process similar to powder metallurgy; fine powders are converted into a "paste" by the addition of 15–25% by weight of a hydrocarbon lubricant (e.g., kerosene); the paste is then shaped into unsintered tapes or sintered thin tubing or wire coating after removal of the lubricant; the aqueous dispersions are mainly used to make coated fabrics or cast films.
- Modified PTFE is made in a similar fashion as the standard polymer except that a small amount of a comonomer is added. The resulting resin has a lower melt viscosity and considerably fewer microvoids than the standard PTFE resin. It can be shaped and welded easier and films and sheets made from modified PTFE have lower permeability and higher dielectric strength [56].
- PTFE micropowders are low-molecular weight PTFE grades used as additives to increase wear resistance, release properties, and reduce the coefficient of friction of a large number of matrix materials [57]. Micropowders are produced by controlled

polymerization of TFE or controlled degradation of high-molecular weight PTFE.

1. Melt-Processible Perfluoroplastics

   The need for highly fluorinated thermoplastic polymers that, unlike PTFE, can be fabricated by the conventional melt-processing methods led to the development of a group of resins that are copolymers of TFE with other perfluorinated monomers. Copolymerization of TFE and HFP produces a polymer that is frequently called fluorinated ethylene–propylene or FEP; copolymerization of TFE and perfluorpropylvinly ether (PPVE) leads to perfluoroalkoxy (PFA) resins and copolymerization of TFE and perfluoromethylvinyl ether (PMVE) to methylfluoroalkoxy (MFA) resins.

2. Other Melt-Processible Fluoroplastics

- Copolymers of ethylene and TFE (ETFE) essentially comprise alternating ethylene and TFE monomeric units. They have an excellent balance of physical, chemical, mechanical, and electrical properties and are easily processed by the usual melt-processing techniques (e.g., extrusion, injection molding, transfer molding, blow molding, and etc.).

  Ethylene tetrafluoroethylene exhibits excellent dielectric properties. Its dielectric constant is low and essentially independent of frequency. The dissipation factor is low, but increases with frequency and also can be increased by cross-linking. Dielectric strength and resistivity are high and unaffected by water.

- Polyvinylidene fluoride (PVDF) homopolymer is a semicrystalline polymer. Its degree of crystallinity can vary from 35 to >70%, depending on the method of preparation and thermomechanical history. Similar to other linear polyolefins, crystalline forms of PVDF fluoride involve lamellar and spherulitic forms [58].

  Polyvinyledene fluoride exhibits a complex crystalline polymorphism, which cannot be found in other known synthetic polymers. There are a total of four distinct crystalline forms: alpha, beta, gamma, and delta. The alpha and beta forms are most common in practical situations. Generally, the alpha form develops in normal melt processing; the beta form is generated under mechanical deformation of melt-fabricated specimens [59].

  Typical values of electrical properties of the homopolymer without additives and treatments are listed in Table 2.11. The values can be substantially changed by the type of cooling and posttreatments, which determine the morphological state of the polymer. Dielectric

TABLE 2.11 Typical Electrical Properties of PVDF Homopolymer

Property		Value	ASTM
Volume resistivity (Ω cm)		1.5×10^{14}	D257
Surface arc resistance (s)		50–60	D495
Dielectric strength[a], short time, (500 V s^{-1})			D149
kV 10^{-3} in.		1.7	
kV mm^{-1}		67	
Dielectric constant[b]	1 kHz	9.54	D150
	100 kHz	9.0	
	1 MHz	8–9	
	1 GHz	3–4	
Dissipation factor	1 kHz	0.005–0.02	D150
	100 kHz	0.04–0.06	
	1 MHz	0.03–0.05	
	1 GHz	0.09–0.11	

[a] In air, using 1-in. (2.54-mm) electrode with 4 × 0.005 in. (100 × 0.125 mm) compression molded disk.
[b] Measured in. air using 0.50 in. (12.5 mm) thick, 2 in. (50 mm) diameter compression molded disks that were water quenched after molding.

constant values as high as 17 have been measured on oriented samples. The structure yielding the high dielectric constant and the complex polymorphism of PVDF also exhibits a high dielectric loss factor. The unique dielectric properties combined with polymorphism of PVDF are the sources of piezoelectric and pyroelectric activity [59, 60]. Polyvinylidene fluoride is among the few semicrystalline polymers that exhibit thermodynamic compatibility with other polymers, in particular with acrylic and methacrylic resins [61]. Commercial grades of PVDF are copolymers of vinylidene fluoride (VDF) with small amounts (typically <6%) of other fluorinated monomers (e.g., HFP, CTFE, and TFE). These exhibit different properties than the homopolymer.

Poly(chlorotrifluoroethylene) (PCTFE) exhibits a reduced tendency to crystallize due to the inclusion of the relatively large chlorine atom into the monomeric unit. Commercial grades include a homopolymer which is used mainly for special applications, and copolymers with small amounts (<5%) of VDF.

- Poly(chlorotrifluoroethylene) exhibits excellent mechanical properties, chemical resistance, and resistance to ionizing radiation, as well as outstanding optical properties (clarity). The weak point of PCTFE is its low thermal stability in the molten state; this requires great care in processing.

- The copolymer of ethylene and chlorotrifluoroethylene (ECTFE) consists of an overall CTFE to ethylene ratio 1:1 and contains PE and PCTFE blocks of <10 mol% each. The ECTFE resins are tough, moderately stiff materials, having a good chemical resistance, outstanding weathering resistance, creep resistance, and resistance to ionizing radiation.
- Terpolymers of TFE, HFP, and VDF (THV Fluoroplastics) possess the unique combination of relatively low processing temperatures, good adhesion to itself and other substrates, excellent clarity and high flexibility. They also exhibit a very good chemical resistance, good resistance to weathering, low friction, and low flammability.
- Terpolymer of HFP, TFE, and ethylene (THE Fluoroplastic) exhibits an exceptional balance of tensile strength, good electrical insulating properties, good light transmission, and a very good dimensional stability [62].
- Poly(vinyl fluoride) (PVF) exhibits excellent resistance to weathering, mechanical properties, chemical resistance, excellent hydrolytic stability, high dielectric strength, and a high dielectric constant [63].

2.12.2 Elastomers

As pointed out earlier, elastomers (often called rubbers, although the term "rubber" is more fitting to the product made from elastomers) are unique materials in that they exhibit a fair amount of elasticity. Besides the elastic recovery, the majority of elastomers possess other useful properties, including low permeability to air, several gases, to water, and steam; good electrical insulation; good thermal insulation; good mechanical properties; and a capability to adhere to a variety of fibers, metals, and plastics.

The majority of elastomers can be compounded, that is, mixed with fillers, plasticizers, and other ingredients to produce materials with improved or specific properties; such compounds (also referred to as formulations), can be designed to meet a wide variety of service conditions.

The base of every such compound is an elastomer (single or a blend of two or more), but other ingredients (e.g., curing agent, activators, and accelerators) are needed to cross-link the elastomer. This cross-linking, also referred to as *vulcanization* or *curing* is responsible for the transition of the thermoplastic compound into thermoset, a much harder, and much more elastic material. Depending on specific requirements, the compound may contain antioxidants, fillers, reinforcing fillers, plasticizers and extenders, tackifiers, colors, blowing agents, internal lubricants, release agents, and so on. Although a great deal is known about the properties of each ingredient and about their interaction, formulation of rubber compounds is to a high degree still an art and requires adequate experience.

Typically, a compound can contain 6–12 ingredients, sometimes even more. The idea is to produce a material, which in the raw state can be processed in a desired way and also yields required physical and mechanical properties after vulcanization. In the following sections, the most common compounding ingredients will be discussed in some detail.

- Curing agents are chemicals that are necessary for vulcanization, which is chemical cross-linking of the polymeric molecules. This causes major irreversible changes in the properties of the elastomeric compound: It becomes stronger, harder, and more elastic. The most common curing agent is sulfur, which can be added in powder form, as a paste, or a masterbatch into the rubber compound. Organic peroxides that work with most general purpose rubber and produce vulcanizates with somewhat different properties are an alternative to sulfur. There are other substances that are used only for specific elastomers and/or specific applications.
- Activators activate the process of vulcanization. The most common activator for general-purpose rubber is zinc oxide, which requires an addition of a fatty acid (e.g., stearic acid).
- Accelerators are chemicals accelerating the curing process. There are many types and their choice is dictated mainly by the manufacturing process and to a lesser degree by the desired physical and mechanical properties.
- Antioxidants protect most of the general-purpose elastomers from damage by oxygen, especially in service at elevated temperatures. There is a large variety of antioxidants and their selection is dictated by the elastomer used and the service conditions.
- Antiozonants protect *diene* elastomers, such as natural rubber, styrene–butadiene rubber, polybutadiene, polyisoprene rubber, against ozone attack and destruction. These substances form a thin layer on the surface of the vulcanizate, which prevents the ozone from reacting with the elastomer. Frequently, an addition of a special wax enhances the effectiveness of the antiozonant. Some waxes alone act as antiozonants for low concentrations of ozone or in cases where the product is in a static application.

- Fillers are used mainly for reducing cost, but also improve processing and modify properties of the final product. Fillers used in rubber are classified as reinforcing and nonreinforcing (inactive). Examples of inactive fillers are clay, whiting, and precipitated calcium carbonate. Reinforcing fillers increase tensile strength, modulus, and hardness much more than the nonreinforcing fillers. Examples of reinforcing fillers are silica fillers.
- Carbon blacks are reinforcing ingredients for elastomers. There are many grades modifying properties and processing of elastomeric compounds. Large amounts of carbon blacks are used in tires, conveyor belting, hose, footwear, and technical molded parts.
- Plasticizers are added to a rubber compound to soften it and often to improve processing. They also reduce hardness, modulus, and tensile strength of a vulcanizate. Most common plasticizers for general-purpose elastomers are oils. For specialty elastomers, other liquid substances (esters, e.g., dibutyl phtalate) are used.
- Colorants include pigments, which are of mineral origin and organic dyes, specifically developed for the use in rubber. Colorants almost always need an addition of titanium dioxide to achieve good color.
- Other additives include processing aids for the improvement of processing, tackifiers, release agents, blowing agents, and so on. Reclaim (special processed rubber) is used for improved extrusion and calendering and for additional cost reduction. Ground rubber scrap is often added to inexpensive compounds to further reduce their cost.

This remarkable combination of useful properties makes them suitable for a great number of engineering applications (seals, shock absorbers, electrical and thermal insulation, elastic functional parts, tapes, profiles, etc.). Other applications are tires, mats, shoe soles, belts, toys, balloons, medical devices, household products, sponge rubber, raincoats, hose, and a multitude of products essential to modern civilization. The vast majority of elastomers, both unfilled and compounded, are processed by vulcanization (curing), which is essentially a cross-linking reaction.

The conventional (thermoset) elastomers are classified as general-purpose grades and special-purpose grades that exhibit such properties as oil-resistance, ozone resistance, heat resistance, chemical resistance, and low permeability. There are two types of rubbers, natural and synthetic. The American Society for Testing and Materials (ASTM) classification of elastomers is found in Table 2.12.

TABLE 2.12 Abbreviations of Common Commercial Elastomers (Rubbers)[a]

Abbreviation	Name of the Rubber
ABR	Acrylate–butadiene rubber
AU	Polyester polyurethane
BIIR	Brominated isobutene–isoprene (bromobutyl) rubber
BR	Polybutadiene rubber
CIIR	Chlorinated isobutene–isoprene (chlorobutyl) rubber
CO	Epichlorohydrin rubber (homopolymer)
CPE	Chlorinated polyethylene
CR	Polychloroprene rubber (neoprene)
CSM	Chlorosulfonated polyethylene (Hypalon)
EAM	Ethylene–vinyl acetate copolymer (EVA)
ECO	Copolymer of ethylene oxide and epichlorohydrin
EPDM	Terpolymer of ethylene, propylene, and a diene (cure site monomer)
EPM	Copolymer of ethylene and propylene
EU	Polyether, urethane
FFKM	Perfluoro-rubber (with a CSM, cure site monomer)
FKM	Fluorocarbon rubber (also referred to as FPM)
FVMQ	Fluorosilicone rubber
FZ	Fluorinated polyphosphazene rubber
HNBR	Hydrogenated NBR (hydrogenated nitrile rubber)
IIR	Isobutene–isoprene rubber (butyl rubber)
IM	Polyisobutylene
IR	Polyisoprene rubber (synthetic NR)
MQ	Silicone rubber with methyl substituent groups on the polymer chain
NBR	Acrylonitrile–butadiene rubber (nitrile rubber)
NR	Natural rubber (polyisoprene)
PMQ	Silicone rubber with both methyl and phenyl substituent groups
PNR	Polynorbornene rubber
PZ	Polyphosphazene rubber
SBR	Styrene–butadiene rubber
T	Polysulfide rubber (Thiokol)
VMQ	Silicone rubber with both vinyl and methyl substituent groups
XLPE	Cross-linked polyethylene
XNBR	Carboxylic–acrylonitrile–butadiene rubber

[a] Based on ASTM D1418: Standard Practice for Rubber and Latices–Nomenclature.

2.12.2.1 Natural Rubber Natural rubber (NR) is produced from latex of the rubber tree (*Hevea brasiliensis*) grown in the tropics where an equatorial monsoon climate prevails. This comprises a large region between 10°S and 8°N latitude in most of the Indonesian archipelago, Malaysia, southern part of Sri Lanka, and several regions in southern India [64]. The latex is tapped and collected from individual trees (see Fig. 2.22) and is processed into either dry rubber or latex. Dry rubber is obtained from coagulated latex by squeezing between steel rolls and then

FIGURE 2.22 Tapping of latex from a rubber tree (Photo Courtesy of J.G. Drobny).

processed by drying. Drying is done by smoking or in hot air. The finished raw rubber is Ribbed Smoked Sheets (RSS), Crepe (air dried), and Technically Specified Rubber (TSR). Currently, most of the dry rubber is TSR, which is supplied in blocks; RSS and Crepe are produced only in small- or medium-size rubber estates. The block rubber consists of rubber granules pressed together [65].

Latex is used for the production of dipped goods, carpet backing, foam, thread, and adhesives. Since the latex obtained from trees has 30–40% solids, it is concentrated to ~60% solids [66]. The most widely used method is centrifuging, an alternative is creaming that is used much less.

Chemically, natural rubber is *cis*-1,4-polyisoprene with a molecular weight ranging from 10^4 to 10^7 and with the polydispersity (M_w/M_n) ranging from 2.5 to 10. The commercial NR contains ~93–94% rubber hydrocarbon; the remainder are lipids, proteins, and other non-rubber materials [67]. Natural rubber is essentially a general-purpose elastomer with some unique properties (e.g., high resilience, tear resistance, low permanent deformation very good mechanical properties, excellent building tack, and good low-temperature properties). It can be easily dissolved in usual aliphatic hydrocarbon and chlorinated solvents, as well as in benzene and toluene. It is softened by milling on rolls (so-called mastication); prolonged milling can reduce it to a very low viscosity. Compounds from NR often contain fillers, reinforcing fillers, plasticizers (hydrocarbon oils) antioxidants, and antiozonants and curatives (vulcanizing additives including activators, sulfur, or sulfur donors and accelerators). Some compounds are cured by means of organic peroxides. Certain products are made from unfilled NR containing only curatives. One unique attribute of natural rubber is its strong tendency to crystallize when being stretched.

2.12.2.2 Synthetic Rubbers Synthetic rubbers (elastomers) are produced by polymerization or copolymerization of monomers that produce long flexible chains.

2.12.2.2.1 General-Purpose Rubbers

Polyisoprene Rubber

Polyisoprene rubber (IR) is synthetic natural rubber made by polymerization of 1, 4-isoprene. The commercial grades available are with 96–97% and 92–93% *cis*-1.4 polyisoprene, the former having better physical properties [68]. General properties are similar to NR with lower strength and lesser building tack. Polyisoprene rubber is mostly used in blends with other diene elastomers.

Styrene–Butadiene Rubber

Styrene–butadiene rubber (SBR) is produced by copolymerization of styrene and butadiene. Commercial products contain 23% styrene, although other grades contain higher levels than that when rain foresting pillars are added to SBR compounds the rubber articles made from them exhibit similar mechanical properties to those of natural rubber, but with greater abrasion resistance and improved heat and aging resistance than those from NR. The chemical properties of cured SBR compounds are similar to those of natural rubber.

Polybutadiene Rubber

Polybutadiene rubber (BR) is a homopolymer of butadiene. The most common high cis grades contain 97–98% *cis*-1,4-butadiene units. Vulcanizates of 100% BR have high abrasion resistance, the highest resilience of all known elastomers and a very low T_g. Polybutadiene rubber is used mainly in blends with NR for tires and special mechanical goods requiring improved resilience, low-temperature flexibility, and reduced heat build-up upon flexing.

Ethylene–Propylene Rubbers

The EPM rubbers are fully saturated copolymers of ethylene and propylene, while EPDM rubbers are copolymers of ethylene with polypropylene with a small amount of diene monomer (e.g., 5-ethylidene norbornene, ENB) and have a greater reactivity for sulfur vulcanization. The ethylene content can vary from 75 to 45% in both grades. The EPM is completely saturated and can be cross-linked only by organic peroxides or by ionizing radiation (e.g., electron beam process). The EPDM can be vulcanized by sulfur, organic peroxides, or ionizing radiation. In general, both EPM- and EPDM rubber articles show a good heat, ageing and chemical resistance, high elasticity, good low temperature, and electrical insulating properties. The typical operating range for continuous service of grades with low ethylene content or vulcanizates cured with peroxides is from −40 to +150°C (−40 to 302°F).

Butyl, Chlorobutyl, and Bromobutyl Rubbers

Butyl (isobutene–isoprene) rubber (IIR) is a copolymer containing mainly isobutene (isobutylene) units with a small percentage of isoprene units. Thus the raw elastomer is a highly unsaturated polymer with just enough double bonds as cure sites for vulcanization. Due to their low level of unsaturation, raw butyl rubbers are not compatible with highly unsaturated rubbers (e.g., NR, IR, SBR, BR, and NBR). Moreover, they cannot be cross-linked with organic peroxides since the peroxides degrade the polymer. To enable the covulcanization with diene rubbers and to increase the cure rate, butyl rubbers are modified with halogens (chlorine, bromine) to obtain chlorobutyl rubber (CIIR) and bromobutyl rubber (BIIR). The process involves reaction of elemental chlorine or bromine, respectively, with IIR dissolved in a light aliphatic hydrocarbon (e.g., hexane) [69]. The prime importance of butyl rubbers is the low gas permeability and a good resistance to heat, ageing, weather, and chemicals. The main disadvantages are low grease and oil resistance, and low curing speed. Also, sometimes low elasticity, low compression- and tension-set can only be regarded as being poor. The IIR is suitable for continuous use at a temperature range from −40 to +120°C (−40 to 248°F).

Polychloroprene Rubber

Polychloroprene (CR) rubbers are homopolymers of chloroprene (2-chloro 1,3-butadiene), with a predominantly trans-1,4 configuration. Because of this high degree of stereoregularity, CR can crystallize on stretching like NR. Certain grades are highly crystalline even at ambient temperature. These are used for specialty adhesives. In general, vulcanizates of chloroprene rubber exhibit outstanding flame resistance, good resistance to ozone, weathering, chemicals and good ageing resistance, average oil resistance, good mechanical properties, and acceptable elastic behavior even at low temperatures. The CR rubber is often referred to as "neoprene", although Neoprene is the brand name of polychloroprene rubber, a product from the DuPont Company.

Articles made from chloroprene rubber exhibit in general outstanding flame-resistance, good resistance to ozone, weathering, chemicals and aging resistance, average oil resistance, good mechanical properties, and acceptable elastic behavior even at low temperatures.

Acrylonitrile–Butadiene Rubber

Acrylonitrile–butadiene rubber (NBR) or simply *nitrile rubber* is a series of copolymers of butadiene and acrylonitrile with different acrylonitrile (ACN) content between 18 and 48%. There are essentially five grades available on the market: with very low ACN (18–20%); low ACN (28–29%); medium ACN (33–34%); high ACN (38–39%); very high ACN (typically 45–48%). The amount of acrylonitrile has a profound effect on the properties of the vulcanizate. The main consequences of the increasing proportion of ACN are increase of swelling resistance in petroleum-based oils and hydrocarbon fuels, decrease of low-temperature flexibility, decrease of rebound resilience, increase of compression set, decrease of gas permeability, and increase of ozone resilience.

Vulcanizates of NBR compounds exhibit poor electrical insulation properties, but are still adequate for sheathing when oil resistance is required. Continuous service temperature range typically from −40 to +100°C (−40 to 212°F) and for intermittent periods they can be as high as 120°C (257°F).

Carboxylated Nitrile Rubber

Carboxylated nitrile rubber (XNBR) is a terpolymer of butadiene, acrylonitrile, and an acidic monomer with a carboxylic group (e.g., acrylic or methacrylic acid, acrylonitrile, and acidic monomer unit) are distributed randomly through the polymer chain. Commercial grades of XNBR are available in grades that differ in ACN and acid content. The acid content varies between 2 and 10%.

The carboxylic groups provide additional cure sites for zinc oxide or zinc peroxide. The XNBR vulcanizates exhibit the general properties of the regular NBR elastomers, but they have a much higher modulus, tensile strength, abrasion resistance, and tear resistance.

2.12.2.2.2 High-Performance Elastomers High-performance elastomers (HPE) are generally distinguished from other elastomers by exceptional chemical, thermal, and mechanical properties. This family represents only a small part of the total demand for elastomers where rubber was traditionally used [70].

- Acrylic rubber (ACM)
- Chlorinated polyethylene (CPE)
- Chlorosufonated polyethylene (CSM)
- Epichlorohydrin (ECO)
- Ethylene/acrylic elastomers (AEM)
- Ethylene /vinyl acetate elastomers (EAM)
- Fluorocarbon elastomers and perfluoroelastomers (FKM/FPM and FFKM)
- Hydrogenated nitrile rubber (HNBR)
- Polyurethane rubber (AU/EU)
- Silicone and fluorosilicone elastomers (VMQ/MQ and FVMQ)

According to ASTM D2000 Heat and Oil Resistance Designation for selected HPE the best performing elastomer in heat resistance is FFKM and ECO is the lowest performer. In oil resistance, FFKM is the highest performer and silicone elastomer the lowest.

Chlorinated Polyethylene Rubber

Chlorinated polyethylene rubber (CM) is produced by reacting PE with chlorine. There are several grades with a chlorine content ranging from ~25–42% by weight. As the chorine content increases, the oil and flame resistance increases, but the low-temperature flexibility and heat resistance are impaired [71]. The main characteristics of vulcanizates from CM [72]:

- High resistance to ozone, weather, and to a great number of corrosive and oxidizing chemicals.
- Very good color stability.
- Good resistance to dry heat (continuous service to ~125°C or 257°F and for peaks up to 150°C or 302°F).

- Fair resistance to mineral oils and aliphatic solvents.
- Low flammability.
- Low compression set at high temperature (if cured by peroxide).
- Rather poor low-temperature flexibility.

Applications: wire and cable sheathing for service in environments including oil, chemical, ozone, and flame; hoses, membranes roll covers, and tank linings

Acrylic Rubbers

Polyacrylate rubbers (ACM) are saturated copolymers of monomeric acrylic esters and monomers with reactive cure sites. The cure site monomers include alkoxyalkyl acrylates, 2-chloroethyl vinyl ethers, vinyl chloroacetates, and allylglycidyl ether.

The saturated backbone provides excellent heat and ozone resistance, the presence of polar side groups give a good oil resistance. Acrylic rubber (ACM) vulcanizates require postcure in an oven with circulating air (typically 6 h at 175°C or 347°F) to complete the cross-linking process and to develop optimum physical properties, especially compression set [73].

Vulcanizates have the following properties: excellent heat resistance (continuous service at 150°C or 302°F, limited exposure up to 175°C or 347°F); high resistance to oxygen and ozone; good resistance to hot industrial oils including those containing sulfur; and relatively poor resistance to water. These rubbers are mainly used in applications where combination of resistance to heat, oils, and oil additives is required.

Chlorosulfonated–Polyethylene Rubber

Chlorosulfonated–PE rubbers (CSM) are produced by reacting PE with a mixture of chlorine and sulfur dioxide gases. The range of chlorine in the elastomer is 24–43% by weight and that of sulfur 1.0–1.5% by weight. This rubber is vulcanized by metal oxides (e.g., magnesium oxide or peroxides).

The properties of products made out of CSM are mainly influenced via the content of chlorine in the CSM and on the compound ingredients. In general, CSM rubber exhibits good mechanical properties (even without reinforcing fillers), outstanding resistance against light, ozone, weathering and oxidants, good flame and heat resistance, average mineral oil resistance, and poor low-temperature characteristics.

Ethylene–Vinyl Acetate Rubber

Ethylene–vinyl acetate rubbers (EAM) are copolymers of ethylene and vinyl acetate with fully saturated backbones. Commercially available EAM rubber grades have vinyl acetate contents ranging from 40 to 70% by weight. The fully saturated backbone provides a very good heat and ozone resistance, the presence of polar vinyl acetate groups markedly improve oil resistance. With the increase of vinyl acetate monomeric units the oil resistance improves, but the electrical insulation properties are impaired. The properties of only peroxide curable EAM rubber are mainly influenced by the concentration of vinyl acetate in the base polymer. The EAM vulcanizates have a good dry-heat resistance (continuous service up to ~125°C or 257°F and for a limited time up to 150°C or 302°F), but rather poor low-temperature flexibility.

Ethylene–Acrylic Rubbers

Ethylene–acrylic rubbers (AEM) elastomers are terpolymers of ethylene and methyl acrylate with a cure site monomer that contains a carboxylic acid group. The presence of the carboxylic acid in the polymer permits the cure with certain bifunctional compounds (e.g., diamines) [74]. A postcure is required for optimum properties (typically 6 h at 175°C or 347°F). Ethylene–acrylic vulcanizates exhibit a very good heat resistance with continuous service up to 150°C (302°F) and 175°C (347°F). Other valuable properties of AEM rubbers are excellent ozone and weather resistance, good low-temperature flexibility, and moderate resistance to mineral oils.

Epichlorohydrin Rubbers

Epichlorohydrin rubbers (CO, ECO, GECO) are saturated polymers of aliphatic polyethers with chloromethyl side chains. Three types of epichlorohydrin rubbers are currently available [75]:

1. Homopolymer of epichlorohydrin (designated as **CO**).
2. Copolymer of epichlorohydrin and ethylene oxide designated as **ECO**.
3. Terpolymer of epichlorohydrin, ethylene oxide, and an unsaturated monomer (allyl glycidyl ether) designated as **GECO**. The unsaturated monomer provides double bonds in the pendent side chains to enable sulfur cure.

The saturation of the backbone provides a high resistance to oxygen and ozone, while the oxygen and chlorine atoms impart polarity to the polymer chains resulting in resistance to swelling in nonpolar and hydrocarbon liquids. The fully saturated grades (CO and ECO) can be cured with diamines, or ethylene thiourea, which react with the chloromethyl groups. The terpolymer can be cured with conventional sulfur, peroxide systems, or ethylene thiourea [78]. The vulcanizates of epichlorohydrin exhibit an outstanding low air permeability and good resistance to swelling in mineral oils and aliphatic hydrocarbons. Their electrical insulation and properties and radiation resistance are generally poor. The temperature range for continuous service depends on the grade. The CO grades can be used within the range from −15 to +130°C (+5 to 266°F) and ECO grades from −30 to +120°C (−22 to 248°F). The major drawbacks are their corrosive effects on metals and their tendency to revert (soften) after a long-term exposure to high temperatures.

Hydrogenated Acrylonitrile–Butadiene Elastomers

Hydrogenated acrylonitrile elastomers (HNBR) are made by selective and controlled hydrogenation of NBR, Depending on the degree of hydrogenation (partially or completely hydrogenated basic polymers), HNBR rubber articles show oil and fuel resistance similar to NBR, but an improved durability against chemicals, heat, weathering, and ozone. Compared to NBR, parts made of HNBR also provide improved abrasion resistance [79]. A postcure in an oven with circulating hot air (usually 3 h at 160°C or 320°F) improves the physical properties of the vulcanizate, particularly compression set. The HNBR vulcanizates are suitable for continuous service in the temperature range from −25 to +150°C (−13 to +302°F) and for short time up to 170°C (358°F).

Polyurethane Rubbers

Polyurethane rubbers (AU/EU) are based on the reaction of diisocyanates with difunctional reactants (e.g., long-chain glycols). In addition short-chain glycols or diamines are used as chain extenders or cross-linkers. The polyols have a large influence on the properties of the final product. Both polyester and polyether polyols are used and the resulting polyurethanes are polyester urethanes (AU) and polyether urethanes (EU). Essentially, there are three ways of processing polyurethane rubbers:

1. Millable rubbers that are processed and subsequently cured using conventional mixing and processing methods of the rubber industry.

2. Liquid rubbers, used for casting, liquid injection molding (LIM) and reaction injection molding (RIM).
3. Thermoplastic polyurethanes that are processed using methods typical for plastics (see Section 2.14.2).

Polyurethane elastomers have outstanding tensile strength, tear, and abrasion resistance, better than all other elastomers. They also exhibit an excellent resistance to degradation by oxygen and ozone because of their unsaturated chains, good resistance to mineral oils and greases, and to effects or high-energy ionizing radiation [80]. Polyether polyurethanes are more resistant to hydrolysis than the polyester-based polymers. Service temperatures range from −35 to 100°C (−31 to +212°F) in dry conditions. Polyurethane as a group are not recommended to be used in contact with glycol-based break fluids, aromatic hydrocarbons, chlorinated hydrocarbons, and ketones.

2.12.2.2.3 Elastomers for Extreme Conditions Elastomers for extreme conditions, that is, for service temperatures >175°C (347°F) and/or aggressive environments include the following [70]: fluorocarbon elastomers, perfluoroelastomers, silicone elastomers, and fluorosilicone elastomers.

The above elastomers have a few things in common: either they have hydrocarbon backbone protected by large atoms (e.g., fluorine or chlorine) or have inorganic atoms in their backbone. The carbon–fluorine bond is very strong and also accounts for chemical stability of fluorocarbons. A comparison of different carbon bond energies is in Table 2.8 (see Section 2.12.1.1.2)

Fluorocarbon Elastomers

The most widely used members of this group are fluorocarbon elastomers (FKM/FPM), which are copolymers of halogenated (mostly fluorinated) monomers (e.g., vinylidene fluoride (VDF), polytetrafluoroethylene (TFE), hexafluoropropane (HFP), perfluoromethylvinyl ether (PMVE), or perfluoropropylvinyl ether (PPVE) and chlorotrifluoroethylene (CTFE). The VDF based fluorocarbon elastomers have been the most commercially successful of fluorocarbon elastomers. Chemical formulas of monomers, used for FKM are in Table 2.13. The first commercially available fluorocarbon elastomer was Kel F, a copolymer of VDF and CTFE, developed in the late 1950s. Since then literally dozens of grades have been developed and marketed. An example of the most

TABLE 2.13 Typical Compositions of Fluorocarbon Elastomers

Designation	Composition
FKM (FPM)	VDF/HFP
	VDF/CTFE
	VDF/HFP/TFE
	VDF/HFP/TFE/CSM[a]
	TFE/P
FFKM (Perfluoroelastomers)	TFE/PPVE/CSM[a]
	TFE/PMVE/CSM[a]

[a] Cure site monomer (e.g., a monomer with nitrile, iodine, or bromine group).

common commercial FKM grades and some of their properties are in Table 2.14. Currently, there are five major manufacturers of FKM.

Another fluorocarbon elastomer is tetrafluoro–propylene rubber (TFE/P) produced by copolymerization of TFE with propylene. The product contains only 53.5% fluorine (compared with typically 65% in FKM elastomers). Vulcanizates from TFE/P have the same thermal stability, but better electrical insulation properties and a different chemical resistance profile [79]. They are resistant to a wide variety of chemicals, including amines, sour oil and gas (containing H_2S), hydraulic and break fluids, petroleum oils and greases, acids, alkali and oxidizing chemicals, and high-energy ionizing radiation. They are not resistant to aromatic hydrocarbons, chlorinated hydrocarbons, ketones, acetic acid and its esters, most organic refrigerants. In general, TFE/P vulcanizates show a somewhat higher compression set than FKM and poor low-temperature flexibility [80].

Perfluoroelastomers

Perfluoroelastomers (FFKM) represent a special subgroup of fluorocarbon elastomers. They are essentially rubbery derivatives of PTFE and exhibit exceptional properties, such as unequaled chemical inertness and thermal stability. Commercial perfluoroelastomers are normally terpolymers of tetrafluoroethylene, perfluorovinyl ether (PMVE or PPVE), and a cure-site monomer (CSM). The structure and content of the perfluorovinyl ether determines the T_g of the resulting polymer. The cure-site monomer

TABLE 2.14 Main Commercial Grades of Fluorocarbon Elastomers

Monomer	HFP	PMVE	CTFE	P	HPFP
VDF	Dai-el® 801 (Daikin) Fluorel® (Dyneon) Tecnoflon® (Solvay Solexis) SKF-26 (KCKK)[a] Elaftor 2000 Series (KCKK) Viton® A (DuPont)		KelF® (Dyneon) SKF-32 (KCKK)[a] Elaftor 2000 Series (KCKK)[a]		Tecnoflon® SL (Solvay Solexis)
TFE		Kalrez® (DuPont)		Aflas® (Asahi Glass) Extreme® (DuPont)	
VDF+TFE	Dai-el® 901 (Daikin) Fluorel® (Dyneon) Tecnoflon® (Solvay Solexis) Viton® B (DuPont)	Viton® GLT (DuPont)			
VDF+TFE+CSM	Viton® GH (DuPont) Elaftor ® 7000 Series (KCKK)[a]				Tecnoflon® T (Solvay Solexis)

[a] Kirovo-Chepetsk Chemical Plant, Russia available at (www.kckk.ru).

is necessary to be included in the perfluorinated carbon backbone to assure the ability to cross-link [79].

Perfluoroelastomers have engineering properties similar to FKM; with useful hardness range from 60 to 95 Durometer A. They have a wide service temperature window, rather skewed to high temperatures. Many perfluoroelastomers have an upper service temperature >300°C (572°F) and a lower service temperature in the −20 to −30°C (−4 to −22°F) range. They are considered the ultimate in performance as to thermal and chemical resistance, especially when the requirement is to handle mixtures of aggressive chemicals and/or elevated temperatures for extended periods of time [81].

One of the weaknesses of FFKM is their higher than normal compression set when compared to other elastomers attributable to the higher T_g and higher degree of crystallinity associated with their composition and structure. Another issue is, that they often exhibit a 50–100% higher coefficient of thermal expansion than FKM materials. This has to be taken into consideration in designs, where FFKM is used.

Silicone Elastomers

Silicones or organosiloxanes have unique properties because of their flexible inert backbone consisting of alternating silicon and oxygen atoms. The free rotation of the silicon and oxygen atoms results in an extremely high degree of flexibility under harsh environmental conditions. By varying the substituents on the silicon atom, the properties of the resulting polymer can be changed significantly. Thus, a methyl group is used for a general-purpose polymer, a phenyl improves resistance to extreme low and high temperatures, and vinyl provides network control, high hardness, and strength. Linear polymers with moderate and high molecular weight are elastomeric materials. The vinyl group is essential for the cross-linking reaction regardless of what curing system is used [82].

Generally, silicone elastomers are classified by their viscosity and the curing systems used

- HTV (high-temperature vulcanizing solid silicone elastomers).
- RTV1 and RTV2 (room temperature vulcanizing 1 and 2 component systems).
- LSR (liquid silicone rubber, two component, high-temperature curing).

High-temperature vulcanizing, also referred to as millable silicone elastomers, have viscosity in the range 40,000–200,000 Pa.s (or Mooney 20–100). They can be cross-linked via functional organic groups attached to the silicone atoms in the backbone. Catalysts for their cross-linking are either organic peroxides or transition metals. Because of their high viscosity and consequently high green strength, they can be processed on standard rubber processing equipment by compression molding, injection molding, calendering, and extrusion.

Liquid silicone rubber, liquid elastomers, or "pumpable" silicones are an alternative to pourable and millable polymers. Their viscosity values are in the range of 100–5000 Pa.s. These are most frequently processed by liquid injection molding (LIM). The LIM process employs an addition cure reaction, which allows fast curing throughout the entire mass of the material. The most commonly used is a platinum catalyst with cure times at elevated temperatures (e.g., 177°C or 350°F) being typically <1 min.

Room temperature vulcanizing systems have viscosity values in the range of 40–100 Pa s and are used mostly as sealants. Room temperature vulcanizing systems with viscosities from 10 to 50 Pa. s are used as potting compounds, cured-in place gaskets, encapsulants of electronic components, and in prototype mold making. The RTV systems can be cross-linked by addition or condensation reactions.

Fluorosilicone Elastomers

The properties of fluorosilicones span the world of silicones and fluorocarbon elastomers, since they contain both a C—F and Si—O bonds with hydrocarbon entities between them. Thus the repeating structure may be generally written as $[R_f\ X\ (CH_2)_n]_x(CH_3)_ySiO_z$, where R_f is a fluorocarbon group, which can be linear or branched, aliphatic, or aromatic, but in commercial products it is either CF_3 or $CF_3(CF_2)_3$. The X can be oxygen or sulfur, for example, chosen to link the fluorocarbon group to the hydrocarbon "spacer" and is not present in commercial products. The length of the hydrocarbon spacer, n, is optimally 2.

Commercially available fluorosilicones are based on polymethyltrifluoropropylsiloxane (PMTFPS) or more accurately poly[methyl (3,3,3-trifluoropropyl siloxane]. In some cases, PMTFPS is copolymerized with polydimethyl siloxane (PDMS) for cost/benefit balance [83]. For the most part, fluorosilicones are compounded using similar ingredients to those used in silicone compounds and are supplied precompounded as a base and are not sold in pure form. Then, they are blended and compounded for specific applications. They are frequently blended with VMQ for less demanding applications and lower cost.

Curing systems for FVMQ are dialkyl peroxides and platinum curing agents. Peroxide cures are slower, but platinum catalysts are much faster offering more flexibility to adjust scorch time. Fluorosilicones can be compression molded, injection molded, calendered, and extruded. Their continuous operation temperature ranges from −60 to +170°C (−76 to 338°F) with short term use up to 230°C (446°F) [84].

2.12.3 Synthetic Fibers

Synthetic fibers are chemically identical to plastics, but special processing methods, such as spinning and often stretching, are used. Examples are rayon, polyamide (Nylon), polyester (PET), and aramid (Kevlar, Nomex). Their applications depend on their specific properties.

2.12.4 Composites

Composites or *reinforced plastics* covered in this section consist essentially of a polymeric matrix combined with a reinforcing agent or agents, which most frequently are fibers. The reinforcement improves strength and stiffness properties, compared to neat resin. Almost every type of thermoplastic or thermosetting resin can be reinforced. *Hybrid composites* combine more than one type of reinforcement; an example is the combination of carbon and glass fibers. The *matrix* of polymeric fiber reinforced composites is made to be either a thermosetting or a thermoplastic resin.

Examples of thermosetting resins for a matrix of composites are unsaturated polyesters, epoxies, phenolics, polyurethanes, thermoset acrylics, melamines, bismaleimides, thermoset polyimides, and silicones.

Examples of thermoplastics used as matrix materials are polyolefins, styrenic polymers and blends, thermoplastic elastomers, fluoropolymers, LCPs, polyphenylenes, ketones, sulfones, acetals, ketones, sulfones, acetals, acrylics, polycarbonates, PVC, and polyacrylates. Details regarding the properties and preparation of composite materials are found in Section 2.13.3.

2.12.5 Other Polymeric Materials

Other forms of polymeric materials are dispersions, latexes, emulsions, and solutions. Some of them, such as latexes and dispersions, are commercial forms of specific polymers. They are used in applications, such as adhesives, coatings, lacquers, and processing aids. Some of them are used as coatings on plastics, paper, textiles, leather metals, and other substrates.

2.13 GENERAL TECHNOLOGICAL CHARACTERISTICS OF POLYMERS

In general, polymers are more sensitive to processing and environmental conditions than traditional engineering materials (steel and other metals, ceramics, glasses, etc.). In spite of that, they offer many advantages, such as a high strength/weight ratio, ease of processing, and chemical resistance.

Polymeric materials are generally perceived as materials that consist of relatively simple macromolecular organic compounds. These compounds can be either manufactured synthetically or transformed from natural products.

2.13.1 Thermoplastics and Thermosets

Thermoplastics consist of physically bound linear or branched macromolecules. They are solids at ambient temperatures and become deformable at elevated temperatures. The process of softening and hardening is fully reversible. Linear macromolecules may arrange themselves parallel to one another and crystalline regions (domains). Usually the crystalline domains exist next to the amorphous regions. Amorphous areas are not ordered. Macromolecules containing both crystalline and amorphous regions are known as semicrystalline (see Fig. 2.9).

Thermosets solidify as a result of chemical reactions between different components (e.g., resins, hardeners, as well as different accelerators and additives). A basic material in a thermoset is a monomer or an oligomer that consists of several monomeric units. Both monomers and oligomers are mostly in the liquid state and exhibit a relatively wide range of viscosities. Chemical cross-linking during processing of thermoset materials leads to an intertwined physical network that can be modified by the material temperature. Typically, cross-linked thermoset materials cannot be fully melted or dissolved. If exposed to solvents, they only swell.

2.13.2 Elastomers

Elastomers, both natural and synthetic, are mostly processed as thermoplastic materials to attain the desired shape and/or dimensions mainly by extrusion or calendering and are subsequently cross-linked. This cross-linking process is referred to as vulcanization or cure. The curing agents are most frequently sulfur, sulfur donors, inorganic oxides (e.g., zinc oxide), or organic peroxides. In most cases, the cure takes place at elevated temperatures (typically in the range from 140 to 200°C or 284 to 392°F) although some systems allow a cure at ambient temperatures. Some materials require additional treatment at elevated temperatures, so-called postcure. Certain elastomers can be cross-linked by ionizing radiation (electron beam or gamma radiations) at ambient temperatures. An elastomer cross-linked by sulfur is shown in Fig. 2.23. Additional details about vulcanization process are shown in Section 2.14.3.2.

Thermoplastic elastomers (TPEs) are thermoplastics that behave similarly to elastomers at ambient temperatures and can be shaped using established melt-processing methods, because of their molecular structure [85]. Most thermoplastic elastomers are essentially phase-separated systems. The only currently known exceptions are Alcryn® (registered trademark of Advanced Polymer Alloys), a single-phase

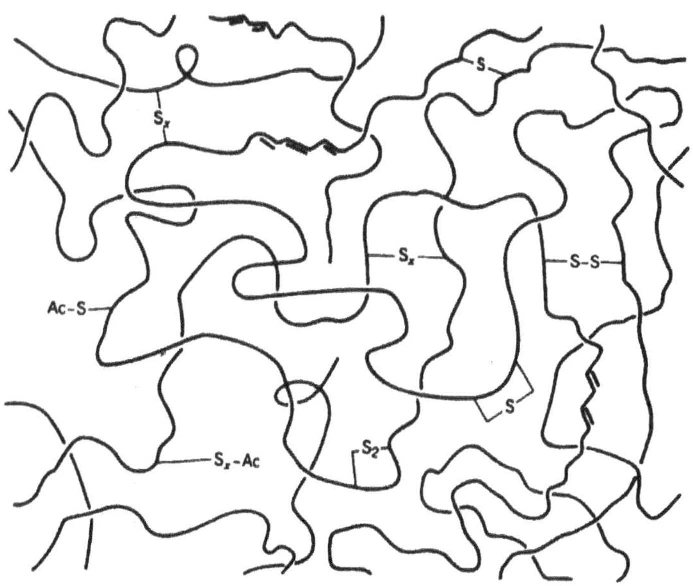

FIGURE 2.23 Cross-link types and chain modifications sulfur cure.

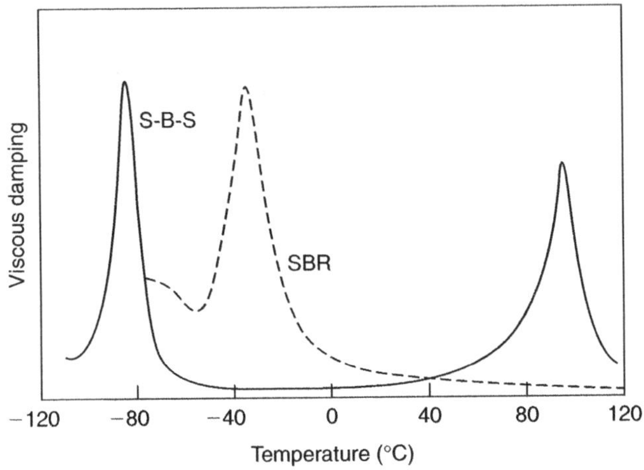

FIGURE 2.24 Glass transition temperatures of a SBS block copolymer and SBR (from viscous damping).

melt-processable rubber (MPR) and materials based on ionomers. Usually, one phase is hard and solid at ambient temperature, the other is an elastomer. Often, the phases are bonded chemically by block or graft polymerization. In other cases, a fine dispersion of the phases is apparently sufficient. The hard phase gives these thermoplastic elastomers their strength and represents the physical cross-links. Without it the elastomer phase would be free to flow under stress and the polymer would be practically unusable. On the other hand, the elastomer phase provides flexibility and elasticity to the system. When the hard phase is melted or dissolved in a solvent, the material can flow and be processed by usual respective processing methods. Upon cooling or evaporation of the solvent the hard phase solidifies and the material regains its strength and elasticity [85]. The individual polymers constituting the respective phases retain most of their characteristics so that each phase exhibits its specific T_g (see Fig. 2.24) or crystalline T_m. These two temperatures determine the points at which the particular elastomer goes through transitions in its physical properties. Examples are styrene–butadiene–styrene rubber (SBS, Kraton), thermoplastic polyurethane (TPU), thermoplastic vulcanizate (TPV), thermoplastic polyolefins (TPO), and copolyester elastomers (COPE). Each of the above classes includes a multitude of elastomeric materials with a variety of properties (hardness, stiffness, processing behavior, transparency, etc.). The morphologies of different types of TPEs are depicted in Figs. 2.25–2.29 [85].

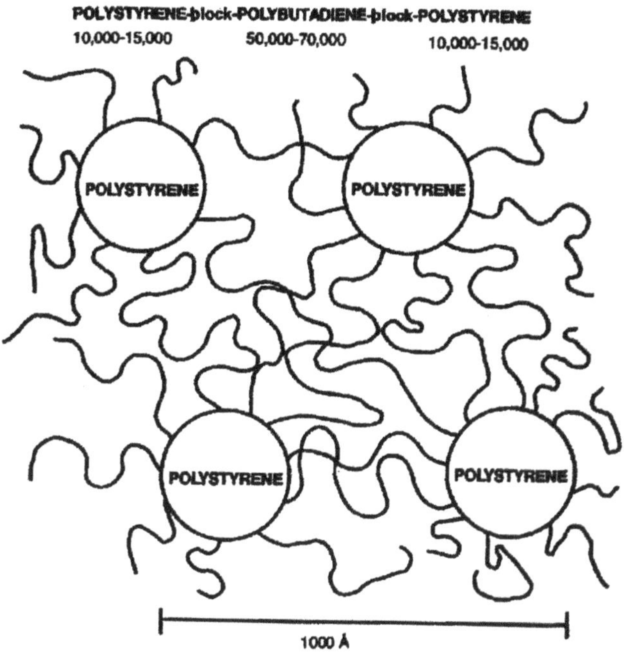

FIGURE 2.25 Morphology of a styrenic block copolymer. (Drobny, J.G., *Handbook of Thermoplastic Elastomers*, William Andrew Publishing, 2007, p. 162)

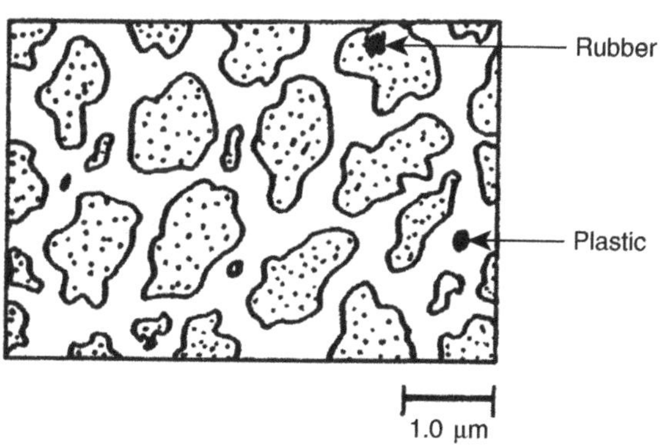

FIGURE 2.26 Morphology of a thermoplastic vulcanizate. (Drobny, J.G., *Handbook of Thermoplastic Elastomers*, William Andrew Publishing, 2007, p. 181)

2.13.3 Composites

The mechanical strength of the composite material depends on the amount, type(s), and arrangement of reinforcement. In general, the strength of the composite material increases with the increased amount of reinforcement. If fibers are used for reinforcement, their

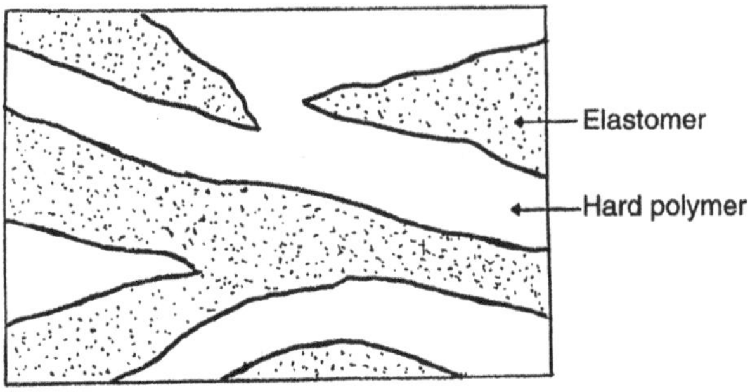

FIGURE 2.27 Morphology of TPO (hard polymer–elastomer blend).

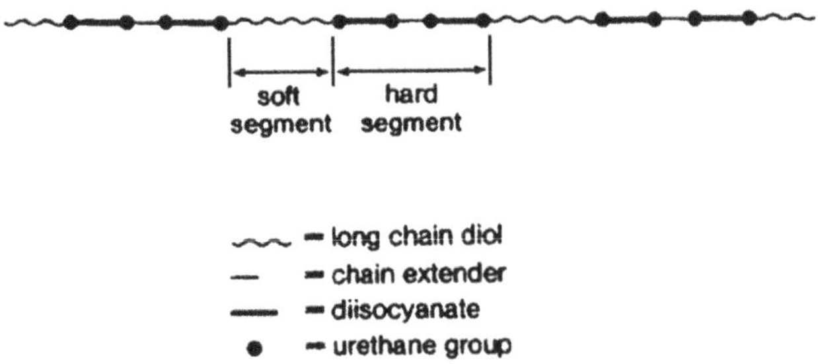

FIGURE 2.28 Schematic representation of TPU composed from diisocyanates, long-chain diol, and chain extender. (Drobny, J.G., *Handbook of Thermoplastic Elastomers*, William Andrew Publishing, 2007, p. 215)

FIGURE 2.29 Model of microstructure of chain conformation in the polyester-polyether block copolymer. (Drobny, J.G., *Handbook of Thermoplastic Elastomers*, William Andrew Publishing, 2007, p. 250)

arrangement is critical. The usual reinforcing fibers are placed in the following fashion to form a *lamina*:

1. *Unidirectional*: The greatest strength is in the direction of the fibers.
2. *Bidirectional*: Two arrangements of the fibers in two direction at certain angles to each other; strength is in each direction of fiber orientation.
3. *Multidirectional (isotropic)*: Strength is uniformly distributed with short (mostly chopped) fibers arranged in all directions or with continuous fibers arranged in various orientations.

The three base laminas are shown in Fig. 2.30.

The most common form in which fiber-reinforced composites are used in structural applications is called a *laminate*. It is obtained by stacking a number of thin layers of fibers and matrix (laminas, shown above) and consolidating them into the desired thickness (see Fig. 2.31). Fiber orientation, as well as the stacking sequence of various layers, can be controlled to generate a wide range of physical and mechanical properties for the composite laminate.

The most widely used reinforcements are glass and carbon fibers. They can be used as *continuous roving* that produces unidirectional reinforcement. Other possibilities are *chopped strands, milled fibers*, and *glass flakes*. In some cases, the fibers are twisted into yarns and the yarns are used in woven or nonwoven fabrics that produce two-directional or multidirectional reinforcement, respectively. Other reinforcement types are aramid fibers or fabrics, polyamide, thermoplastic polyester,

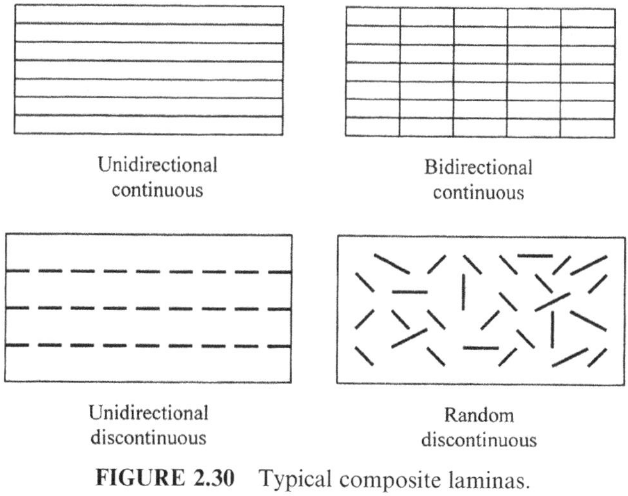

FIGURE 2.30 Typical composite laminas.

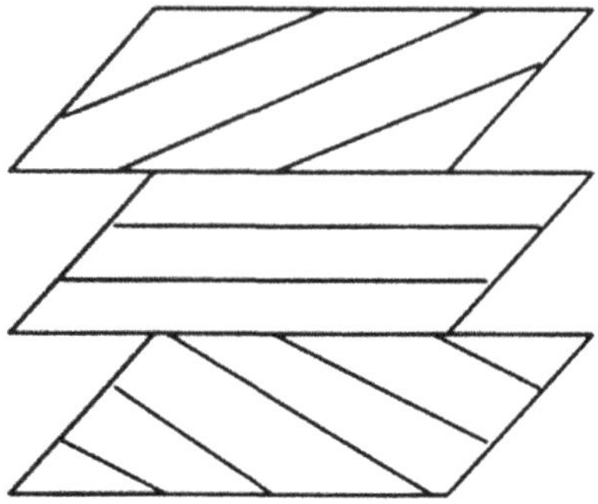

FIGURE 2.31 Example of a laminate.

ultrahigh molecular weight/polyethylene (UHMWPE). When two or more reinforced fibers or fabrics are used in a common polymeric matrix, the resulting composite is referred to as *hybrid*. For example, combining carbon and glass fiber reinforcement create a combination of high performance of carbon and low cost of glass. Reinforcing fillers include mica platelets, fibrous and finely divided wollastonite, and solid glass microspheres (beads).

2.14 PROCESSING OF POLYMERS

Some polymers are used as supplied, some need to be first mixed with a variety of additives. The mixing methods for thermoplastics, thermoset plastics, and elastomers differ in equipment used and the mixing procedures. As for processing there are also significant differences between thermoplastics, thermosets, and conventional elastomers (rubber). Thermoplastic elastomers are processing by melt-processing methods essentially identical to those for thermoplastics. The processes used for the manufacture of reinforced composite materials are somewhat different from the technologies used for the unreinforced products.

2.14.1 Mixing

In many cases, polymeric solids are melted and blended or mixed in the molten stage. Frequently, various additives (e.g., powders, waxes, resins, pigments, reinforcing agents, liquid or solid plasticizers, antidegradants, colorants, or other specialty chemicals) are added in the mixing process. Mixing in the molten stage requires specialized equipment, (e.g., batch mixers or continuous mixers) and consumes a large amount of energy.

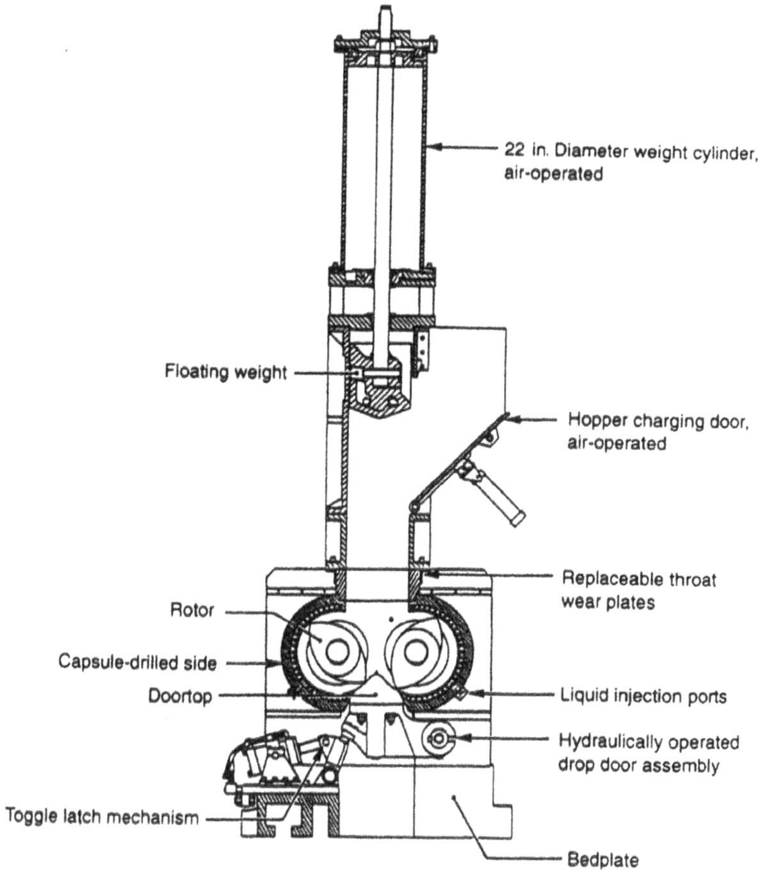

FIGURE 2.32 Schematic of an internal mixer. (Courtesy of Farrel Corporation)

The finished mix is often called a *compound* or *formulation* and can be in the form of pellets, sheets, slabs, or ropes, depending on the form required for the next step in the manufacturing process. Although there are many types of mixing machines, the most widely used are the internal mixers (see Fig. 2.32), continuous mixers (see Fig. 2.33) and certain types of extrusion machines. Conventional elastomers are mixed in similar equipment with mixing procedures being somewhat different.

2.14.2 Melt Processing

Polymer processing most frequently involves melting the polymeric mass and shaping it into a desirable product. Such methods are often referred to as *melt-processing methods*. Most thermoplastics are not subjected to cross-linking, which means that they can be remelted and shaped again.

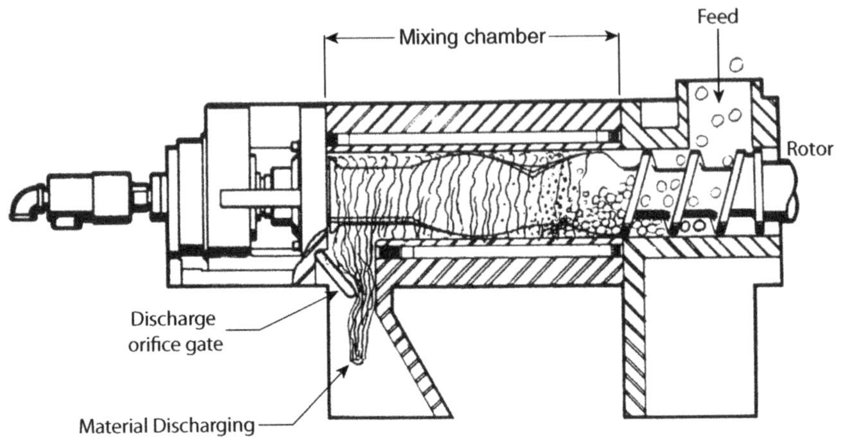

FIGURE 2.33 Schematic of a continuous mixer.

Thus production scrap or used parts can be utilized again, at least as a partial replacement of virgin materials.

Examples of melt-processing methods are extrusion, injection molding, compression molding, rotational molding, blow molding, melt spinning of fibers, calendering, film and sheet extrusion or casting, as well as the blown film process. Certain polymeric materials are available in latex form (i.e., in an aqueous dispersion). Some of them can be dissolved in suitable solvents; in such cases these liquid systems (lattices or solutions) are processed by coating, dipping, and fiber spinning. Many polymers are first processed by being mixed with a variety of additives, fillers, and plasticizers.

Thermosets are most frequently first shaped and only then cross-linked. In some cases, shaping and cross-linking occurs simultaneously (e.g., compression molding). As pointed out earlier, conventional elastomers, as well as other thermosetting polymers (often compounds), are cross-linked by a specific reaction involving substances reacting with the polymer. It may be peroxides, sulfur, amines, isocyanates, and the reaction is carried out most frequently at elevated temperatures, typically at 150°C or 302°F and higher. However, some materials (e.g., epoxy resins or polyurethane systems) react at room or slightly elevated temperatures. Cross-linking is covered in more detail in Section 2.14.3.

2.14.2.1 Extrusion Extrusion is one of the widely used processes for fabricating parts from thermoplastic elastomers. Unlike injection molding (Section 2.14.2.2), extrusion usually results in a semifinished or intermediate article that needs further processing to arrive at an end product. The principal end products made by extrusion are wire insulation, tubing, profiles, film, and sheet.

The flow behavior of the molten polymer is of critical importance in the extrusion of most thermoplastics. Such materials must be processed below the velocity at which *melt fracture* occurs, referred to as the *critical shear rate*. Melt fracture in molten plastics takes place when the velocity of the resin in flow exceeds the critical velocity, the point where the melt strength of the polymer is surpassed by internal stresses. Parts extruded in a process that exceeds the critical shear rate have poor quality. Typical symptoms of melt fracture include a rough (*shark skinning*) and a frosty or cloudy surface.

Extrusion is a continuous process that involves forming a product (*extrudate*) in two dimensions. These x–y dimensions determine the cross-sectional form of the extrudate, and can be almost unlimited in scope, ranging from a simple tube to a very complex extruded profile. The third (z) dimension is the length of the extrudate. In principle, it can be infinite. In fact, it is limited by practical considerations governing winding, reeling, storage, and transport. The essential point is that extrusion always produces an object of constant cross-section.

The product cross-section is formed in a die; the extrusion process consists of heating a thermoplastic above its melt temperature and forcing it through the die. The heating and pressurizing device involves one or more screws operating in a heated barrel, and is known as an *extruder*. Downstream of the die, the extrudate is calibrated, cooled, and packaged by an array of ancillary devices, including vacuum calibrators, air cooling chamber, water tanks, cooling rolls, haul-offs, cutters, and winders.

The function of the extruder is to heat the plastics material to a homogeneous melt and to pump it through the extrusion die at a constant rate. Because the extrusion of plastics is a continuous process, the melt preparation device must be capable of a constant output. Thermoplastic extrusion depends almost entirely on the rotating screw as a melt delivery device. Thermoplastics are generally characterized by low thermal conductivity, high specific heat, and high melt viscosity, so the preparation of a uniform melt and its delivery at adequate pressure and a constant rate pose considerable challenges. These have been countered by the introduction of various types of extruders.

The principle variants are the single- and twin-screw types. Of these, the single-screw extruder (Fig. 2.34) is by far the most popular. The twin-screw extruder may have parallel or conical screws, and these screws may rotate in the same direction (*corotating*) or in opposite directions (*counterrotating*).

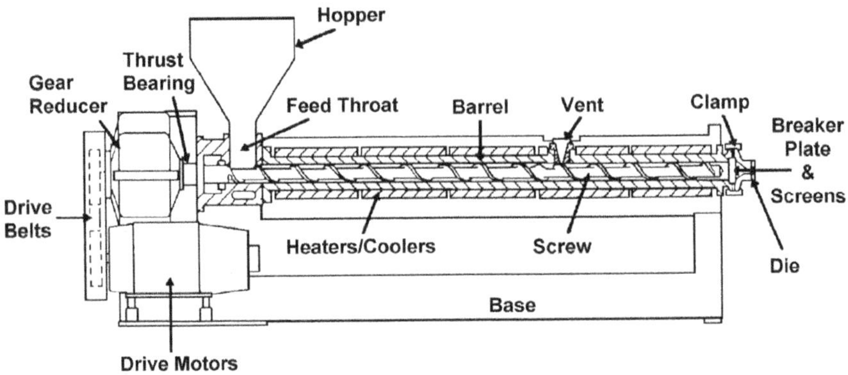

FIGURE 2.34 Schematic of as single-screw extruder.

The single-screw extruder consists essentially of a screw that rotates in an axially fixed position within the close-fitting bore of a barrel. The screw is motor-driven through a gear reduction train, and is supported by a thrust bearing that opposes the force exerted on the plastic melts. A helical flight on the screw provides the drag-induced conveying motion that develops this force. The barrel is equipped with both heating and cooling means, and its downstream end is provided with an attachment device for a shaping die that determines the cross-section of the extruded product. The upstream or inlet end of the barrel is equipped with a feed throat or an aperture in the barrel wall where a plastics material is input, generally in the form of granules or pellets. During its passage along the helical screw flight, this material is heated by a combination of conducted heat received from the barrel and mechanical shear heat derived from the mixing and kneading action of the screw. The output rate of the extruder is a function of screw speed, screw geometry, and melt viscosity.

Depending on the intended use, the corresponding extrusion method is used. Each method requires specific tooling and process conditions. Often the method and the polymer grade have to be matched to assure a satisfactory product. The most common extrusion methods are film and sheet extrusion, pipe and tube extrusion, wire coating, and profile extrusion.

The process of coating wire and cable by extrusion has been around for quite some time. Molten plastic is extruded into a *cross-head die* through which the wire being coated is passed (see Fig. 2.35). After leaving the die, the coated wire is cooled in air and/or in a water bath, while it is continuously tested for spark and *concentricity* (roundness of the coating). In this process, *primary insulation* is defined as a metal wire directly coated with a plastic to isolate the metal electrically. Jacketing (or sheathing) is referred to covering one wire or a group of wires with a

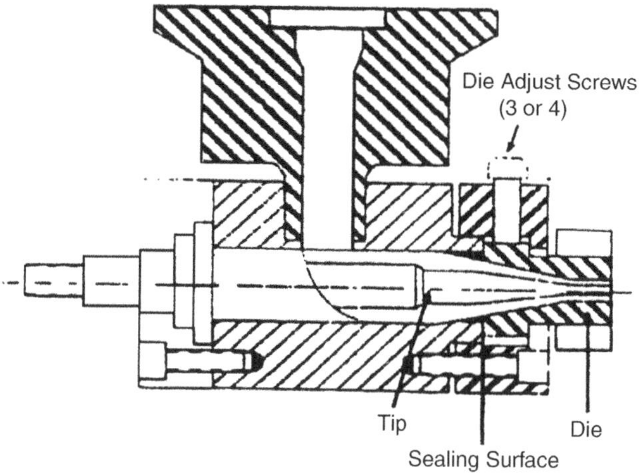

FIGURE 2.35 Schematic of a cross-head extrusion die. (Drobny, J.G., *Handbook of Thermoplastic Elastomers*, William Andrew Publishing, 2007, p. 62)

plastic coating or jacket for nonelectrical protection. Jackets are usually put on primary wires. Rubber compounds are extruded in a similar fashion, but the extruders have a different design. In addition, the downstream operations are different (see Section 2.14.3.2).

2.14.2.2 Injection Molding A significant portion of thermoplastics and elastomers is processed by means of injection molding. The process produces a complex finished part in a single rapid and automatic operation. This distinguishes plastics injection molding from most other manufacturing processes, although there are parallels with metal casting, particularly with die casting. Generally speaking, a whole series of forming, joining, and finishing operations would be needed to replicate an injection molded article in other materials using different manufacturing methods. It is the single operation that makes injection molding economically viable despite the high cost of machinery and molds. Injection molding machines and molds are very costly because of the high pressures required when injecting most melts (thermoplastics, in general) and the complexity of the process controls. The ability to produce a completely finished part at high speed balances the equation with high machine and mold costs and makes injection-molded articles highly cost effective. Injection molding of thermoplastics require rapid cooling of the molded part before it is removed from the mold. Injection molding of thermosets differs in that the finished product is cross-linked in the mold and does not require cooling before its removal from the mold. Example of the design of an injection mold is in Fig. 2.36.

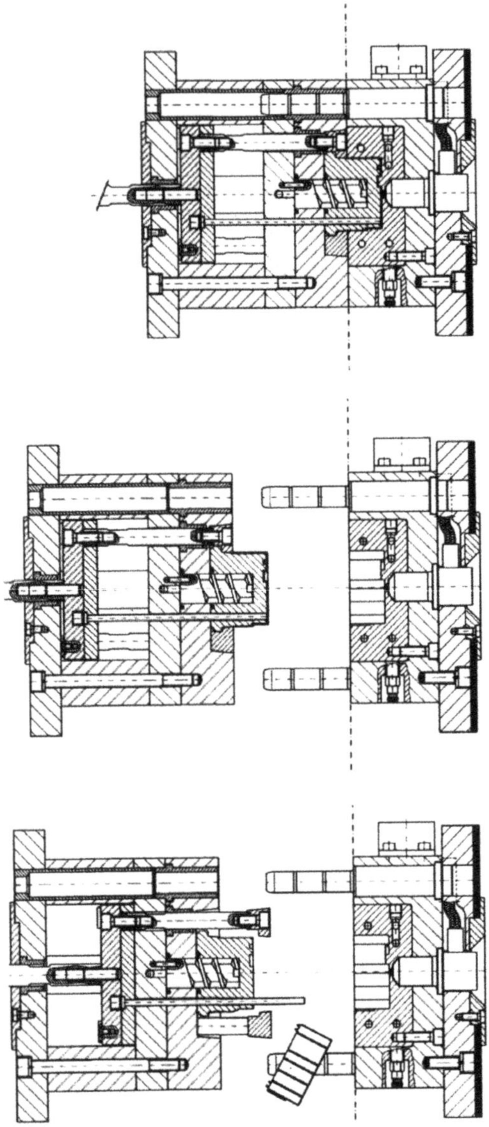

FIGURE 2.36 Injection molding sequence.

2.14.2.3 Compression Molding Compression molding was among the first to be used to form plastics. It involves four steps:

1. Preformed blanks, powders, or pellets are placed in the bottom section of a heated mold or die.
2. The other one-half of the mold is lowered and pressure is applied.

3. The material softens under heat and pressure, flowing to fill the mold. Excess material is squeezed from the mold. If it is a thermoset plastic, cross-linking occurs in the mold. In thermoplastic molding, the part is hardened under pressure by cooling the mold.
4. The mold is opened and the part is removed.

As pointed out above, for thermoplastics, the mold is cooled before removal so the part will not lose its shape. Thermosets may be ejected while they are hot and after curing is complete. This process is slow, but the material moves only a short distance to the mold and does not flow through gates or runners like in injection molding. Only one part is made from each mold.

In the simplest form of compression molding, a molding powder (or pellets, which are also sometimes called molding powder) is heated and, at the same time, compressed into a specific shape. In the case of a thermoset material, the melting must be rapid since a network starts to form immediately and it is essential for the melt to fill the mold completely before solidification progresses to the point where flow stops. Commercial molding machines use high pressure and temperature to shorten the cycle time for each molding. The molded article is pushed out of the cavity by the action of ejector pins that operate automatically when the mold is opened. An example of a compression mold is shown in Fig. 2.37.

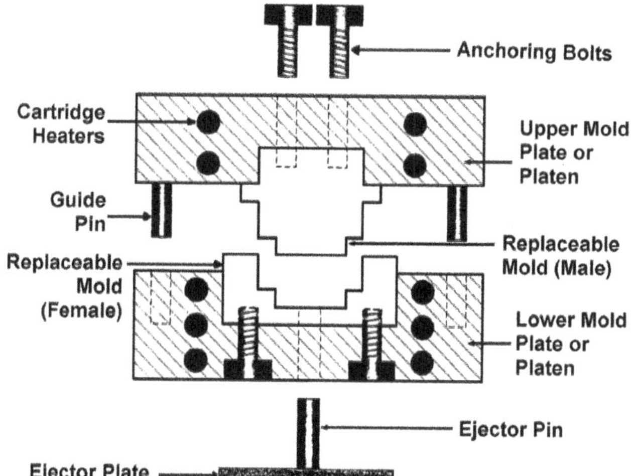

FIGURE 2.37 Example of a compression mold.

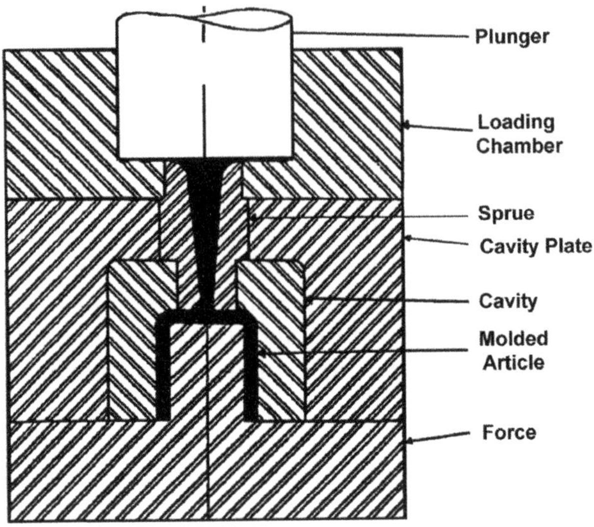

FIGURE 2.38 Example of a transfer mold.

2.14.2.4 Transfer Molding Transfer molding was originally developed for fabricating parts from thermosetting polymers. It is a combination of injection and compression molding processes in which the polymer is melted in a pot, transferred into a mold, and compressed into the designated shape. Transfer molding differs from compression molding in that the plastic is heated in a separate chamber, but compressing is carried out in a conventional compression press. Cooling hardens thermoplastics as opposed to thermosetting resins that are hardened by a cross-linking reaction. The mold design is modified to accommodate the melting pot or the heated mold. An example is shown in Fig. 2.38.

2.14.2.5 Blow Molding Blow molding is a process for converting thermoplastics into hollow objects. Like injection molding, the process is discontinuous or batchwise in nature, involving a sequence of operations that culminates in the production of a molding. This sequence or cycle is repeated automatically or semiautomatically to produce a stream of molded parts. An example of a blow-molding process is shown in Fig. 2.39.

The blow-molded parts are formed in a mold that defines only the external shape. As the name implies, the inner shape is defined by fluid pressure, normally compressed air. In this respect, blow molding differs radically from many molding processes where both inner and outer forms are determined by mold members. The major advantage is that the

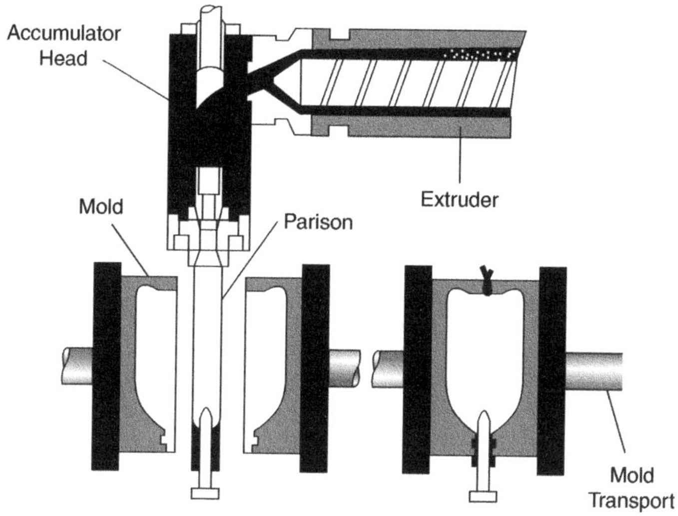

FIGURE 2.39 Example of a blow-molding process.

inner form is virtually free of constraints because there is no core to extract. The main drawback is that the inner form is only indirectly defined by the mold, so high precision and independent internal features are impossible. This has a bearing on wall thickness, which can never attain the consistency and accuracy of a full-mold process (e.g., injection molding).

2.14.2.6 Rotational Molding Rotational molding (also called *rotomolding* or *rotational casting*) is a process for manufacturing seamless hollow plastic parts for a variety of applications ranging from liquid storage tanks to containers of various shapes. Consumer products (e.g., furniture, playground equipment, and toys) are made by this technique. Rotational molding produces dolls, doll parts, sit-in and sit-on toys, and various sizes of heavy-wall balls.

The past decade has seen a large growth in the applications of rotational molding and lining technology. Along with this growth, interest has increased in understanding the complex technical aspects of this technique. Rotational lining is employed when the objective is to coat the interior of a metallic part, instead of manufacturing a plastic part.

Rotational molding (Fig. 2.40) involves the external heating of a thin-walled hollow metal mold containing a polymer powder. Heating occurs while the mold is rotated multiaxially. The powder melts and coats, or sinters, onto the interior wall of the cavity. The mold is then cooled,

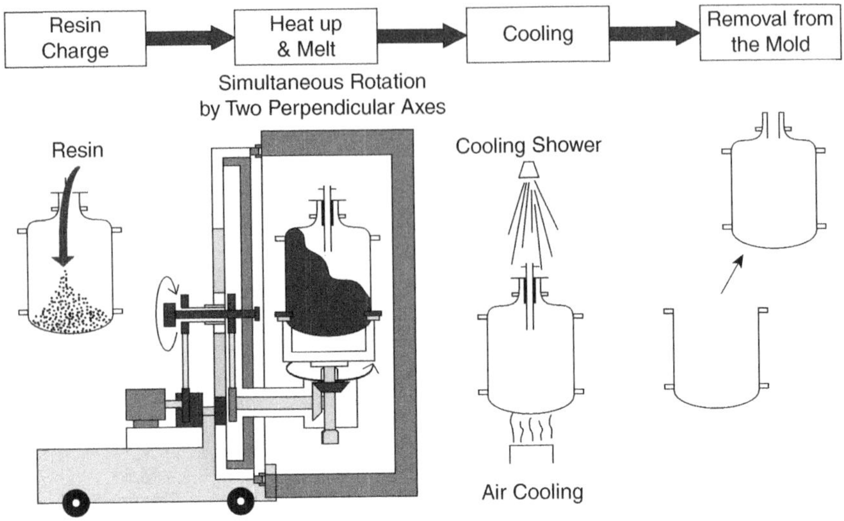

FIGURE 2.40 Example of a rotational-molding process.

which allows the part to solidify and crystallize in the case of thermoplastics. Finally, the part is removed and the mold is charged to repeat the cycle.

2.14.2.7 Thermoforming Thermoforming is a process used to convert thermoplastic sheets and films into shell forms. In the simplest form, a heated thermoplastic sheet is clamped over a mold cavity and drawing a vacuum in the cavity. This process causes the atmospheric pressure to press and plastically deform the sheet into the mold cavity, where it is cooled to retain the formed shape (see Fig. 2.41a). This is known as *vacuum forming*. Another method, *pressure forming* (Fig. 2.41b) is using compressed air instead of vacuum. In this method, compressed air is used to force the preheated sheet into the cavity. *Matched-mold (die) forming* uses closed (compression) mold; the pressure is applied to the male mold half by a power press, forcing the preheated sheet into the female mold half. Today thermoforming is a sophisticated technology using specialized equipment capable of producing many different shapes at very high production rates [85].

2.14.3 Processing of Thermosets

2.14.3.1 Processing of Thermoset Resins Certain polymers and oligomers (low molecular weight polymers or prepolymers) are processed by cross-linking or a combination of polymerization and cross-linking. There are several methods of cross-linking. Typical thermosets

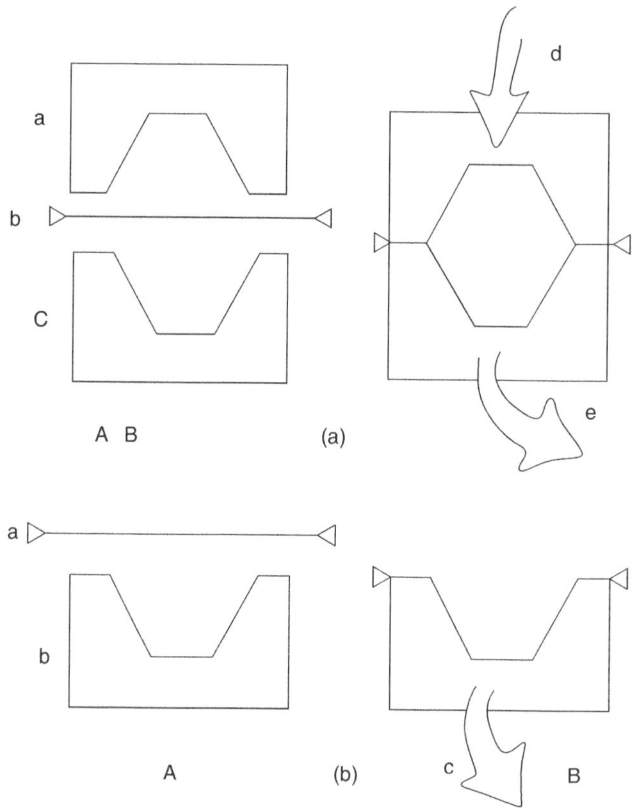

FIGURE 2.41 Examples of thermoforming. (a) Vacuum forming, (b) pressure forming.

come in liquid forms (liquids with different viscosities) or can be easily melted and then reacted to form the required product. Examples are phenolic, melamine and epoxy resins, polyurethanes, and certain acrylic resins. In most cases, heating of the reaction mixture is used to promote cross-linking. Other method to convert liquid oligomers and/or monomers is to expose them to UV or electron beam (EB) radiation, which causes complete polymerization and/or cross-linking. The selection of the processing method of the above materials depends on the type of resin. It may be casting, coating, molding, and so on.

2.14.3.2 Processing of Conventional Elastomers (Rubber)

Mixing of conventional elastomers such as carried out in internal mixers (an example is shown in Fig. 2.32) in batches or continuously (an example of equipment is shown in Fig. 2.33). Only certain specialty materials requiring particular care (e.g., colored compounds and compounds for medical products) are mixed on roll mills (two steel rolls placed horizontally with internal heating or cooling). As mentioned earlier, rubber compounds generally contain many more ingredients than thermoplastics (e.g., fillers, plasticizers, curing agents, antioxidants, and in some cases antiozonants). Because of the presence of curing agents (sulfur, sulfur-donors or organic peroxides), the temperature control of the mixing process is critical for preventing premature curing (scorch) of the compound. That would turn the material into scrap.

Extrusion of rubber compounds is also greatly different from thermoplastics. Rubber extruders are considerably shorter than those for melt processing with a smaller L/D screw ratio. Normally, the stocks are fed into the machine hot, but often special longer cold-feed extruders are used. Cross-head extrusion dies (Fig. 2.35) also can be used with extruders for conventional rubber.

Calendering of rubber compounds is a frequently used process for preparing sheets of varied gauges and for coating fabrics. Most frequently three- and four-roll calenders are used in production setting (examples are shown in Fig. 2.42).

Vulcanization or curing (essentially cross-linking of the elastomer or elastomer blend) is in most cases carried out at elevated temperatures (typically 140–200°C, or 284–392°F) (see also Section 2.13.2). As pointed out earlier, there are several chemistries involved in vulcanization. A great majority of general purpose and diene rubbers are vulcanized by

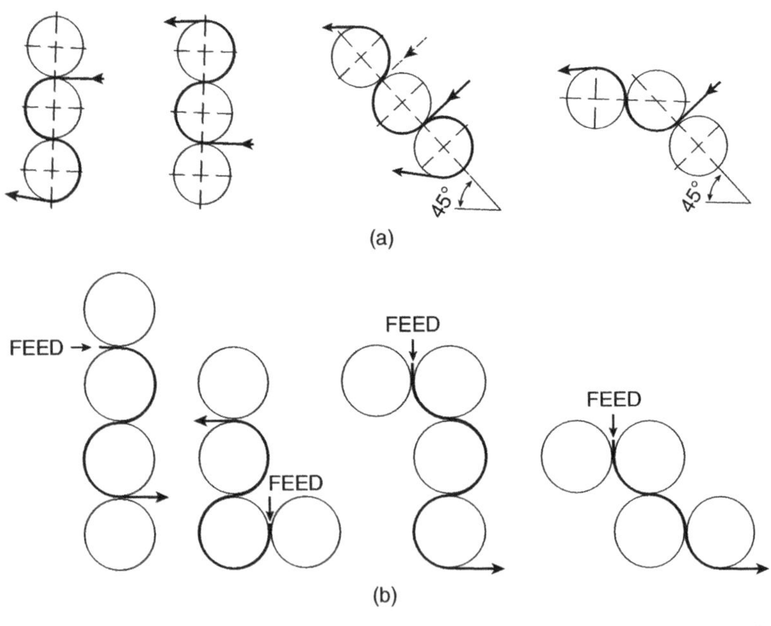

FIGURE 2.42 Examples of calenders: (a) A three-roll calender and (b) a four-roll calender.

the use of sulfur. The chemical reaction involves the formation of cross-links composed of two, three, or four sulfur atoms, between sites of unsaturation on adjacent chains. Typically 2.5 to 3 parts by weight of sulfur per 100 parts by weight of elastomer (phr) is used to obtain usable rubber materials with elongation up to 700%. An addition of 30–32 phr of sulfur produces *ebonite* a highly cross-linked material, which is hard and exhibits a degree of thermoplasticity. *Sulfur donors* are compounds that release sulfur when heated and create short sulfur cross-links. Other compounds used to cure (cross-link) elastomers are organic peroxides, metal oxides, and certain organic compounds used with specific elastomers (e.g., FKM and silicone elastomers).

Many rubber products are cured in compression, transfer, and injection molds; the molding cycles are considerably longer than cycles for molding of thermoplastics or thermoplastic elastomers. Other methods of vulcanization are using a hot air oven, steam heated autoclaves, rotary presses, continuous procedures (e.g., in molten salts or by irradiation by an electron beam). Certain rubber compounds require a relatively long postcuring procedure.

2.14.4 Processing of Reinforced Composites

As pointed out in Section 2.12.4, reinforced composites consist of a polymeric matrix and various reinforcing agents. In addition, in some types of these materials, additional fillers and special additives are used. The processes used require some specific components (e.g., preforms, prepregs, and compounds (Section 2.14.4.1), which are produced before they are used in the specific process.

2.14.4.1 Preforms, Prepregs, and Compounds A *preform* is an arrangement of fibers specially prepared for molding. Preforms are preshaped reinforcements that are combined with resin during molding. They are placed into the mold cavity between the male and female dies used in compression, resin transfer, or reaction injection molding.

A *prepreg* is a fiber–resin combination specially prepared prior to molding. Thermosetting prepregs are made by combining liquid resin with woven and/or continuous fiber reinforcement. Through controlled chemical reaction, the resin is brought into its B-stage, which is a tack-free condition. Prepregs are cured with heat and often are stored refrigerated to prevent premature cure. Thermoplastic prepregs are prepared in a somewhat different fashion. Because a catalyst is not needed to initiate cure, B-staging is not needed. Heat and pressure are used to consolidate the reinforcing fibers and the thermoplastic resin.

There are two kinds of compounds used for the manufacture of specific composite materials:

1. *Sheet Molding Compound* (SMC) is an integrated ready-to-mold composition of glass fibers, resin, and filler. This compound is made by metering a resin paste onto a thin plastic carrier film. The resin typically is a vinyl ester, an epoxy, phenolic, or polyimide resin. The compound is made by chopping continuous fibers onto the resin paste as it is conveyed on the film and the glass and resin mixture is covered with an additional layer of resin on a second carrier film. After maturing, the carrier films are removed and the sheet is used for placement between the top and bottom halves of a compression mold and heated to obtain the finished part.
2. *Bulk Molding Compound* (BMC) is a mixture of chopped glass fibers and resins with necessary fillers and additives. The premixed material has a dough-like consistency and is supplied in bulk form or as an extruded rope. The compound can be processed by compression, transfer, or injection molding.

Fillers and additives are used in both thermosetting and thermoplastic matrix materials to change the performance of the resin system, processing behavior, or material cost. A filler can be a simple extender (e.g., wood flour or crushed nut shells). Mineral fillers (e.g., calcium carbonate and talc) have an effect on the surface appearance, heat deflection temperature, and cost. Alumina trihydrate is one of the additives improving flame resistance of the matrix. Conductive fillers increase static dissipation and conductivity (see Section 3.1.3.5).

Colorants provide integral color that will not chip or fade. Special thermoplastic additives in the thermosetting matrix reduce the shrinkage occurring during the curing process.

2.14.4.2 Manufacturing Processes Essentially, there are six typical processes to manufacture products from fiber reinforced composites:

1. *Open molding* includes lay-up, bagging and autoclaving, and spray-up.
   - Lay-up is the simplest method to produce laminated composites. The mold is typically a female mold cavity. The reinforcement (e.g., mat, woven roving, fabric, or combination of these) usually is placed by hand into the mold and liquid resin is applied (Fig. 2.43). This procedure continues until the desired wall thickness is obtained. To prevent voids in the resinated layers

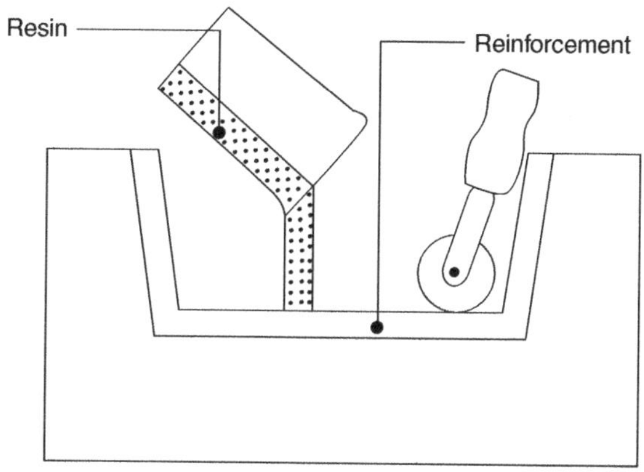

FIGURE 2.43 Schematic representation of composite lay-up. (*Plastics Engineering Handbook of the Society of The Plastics Industry 5th Edition*" (Berins, M.L. Ed.), Van Nostrand Reinhold, 1991, p. 519)

of reinforcement, the entrapped air is worked out by rollers or squeegees. The laid-up parts are usually cured at ambient temperatures, but heat and accelerators can shorten the cure time. This method is mainly used for the manufacture of large structures.

- Bagging and autoclaving can be used to improve a laid-up laminate. In vacuum bagging, a mold release film is placed over the reinforcement–resin composition. The film edges are sealed and vacuum is drawn from within. Atmospheric pressure works the film toward the laminate to create a smoother surface, to eliminate voids, and force out entrapped resin. In pressure bag molding, a deflated bag is placed on top of the film. An airtight seal is placed over the bag, and the bag is inflated. An advanced modification of bagging is to place the laminate and the mold into an autoclave. The autoclave is pressurized to ~80 psi (0.55 MPa). In some cases, the autoclave is heated to accelerate the cure of the resin.
- Spray-up is a mechanized version of lay-up. A spray gun simultaneously sprays resin and chops a continuous glass strand to specific length fibers Fig. 2.44). Chopped fibers enter the resin spray stream and the combined material is applied to the mold.

2. *Compression molding* is accomplished by matched male and female dies. The charge is placed into the cavity and the mold is closed and held under heat and pressure. The mold charge may be a resin impregnated preform, prepreg, sheet molding compound on a bulk-molding compound.

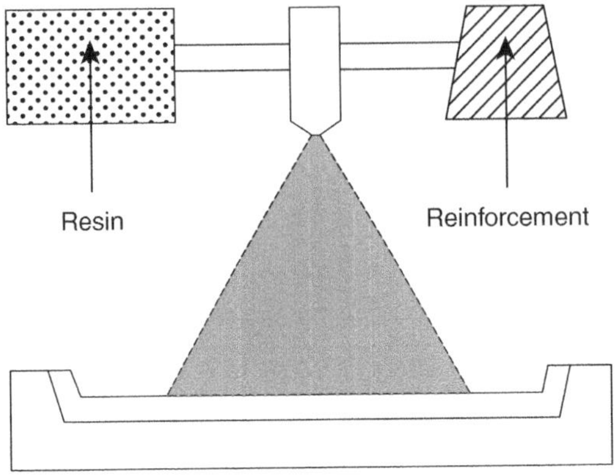

FIGURE 2.44 Schematic representation of composite spray-up. (*Plastics Engineering Handbook of the Society of The Plastics Industry 5th Edition*" (Berins, M.L. Ed.), Van Nostrand Reinhold, 1991, p. 520)

3. *Transfer molding* is related to compression molding. The reinforced thermoset compound is first heated in a chamber or pot separated from the matched dies. Then a plunger transfers the heated material into the closed mold.
4. *Injection molding* of reinforced thermoplastics is very much like unreinforced thermoplastics. The compound is fed from a hopper into a heated cylinder where the resin softens to a flowable polymer melt. During a mold cycle, a rotating screw-ram injects the melt into the mold cavity. The mold is cooled so that the melt can solidify. In injection molding of reinforced *thermosets*, the injection cylinder is cooled instead of heated, and the mold is heated instead of being cooled.
5. *Reaction injection molding* (RIM) involves bringing the liquid resin and additives together in a mixing head and then rapidly injecting it into a closed mold. This process is suitable for reactive materials (e.g., mixtures of polyol and isocyanates). These materials react in the mold to effect rapid polymerization and cure. There are two additional versions of this process, namely, *reinforced reaction molding (RRIM)* that incorporates the reinforcing fibers (milled glass) or glass flakes in the material that is injected from the mixing head (Fig. 2.45). Another version is *structural reaction injection molding (SRIM)*. In this process, the reinforcing medium is first placed into the open mold and then the reactive mixture is injected after the mold closes. The reinforcement typically is a thermoformed mat or fiber-directed preform. In r*esin transfer molding*, a dry reinforcement material that has been cut and/or shaped is

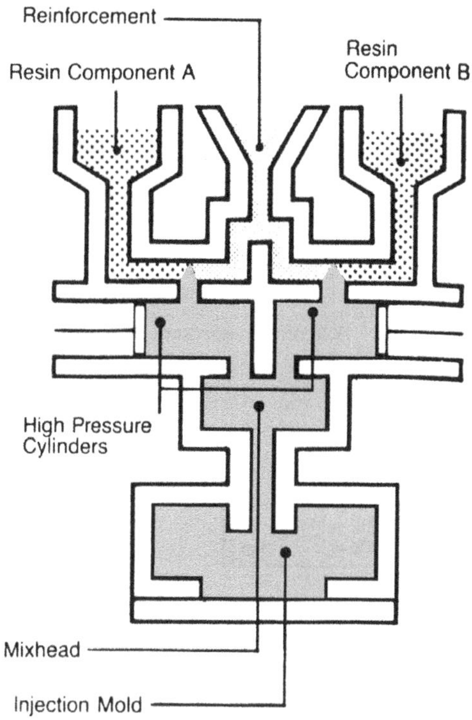

FIGURE 2.45 Schematic representation of reinforced reaction injection molding. (*Plastics Engineering Handbook of the Society of The Plastics Industry 5th Edition*" (Berins, M.L. Ed.), Van Nostrand Reinhold, 1991, p. 522)

placed into a prepared mold cavity. Once the mold has been closed and clamped shut, resin is injected into the mold cavity, where it flows through the reinforcement preform. When all the required resin is transferred into the mold, the cure starts. The cure can take several minutes to several hours, depending on the resin system. The molded cured parts are removed from the mold and may require a postcure to complete the resin reaction.

6. *Continuous Processes*

   - Pultrusion (Fig. 2.46) is a continuous process in which fiber reinforcements that have been impregnated with a solution containing resin and other additives are pulled through a heated die [86]. The reinforcements entering the die are saturated by the resin solution and are solid when exiting the die. The pultrusion process can produce solid, open-sided, and hollow shapes that can be cut to desired lengths.

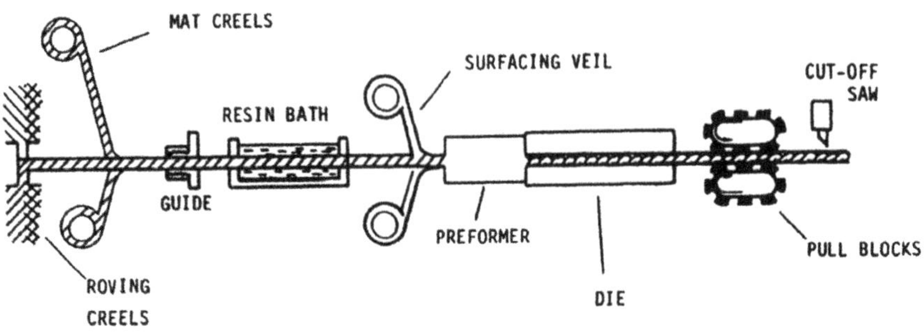

FIGURE 2.46 Schematic representation of pultrusion.

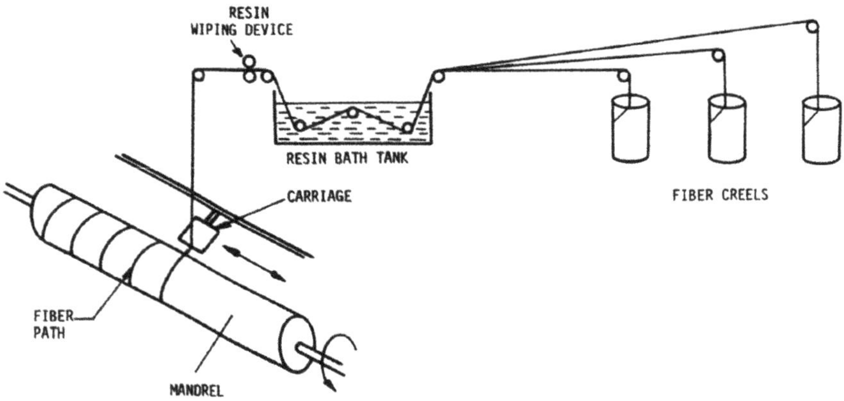

FIGURE 2.47 Schematic representation of filament winding.

- Filament winding (Fig. 2.47) is a process by which a filamentary yarn or row is first wetted by a resin and then is uniformly regularly wound about rotating mandrel. The finished pattern is cured and the mandrel is removed. Typical materials used in filament winding include fiberglass, carbon, or aramid fibers coupled to polyester, vinyl ester, or epoxy resin. The mandrels used can be [87]: metal mandrels, expandable mandrels (frequently made from rubber), and single-use mandrels.

2.15 ADHESION AND ADHESIVES

The majority of polymeric materials can be bonded to other materials by adhesives. Some of them can be joined without any surface treatment; others require surface treatment to achieve a sufficient adhesion.

2.15.1 Principles of Adhesive Bonding

An adhesive is a substance capable of holding materials together by surface attachment. There are many types of adhesives available; they may be classified in several ways, for example, by the mode of application, chemical composition, suitability for certain substrates, and cost.

Essentially, an adhesive must be applied to the *substrate* (or *adherend*) in a fluid form to wet the surface completely and leave no voids, even if the surface is rough. To meet these requirements, it has to have a low viscosity at the time of application. In order to create a strong bond, the adhesive layer has to exhibit a high *cohesive strength*, that is, it has to set. This usually happens by removal of the solvent or water, by solidification, crystallization, polymerization, and/or cross-linking.

Wetting of a surface and spreading of a liquid depends primarily on the nature of the surface of the substrate and on the *surface tension* of the liquid. The wettability of a solid surface depends on its *surface energy*, expressed by its *critical surface tension*. The critical surface tension indicates the maximum surface tension permissible in a liquid (i.e., adhesive) for a particular substrate. To assure spreading and wetting, the liquid adhesive should have a surface tension no higher than the critical surface of tension of the substrate (i.e., solid adherend). When the substrate is organic and not too polar, *solubility parameter* (δ) is a quantity useful in the selection of a suitable adhesive to it. The solubility parameter is defined [88] as:

$$\delta = \left(\frac{\Delta E}{V}\right)^{1/2} \tag{2.27}$$

where ΔE is the energy of vaporization, and V is the molar volume (see Section 2.6.1). The term $\Delta E/V$ is referred to as *cohesive energy density*. The values of solubility parameter and critical surface tension are related [89]. Examples for selected polymers are shown in Table 2.15.

The degree of wetting and spreading is quantitatively expressed in terms of the *contact angle* [90]. Contact angle (θ) is the angle (in the liquid) between the air–liquid and liquid–solid interface (see Fig. 2.48), resulting as a mechanical equilibrium of a drop resting on a plane solid surface under the action of three surface tensions: γ_{LV} at the interface of the liquid and vapor phases, γ_{SL}, at the interface of the solid and the liquid, and γ_{SV} at the interface of solid and vapor, hence,

$$\gamma_{SV} - \gamma_{SL} = \gamma_{LV} \cdot \cos\theta \tag{2.28}$$

TABLE 2.15 Critical Surface Tension Values of Various Polymeric Solids at 20°C

Polymer Solid	Critical Surface Tension, $\gamma_{c.}$ (mN m^{-1})
Polyhexafluoropropylene	16.2
Polytetrafluoropropylene	18.5
Polytrifluoroethylene	22.0
Polyvinylidene fluoride	25.0
Polyvinyl fluoride	28.0
Polychlorotrifluoroethylene)	31.0
Polyethylene	31.0
Polystyrene	33.0
Poly(vinyl alcohol)	37.0
Poly(methyl methacrylates)	39.0
Poly(vinyl chloride)	39.0
Poly(vinylidene chloride)	40.0
Poly(ethylene terephtalate)	43.0
Poly(hexamethylene adipamide)	46.0

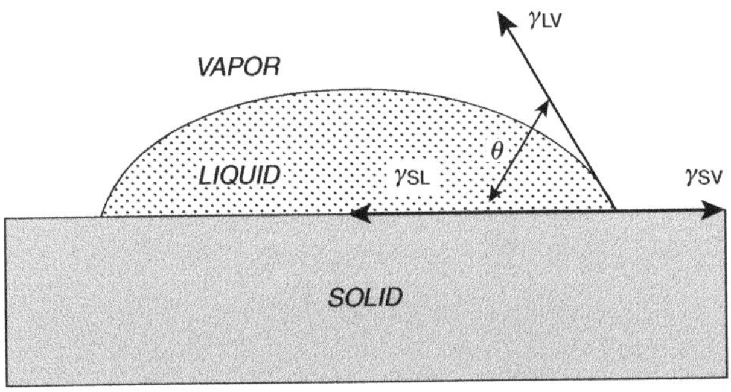

FIGURE 2.48 Contact angle.

From this relation, the critical surface tension, γ_c, is derived for an individual substrate by measuring contact angles of a homologous series of liquids with known surface tension values on it, and plotting the results, as shown in Fig. 2.49 (often referred to as the *Zisman plot*). Extrapolating to a zero contact angle (i.e., cos $\theta = 0$) and extending the intercept to the surface tension axis gives the critical surface tension of the solid surface [91].

If the contact angle is near zero degrees, the liquid spreads and wets the substrate completely. Larger values of the contact angle indicate a poorer wetting. An extreme case is when θ is 180°; then there is absolutely

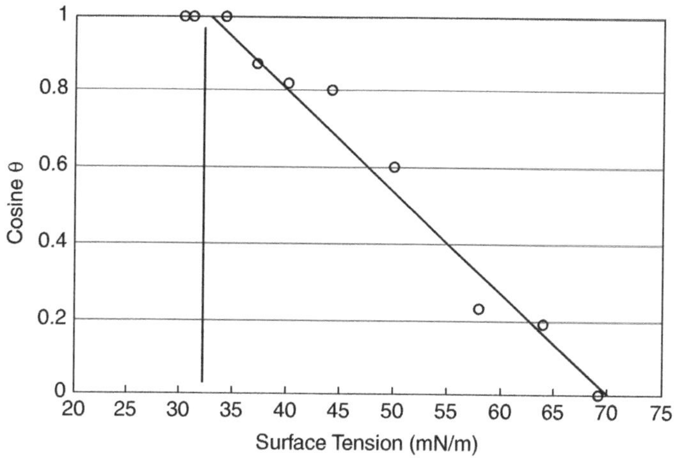

FIGURE 2.49 Zisman plot. (Drobny, J.G., *Handbook of Thermoplastic Elastomers*, William Andrew Publishing, 2007, p. 142)

no wetting of the substrate. Detailed discussions of wetting and spreading of liquids on solids are shown in [90 and 91].

2.15.2 Methods of Surface Treatment

The surfaces to be bonded, regardless of the bonding method or the type of adhesive, have to be clean. Often it is sufficient to wash them with solvents to remove contaminants (e.g., oils), but sometimes grinding the surface cleans it and the roughness improves the bonding by virtue of providing a greater surface area.

If the surface of the polymer to be bonded by adhesive has too low of a surface energy, it has to be treated in order to increase it. The surface treatment energy, improves wetting of the surface by the adhesive and the bond strength. The methods used most frequently are flame treatment, corona treatment, and plasma treatment.

- Flame treatment is a commercial process to render polyolefins and polyethylene terephtalate receptive to adhesives, coatings, and printing inks. The polymer article (e.g., film) is passed over an oxidizing flame formed by an oxygen-rich (relative to stoichiometry) mixture of hydrocarbon gas. The gas flame contains excited fragments and species [e.g., atomic oxygen (O), NO, OH, and others] that can abstract hydrogen from the surface of the polymer, which is then replaced by oxygenated functional groups (mostly —C=O and —OH).

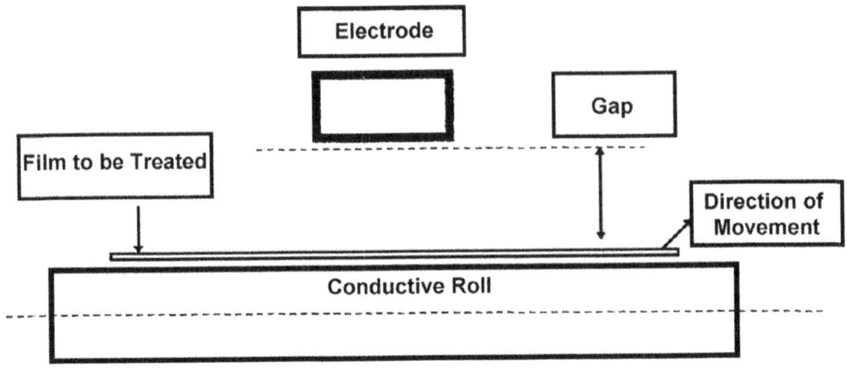

FIGURE 2.50 Conceptual schematic diagram of a film corona treatment system. (Drobny, J.G., *Handbook of Thermoplastic Elastomers*, William Andrew Publishing, 2007, p. 143)

- Corona discharge treatment takes place at atmospheric pressure in contrast to low-temperature (cold) plasma (see below) that requires vacuum. Corona is a stream of charged particles (e.g., electrons and ions) that is accelerated by an electric field. It is generated when a space gap filled with air or other gas is subjected to a sufficiently high voltage to set up a chain reaction of high-velocity particle collisions with neutral molecules resulting in the generation of more ions. The treated surface has a higher surface energy and therefore can be wetted readily. One of the widely used configurations (Fig. 2.50) uses corona discharge produced by high-frequency high-voltage alternating current [92]. The main parameters for the control of this treatment process include the voltage, width of the air gap, film–web speed, and the width of the electrodes. Most machines allow treatment on one side of the web (see Fig. 2.51) and require two passes for two-sided treatment.
- Plasma (glow discharge) is produced by exciting a gas with electrical energy (Fig. 2.52). It is a collection of charged particles containing positive and negative ions [93]. Other types of fragments (e.g., free radicals, atoms, and molecules) may also be present. Plasma is electrically conductive and is influenced by a magnetic field. It is intensely reactive, which is the reason why it can modify surfaces of plastics [94]. It can be used to treat parts to make their surfaces harder, rougher, more wettable, and easier to adhere to. Plasma treatment can be carried out on a variety of plastic parts and even on powder additives like pigments and fillers.

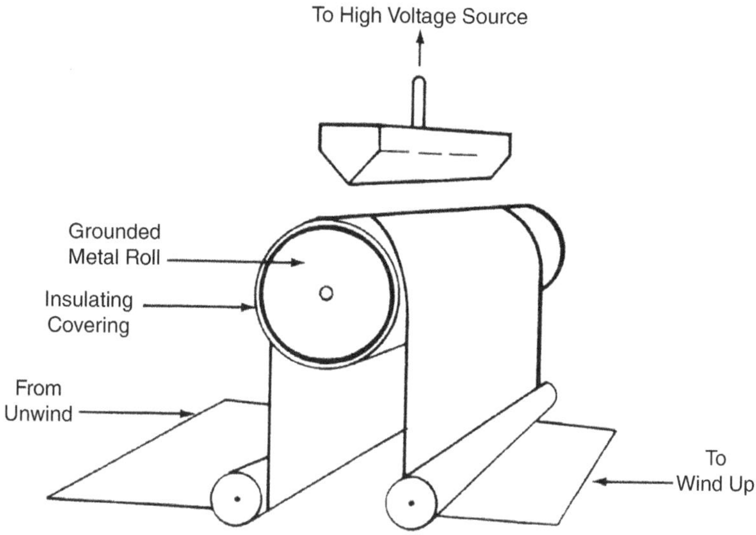

FIGURE 2.51 Typical single-side corona treatment unit.

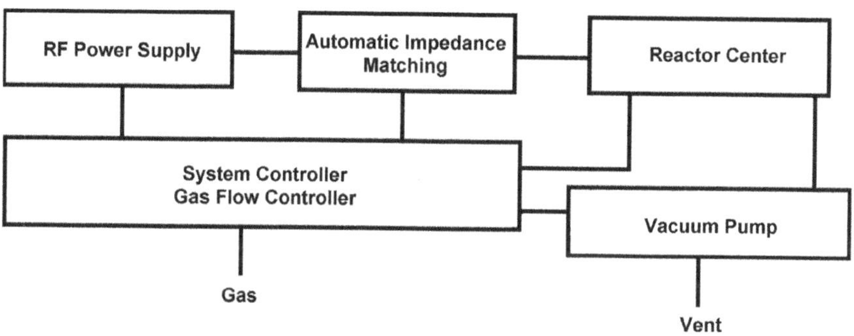

FIGURE 2.52 Schematic diagram of plasma treatment.

- The plasma used for treating surfaces of materials is called *cold plasma*, which means its temperature is about room temperature. Cold plasma is created by introducing the desired gas into a vacuum chamber, followed by rf (13.56 MHz) or microwave (2450 MHz) excitation of the gas [95]. As pointed out earlier, the energy dissociates the gas into electrons, ions, free radicals, and metastable products. Practically any gas may be used for plasma treatment, but oxygen is the most common. The electrons and free radicals created in the plasma collide with the polymer surface and rupture covalent bonds thus creating free radicals on the surface of the polymer (Fig. 2.53).

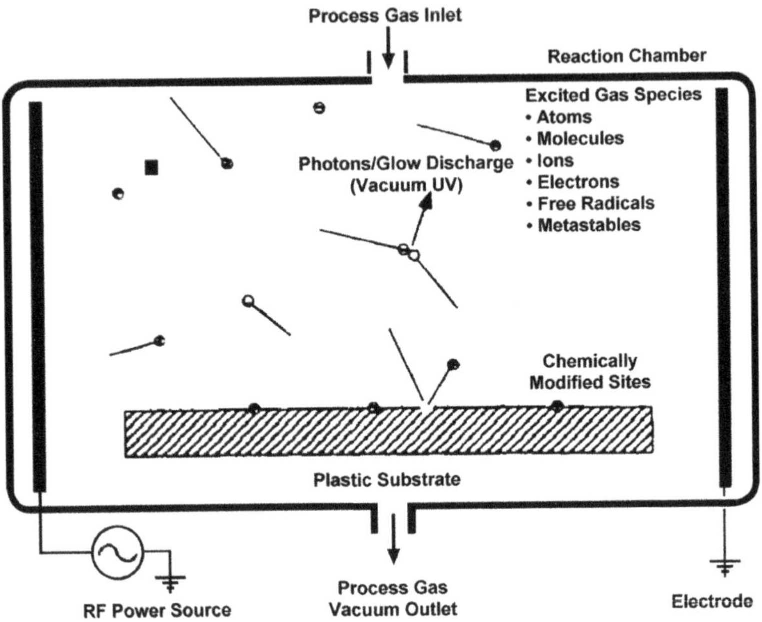

FIGURE 2.53 Schematic diagram of the surface modification of plastics in a gas plasma reactor. (Ebnesajjad, S., *Surface Treatment of Materials for Adhesion Bonding*, William Andrew Publishing, 2006, p. 130)

Other methods of surface treatment include *etching and oxidation* using chromic acid and other strong oxidizing compounds; *thermal treatment*, which is done by a blast of hot air (~500°C or 932°F) [96, 97]; and *surface grafting* that is performed by irradiating the substrate with an EB in the presence of acrylic acid [98]. *Laser treatment* is done by XeCl, ArF, and KrF lasers and introduces oxygenated functional groups to the surface and removes surface contaminants [99].

Primers are reactive compounds that are applied to the substrate surface before the application of the adhesive. They form a chemical bridge between the substrate and adhesive through multifunctional groups that react preferentially with either adhesive or substrate [100]. Methods of surface treatment of materials for adhesive bonding are covered thoroughly in [101 and 102].

2.15.2.1 Evaluation of the Level of the Surface Treatment There are several ways to test the level of any of the above treatments. One common method involves using solutions made from two chemical compounds that produce liquids (dyne liquids) with surface tensions in the range 30–70 dyn cm^{-1}. The test consists of placing droplets of the various dyne liquids on the treated surface and observing the spreading of the drops in 2 s. Successive liquids with different surface tensions allow narrowing of the surface tension range of the web. This method is subjective, but it provides a rapid means of assessment of the treatment level. There are also pens operating in a similar fashion as the dyne liquids. A more quantitative approach is the measurement of contact angle, which decreases with the increase of the treatment level [103].

2.15.3 Types of Polymeric Adhesives

As pointed out earlier, polymer-based adhesives can be classified in several different ways. They may be liquids of different viscosity, represented by *solvent-based adhesives*, *aqueous dispersions* or *emulsions*, and *latexes*, *monomers*, or *oligomers*. In all cases, the main component is an organic compound, such as a polymer (dispersions, emulsions, and latexes) or a monomer or oligomer that forms the adhesive film. Polymers used in adhesives are mainly thermoplastics or elastomers that often are combined with tackifying resins. Monomers and oligomers have to polymerize in order to form a strong enough film. In some cases, all these adhesive substances may be cross-linked in order to further increase the cohesive strength, heat resistance, and chemical resistance of the bond.

- Contact adhesives are mainly solvent- or water-based adhesives, most frequently containing elastomers combined with tackifiers, and exhibit an aggressive tack. Once the components coated with contact adhesives are joined, the bond is strong and permanent.
- Hot-melt adhesives are 100% nonvolatile materials, that is, containing no solvents, water, or other volatile carriers. They are solid at room temperature, but fluid at elevated temperatures. Hot-melt adhesives are applied in the molten state with special equipment, often as beads or stripes. After application, bonding and cooling, hot-melt adhesives return to the solid state and develop their ultimate strength.
- Heat activated assembly adhesives are solution-based adhesives that are applied to the surfaces of two substrates and dried. Then, prior to assembly, the coated surfaces are heated to a sufficiently high temperature, and then pressed together.
- Two-part adhesives consist of two components separately packaged, which are mixed just prior to use. After these two components are mixed, they start to react. As the reaction proceeds, the adhesive layer becomes stronger. The reaction is most widely a polymerization

(polyaddition, polycondensation, etc.). The reaction may be completed in a short time (minutes), but it may takes hours or days, depending on the system.

- Moisture-cure adhesives are based on compounds that cure by exposure to moisture. They are kept in impermeable packaging until they are applied onto the substrate. The reaction usually takes days to be completed since it depends on diffusion of moisture through the applied layer.
- Anaerobic adhesives are single-component liquids that can be stored at ordinary temperatures in the presence of oxygen and then polymerize rapidly to solids on being confined between two surfaces that exclude air.

As for the chemical nature, there are many polymers used in adhesives. A detailed list is beyond the scope of this book, so only the most widely used polymers, monomers, and oligomers are listed (below).

- Elastomers are widely used in solvent-based, water-based (latexes), contact, pressure-sensitive, and hot-melt adhesives. Elastomer-based adhesives may contain a wide variety of component materials (resins, tackifiers, fillers, plasticizers, antioxidants, curing agents, etc.). Solvent-based adhesives are prepared by simply dissolving the rubber compound in a suitable volatile solvent. Elastomer-based hot melts are generally mixtures of the base elastomer and a tackifying resin and other ingredients that affect its melting and melt viscosity. No volatile ingredients are present in hot melts, as pointed out earlier. Latex-based elastomeric adhesives contain a variety of other materials (e.g., protective colloids, emulsifiers, surfactants, thickeners, and emulsion stabilizers).

  The following elastomers are typically used in most commercial adhesives: Natural rubber, polyisoprene, styrene–butadiene rubber, polybutadiene rubber, acrylonitrile rubber, polychloroprene, acrylate rubber, polyurethane rubber, polysulfide rubber, silicone rubber, carboxylic elastomers, and copolymers of ethylene and vinyl acetate.
- Thermoplastic elastomers (e.g., SBS, SIS, and TPU) are most widely used in hot-melt, pressure-sensitive, contact, and heat activated adhesives alone or in blends with other elastomers and resins.
- Epoxy resins are mostly employed in two-part adhesive systems with different curing time and with different bond characteristics.

- Polyurethane- and isocyanate-based adhesives: Certain types of polyurethane adhesives are two-part systems, consisting of a polyol (often an oligomeric polyol) and a polyisocyanate, which are kept separately and mixed only prior to use. The bond strength develops rather rapidly. Other polyurethane adhesive systems include thermoplastic polyurethanes used in hot melts, polyurethane aqueous dispersions, powders, and films. Isocyanates alone are used as primers [104] or as cross-linking agents in certain elastomeric adhesives prepared with thoroughly dried inert solvents [104].
- Silicones: Silicones can be used in two-part systems and also as single-package moisture curing adhesives and sealants.
- Cyanoacrylates: Cyanoacrylate adhesives form a strong adhesive bond rapidly, typically in <1 min. This is the result of an anionic polymerization initiated by weakly basic species (e.g., water or alcohol) at ambient temperatures [105].
- Other polymers used for adhesives are ethylene–vinyl acetate copolymers, polyvinyl acetal, polyamides, polyimides, polybenzimidazoles, polybenzotriazoles, and others used for specialized applications. The reader can find more on the subject in [106]. Special conductive adhesives are used mainly in electronic applications (Chapter 5).

2.15.4 Application of Adhesives

Most of the adhesives are applied in liquid form. The commonly used methods of application are brushing, spraying, dipping, curtain coating, application by roll, and application of a bead or stripe of a hot melt.

2.15.5 Formation of Adhesive Bonds

When an adhesive is applied as a solution in a volatile solvent or as an aqueous dispersion or emulsion, the solvent or water has to be evaporated. In most cases, this happens over a relatively short time at ambient temperature. Sometimes the drying time can be reduced by applying hot air or to the surface of the parts to be joined or by using infrared heaters.

The assembly of the joint has to take place while the adhesive layer has a sufficient tack. So each type of adhesive has a specific *open time*. Open time or *bonding range* is the maximum time two surfaces coated with an adhesive may be allowed to dry (or cool down if applied in the molten state) before bonding and still have satisfactory adhesion. It depends on the specific application and can vary from <1 min to >24 h.

The adhesive joint can be formed by pressing the two adherends together. If the adhesive has a high tack (e.g., a contact adhesive), no additional pressure is required. Similarly, the bond formed by hot melts a brief application of pressure until the melt solidifies. In other cases, pressure has to be applied for a certain period of time by placing the joint between platens of a press or into a specialized fixture that holds the joined parts together until the bond is sufficiently strong. The same applies to adhesive bonds that become stronger due to a chemical reaction. Occasionally, heating of such a joint is used to reduce the time for a sufficiently strong bond or to achieve considerably higher bond strength.

2.15.6 Measurement of the Bond Strength

There are many tests of measuring the strength of the bond between an adhesive and a substrate. They are used for five main purposes [107]:

1. To check the quality of an adhesive to see if it falls within well-defined limits.
2. To determine the effectiveness of the surface treatment.
3. To gather data for the prediction of the performance of the adhesive joint.
4. To select an adhesive from a group for a specific application.
5. To evaluate the effect of aging on the performance of the joint.

The most common test configurations are tensile, shear, and peel.

The most commonly used tests for a quick assessment of adhesion between two rigid substrates are shear tests. The simplest test configuration is the *lap shear test* using two principal forms of test pieces, the single-lap joint (Fig. 2.54) [108], and the double-lap joint (Fig. 2.55) [109]. The advantage of the lap joint is that it can be manufactured easily and quickly tested. After the test piece is ready for the test, it is pulled apart in the axis of the bond and the failure load is then calculated. The test results are of restricted value in the design of adhesively bonded components, because of its complex stress distribution [110]. However, in

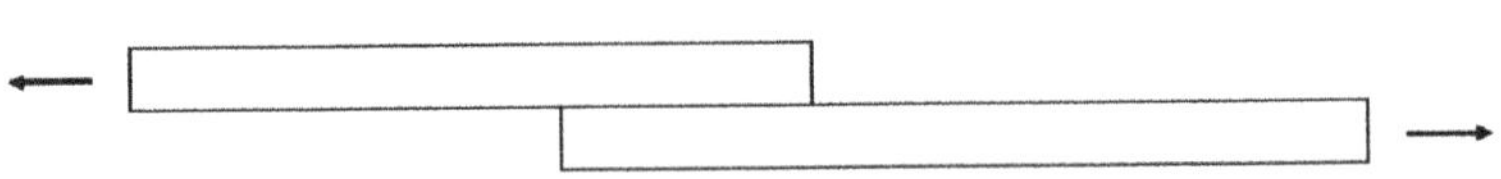

FIGURE 2.54 Single-lap joint.

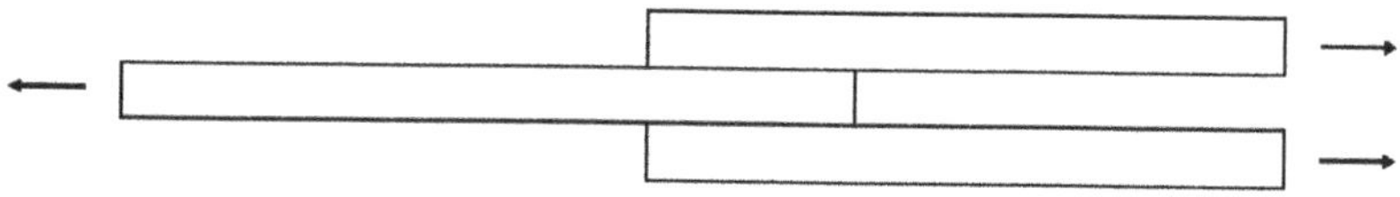

FIGURE 2.55 Double-lap joint.

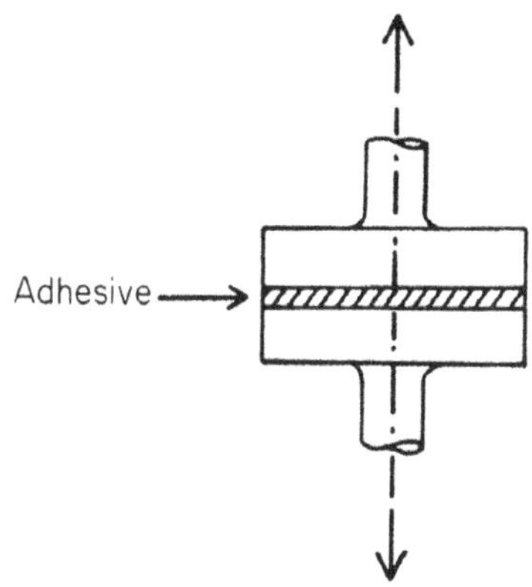

FIGURE 2.56 Adhesive tensile test.

spite of that, this test provides a rapid assessment of the shear stress of the adhesive and may serve as a good method of screening adhesives for a specific application.

The test piece for the adhesive *tensile test* can take several forms; one simple configuration consists of two cylinders (Fig. 2.56) that are bonded together at their ends. The joint is loaded to failure at right angles to the plane of the adhesive and the failure stress is determined by dividing the failure load by the loaded area [111]. The tensile test pieces are widely used for testing the strength of bonds of rubber and plastics to metals.

Peel tests are generally used for elastomeric adhesives. A typical version is the *T-peel test*, in which two strips of substrate bonded together face to face are peeled and the force required to pull them apart is recorded. The peeling can be done at a 90° angle, typically for two flexible substrates (Fig. 2.57) [112] or at a 180° angle for peeling of a flexible substrate from a rigid substrate (Fig. 2.58) [113]. Other adhesive tests (e.g., wedge test, cleavage strength test, impact strength test, and fatigue strength test) [106] are designed for specific applications and service conditions.

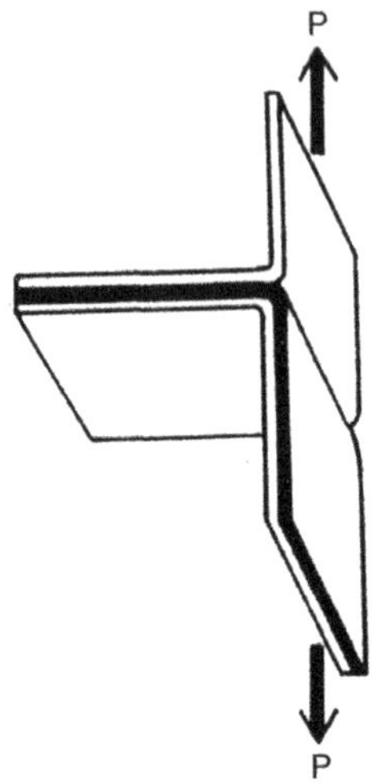

FIGURE 2.57 A 90° test.

FIGURE 2.58 A 180° test.

FURTHER READINGS

Cowie, J.M.G. and Arrighi, V., *Polymers: Chemistry and Physics of Modern Materials*, 3rd ed., CRC Press, Boca Raton, FL, 2008.

Strobl, G., *The Physics of Polymers*, 3rd ed., Springer-Verlag. Berlin, Heidelberg, 2007.

Flory, P.J., *Principles of Polymer Chemistry*, Cornell University Press, NY, 1953.

Mark, J. et al, *Physical Properties of Polymers*, 3rd ed, Cambridge University Press, Cambridge, UK, 2004.

Ehrenstein, G.W., *Polymeric Materials*, Carl Hanser Verlag, Munich, 2001.

Mallick, P.K., *Fiber Reinforced Composites*, 3rd ed., CRC Press, Boca Raton, FL, 2007.

Handbook of Elastomers (Bhowmick, A.K. and Stephens, H.L., Eds.), Marcel Dekker, New York, 2001.

Drobny, J.G., *Technology of Fluoropolymers*, 2nd ed., CRC Press, Boca Raton, FL, 2009.

Ebnesajjad, S. *Fluoroplastics, Volume 1 Non-Melt Processable Fluoroplastics*, William Andrew Publishing, Norwich, NY, 2000.

Polymeric Foams: Science and Technology (Khemani, K.L., Ed.), ACS Symposium Series **669**, American Chemical Society, Washington, DC, 1997.

Wicks, Jr., Z. W., Jones, F.N., and Pappas, S.P., *Organic Coatings: Science and Technology*, Volume **II**: Applications, properties and Performance, Wiley-Interscience, New York, 1994.

Poole, C. P., Jr., and Owens, F. J., *Introducing* to *Nanotechnology*, John Wiley & Sons, Inc., Hoboken, NY, 2003.

Resins for Coatings: Chemistry, Properties and Applications (Stoye, D. and Freitag, W., Eds.), Hanser Publishers, Munich, 1996.

Engineering with Rubber (Gent, A.N., Ed.), Carl Hanser Verlag, Munich, 1992.

Birley, A.W., Haworth, B., and Batchelor, J., *Physics of Plastics: Processing, Properties and Materials Engineering*, Carl Hanser Verlag, Munich, 1992, p. 275.

Polymeric Foams (Klempner, D. and Frisch, K.C., Eds.), Carl Hanser Verlag, Munich, 1995.

Gedde, U.W., *Polymer Physics*, Chapman & Hall, London, 1995.

Birley, A.W., Haworth, B., and Batchelor, J., *Physics of Plastics: Processing, Properties and Materials Engineering*, Carl Hanser Verlag, Munich, 1992.

REFERENCES

1. Ehrenstein, G.W., *Polymeric Materials: Structure-Properties-Applications*, Hanser Publishers, Munich, 2001, p. 14.
2. Strobl, G., *The Physics of Polymers: Concepts for Understanding Their Structures and Behavior*, Third Edition, Springer-Verlag, Berlin, Heidelberg, 2007, p. 2.
3. Cowie, J.M.G. and Arighi, V., *Polymers: Chemistry and Physics of Modern Materials*, 3rd ed., CRC Press, Boca Raton, FL, 2008, p. 9.
4. Stoye, D. and Freitag, W., *Resins for Coatings: Chemistry, Properties and Applications*, Carl Hanser Verlag, Munich, 1996, p. 29.

5. Cowie, J.M.G. and Arighi, V., *Polymers: Chemistry and Physics of Modern Materials*, 3rd ed., CRC Press, Boca Raton, FL, 2008, p. 15.
6. Cowie, J.M.G. and Arighi, V., *Polymers: Chemistry and Physics of Modern Materials*, 3rd ed., CRC Press, Boca Raton, FL, 2008, p. 279.
7. Cowie, J.M.G. and Arighi, V., *Polymers: Chemistry and Physics of Modern Materials*, 3rd ed., CRC Press, Boca Raton, FL, 2008, p. 281.
8. Cowie, J.M.G. and Arighi, V., *Polymers: Chemistry and Physics of Modern Materials*, 3rd ed., CRC Press, Boca Raton, FL, 2008, p. 282.
9. *Resins for Coatings: Chemistry, Properties and Applications* (Stoye, D. and Freitag, W., Eds.) Carl Hanser Verlag, Munich, 1996, p. 30.
10. Bagley, E.B. and Chen, S.A., *J. Paint. Techn.*, **41**, 494 (1969).
11. Hansen, C.M. and Skaadrup, K.J., *J. Paint, Techn.*, **39**, 505 (1967).
12. Hansen, C.M., *J. Paint, Techn.*, **39**, 104 (1967).
13. Brandrup, J. and Immergut, E.H., *Polymer Handbook*, 3rd ed., John Wiley & Sons, Inc., New York, 1989.
14. *Resins for Coatings: Chemistry, Properties and Applications* (Stoye, D. and Freytag, W., Eds.), Hanser Publishers, Munich, 1996, p. 31.
15. Wicks, Jr., Z.W., Jones, F.W. and Pappas, S.P., *Organic Coatings: Science and Technology*, Vol. **1**, John Wiley & Sons, Inc., New York, 1992, p. 26.
16. Turska, E., *J. Pract. Chem.*, **313**, (3) 387 (1971).
17. Kuhn, W., *Kolloid Z.*, **68**, 2–15 (1934).
18. Guth, E, and Mark, H.F., *Monatsh. Chem.*, **65**, 93–121 (1935).
19. Wicks, Z.W., Jones, F.W. and Pappas, S.P., *Organic Coatings: Science and Technology*, Vol. **1**, John Wiley & Sons, Inc., New York, 1992, p. 27.
20. Einstein, A., *Ann, Phys. Ser.* IV, **19**, 289–306 (1906); Einstein, A., *Ann. Phys.*, **24**, 591–592 (1911).
21. Elias, H.-G., Bareiss, R, and Watterson, J.G., *Adv. Polym. Sci.*, **11**, 111 (1973).
22. Meyerhoff, G., *Fortschr. D. Hochpolym-Forsch.*, **3**, 59–105 (1961).
23. Hildebrand, J.H. and Scott, R.L., *The Solubility of Nonelectrolytes*, Reinhold, New York, 1959; Hildebrand, J.H. and Scott, R.L., Regular Solutions, Prentice Hall, Englewood Cliffs, NJ, 1962.
24. Huggins, M.L., *J. Phys Chem.*, **46**, 151 (1942); Huggins, M.L., *J. Am. Chem. Soc.*, **64**, 1712 (1961).
25. Tompa, H., *Trans.Farad. Soc.*, **45**, 1142 (1949).
26. Scott, R.L., *J. Chem. Phys.*, **17**, 279 (1949).
27. Lipatov, Y.S. and Nestorov, A. E., *Thermodynamics of Polymer Mixing*, Technomic Publishing, Lancaster, PA, 1997.
28. *Engineering with Rubber* (Gent, A.N., Ed.), Carl Hanser Verlag, Munich, 1992, p. 69.
29. Strobl, G., *The Physics of Polymers: Concepts for Understanding Their Structures and Behavior*, 3rd ed., Springer-Verlag Berlin, Heidelberg, 2007, p. 223.

30. Birley, A.W., Haworth, B., and Batchelor, J., *Physics of Plastics: Processing, Properties and Materials Engineering*, Carl Hanser Verlag, Munich, 1992, p. 275.
31. Birley, A.W., Haworth, B., and Batchelor, J., *Physics of Plastics: Processing, Properties and Materials Engineering*, Carl Hanser Verlag, Munich, 1992, p. 276.
32. Birley, A.W., Haworth, B., and Batchelor, J., *Physics of Plastics: Processing, Properties and Materials Engineering*, Carl Hanser Verlag, Munich, 1992, p. 277.
33. *Handbook of Plastic Foams: Types, properties, Manufacture and Applications*, (Landrock, A.H., Ed.). Noyes Publications, Park Ridge, NJ, 1995, p. 1.
34. *Polymeric Foams* (Klempner, D. and Frisch, K.C., Eds.), Carl Hanser Verlag, Munich, 1995, p. 1.
35. *Polymeric Foams: Science and Technology* (Khemani, K.L., Ed.), ACS Symposium Series **669**, American Chemical Society, Washington, DC, 1997, p. 1.
36. *Polymeric Foams: Science and Technology* (Khemani, K.L., Ed.), ACS Symposium Series **669**, American Chemical Society, Washington, DC, 1997, p. 2.
37. Semerdjiev, S., *Introduction to Structural Foams*, Society of Plastics Engineers, Brookfield Center, CT, 1982, p. 1.
38. John, B. and Nair, C.P.R., *Update on Syntactic Foams*, Smithers Rapra, 2010, p. 2.
39. Erikson, R., "Foams on the Cutting Edge", *Mech. Eng. Mag.*, January 1, 1999.
40. Karthikeyan, C.S. and Sankartan, S., *J. Reinforced Plastics Composites*, **20** (11), 982 (2001).
41. John, B. and Nair, C.P.R., *Update on Syntactic Foams*, iSmithers Rapra, Shaubury, Shrewsbury, Shropshire, UK, 2010, p. 17.
42. Bruneton, E., Tallaron, C., Nanlin, N.G and Consculluela, A., *Carbon*, **40** (2), 1919 (2002).
43. John, B. and Nair, C.P.R., *Update on Syntactic Foams*, iSmithers Rapra, Shaubury, Shrewsbury, Shropshire, UK, 2010, p. 13.
44. Puterman, M., Narkis, M. and Kenig, S. *J. Cellular Plastics*, **16** (4), 223 (1980).
45. Narkis, M., Puterman, M., and Kenig, S., *J. Cellular Plastics*, **16** (6) 326 (1980).
46. Ehrenstein, G.W., *Polymeric Materials: Structure-Properties-Applications*, Hanser Publication, Munich, 2001, p. 89.
47. Gedde, U.W., *Polymer Physics*, Chapman & Hall, London, 1995, p. 111.
48. Gedde, U.W., *Polymer Physics*, Chapman & Hall, London, 1995, p. 127.
49. Ehrenstein, G.W., *Polymeric Materials: Structure-Properties-Applications*, Hanser Publication, Munich, 2001, p. 138.

50. "World plastics resin consumption", Plastics News, *FYI-Materials*, January 7, 2008, available at www.plasticsnews.com.
51. Ehrenstein, G.W., *Polymeric Materials: Structure-Properties-Applications*, Hanser Publishers, Munich, 2001, p. 2.
52. "The Compelling Facts about Plastics", *PlasticsEurope*, January 2008, available at www.plasticseurope.org.
53. *Plastics Engineering Handbook of the Society of the Plastics Industry*, 5th ed. (Berins, M., Ed.), Van Nostrand Reinhold, New York, 1991, p. 64.
54. Drobny, J.G., *Fluoropolymer Technology*, 2nd ed., CRC Press, Boca Raton, FL, 2009, p. 1.
55. Drobny, J.G., *Fluoropolymer Technology*, 2nd ed., CRC Press, Boca Raton, FL, 2009, p. 9.
56. Scheirs, J., *Fluoropolymers: Technology, Markets and Trends*, Rapra Technology Ltd., Shaubury, Shrewsbury, Shropshire, UK, April 2001, p. 9.
57. Drobny, J.G., *Fluoropolymer Technology*, 2nd ed., CRC Press, Boca Raton, FL, 2009, p. 12.
58. Drobny, J.G., *Fluoropolymer Technology*, 2nd ed., CRC Press, Boca Raton, FL, 2009, p. 44.
59. Bloomfield, P.E. et al., *U.S. Naval Res. Rev.* **31** (1), 1 (1978).
60. Robinson, A.L., *Science*, **200**, 137 (1978).
61. Drobny, J.G., *Fluoroplastics* Rep. 184, Vol. **16**, No. 4, (2006), Rapra Technology Ltd., Shaubury, Shrewsbury, Shropshire, UK, p. 21.
62. Drobny, J.G., *Fluoroplastics* Report 184, Vol. **16**, No. 4, 2006, Rapra Technology Ltd., Shaubury, Shrewsbury, Shropshire, UK, p. 25.
63. *Tedlar® Technical Information*, Publication H-9905, E.I. duPont de Nemours & Co., Wilmington, DE, 1993.
64. *Natural Rubber: Agromanagement and Crop Processing* (George, P.J. and Jacob, C.K., Eds.), Rubber Research of India, Kottayam, Kerala, 2000. p. 97.
65. *Handbook of Elastomers*, 2nd ed. (Bhowmick, A.K. and Stephens, H.L., Eds.), Marcel Dekker Inc., New York, 2001, p. 41.
66. *Handbook of Elastomers*, 2nd ed (Bhowmick, A.K. and Stephens, H.L., Eds.), Marcel Dekker Inc., New York, 2001, p. 46.
67. Nagdi, K., *Rubber as an Engineering Material: Guideline for Users*, Hanser Publishers, Munich, 1993, p. 98.
68. Nagdi, K., *Rubber as an Engineering Material: Guideline for Users*, Hanser Publishers, Munich, 1993, p. 100.
69. Nagdi, K., *Rubber as an Engineering Material: Guideline for Users*, Hanser Publishers, Munich, 1993, p. 108.
70. Drobny, J.G., "High-Performance Elastomers: A Review," Paper 29, Rubber Mini Expo™, 178th Technical Meeting, October 14–16, 2008, Louisville, KY.

71. Nagdi, K., *Rubber as an Engineering Material: Guideline for Users*, Hanser Publishers, Munich, 1993, p. 120.
72. Nagdi, K., *Rubber as an Engineering Material: Guideline for Users*, Hanser Publishers, Munich, 1993, p. 121.
73. Nagdi, K., *Rubber as an Engineering Material: Guideline for Users*, Hanser Publishers, Munich, 1993, p. 126.
74. Nagdi, K., *Rubber as an Engineering Material: Guideline for Users*, Hanser Publishers, Munich, 1993, p. 128.
75. Nagdi, K., *Rubber as an Engineering Material: Guideline for Users*, Hanser Publishers, Munich, 1993, p. 146.
76. Nagdi, K., *Rubber as an Engineering Material: Guideline for Users*, Hanser Publishers, Munich, 1993, p. 148.
77. Röthemeyer, F. and Sommer, F., *Kautschuktechnologie: Werkstoffe-Verarbeitung-Produkte*, Carl Hanser Verlag, Munich, 2001, p. 121 (In German).
78. Nagdi, K., *Rubber as an Engineering Material: Guideline for Users*, Hanser Publishers, Munich, 1993, p. 145.
79. Drobny, J.G., *Fluoropolymer Technology*, 2nd ed., CRC Press, Boca Raton, FL, 2009, p. 96.
80. Scheirs, J., *Fluoropolymers: Technology, Markets and Trends*, Rapra Technology Ltd., Shawbury, Shrewsbury, Shropshire, UK, April 2001, p. 94.
81. Scheirs, J., *Fluoropolymers: Technology, Markets and Trends*, Rapra Technology Ltd., UK, April 2001, p. 99.
82. *Rubber Technologist's Handbook*, Vol. **2** (White, J., De, S.K. and Naskar, K., Eds.), iSmithers/Rapra, Shawbury, Shrewsbury, Shropshire, UK, 2009, p. 382.
83. Maxson, M., Norris, A.W., and Owen, M.J., in *Modern Fluoropolymers* (Scheirs, J., Ed.), John Wiley & Sons, Ltd., Chichester, UK, 1997, p. 360.
84. Scheirs, J., *Fluoropolymers: Technology, Markets and Trends*, Rapra Technology Ltd., Shawbury, Shrewsbury, Shropshire, UK, April 2001, p. 110.
85. Drobny, J.G., *Handbook of Thermoplastic Elastomers*, William Andrew Publishing, Norwich, NY, 2007.
86. *Plastics Engineering Handbook of the Society of the Plastics Industry*, 5th ed. (Berins, M., Ed.), Van Nostrand Reinhold, New York, 1991, p. 523.
87. *Composite Materials Technology: Processes and Properties* (Mallick, P.K. and Newman, S., Eds.), Hanser Publishers, Munich, 1990, p. 196.
88. Hildebrand, J.H., and Scott, R.L., *The Solubility of Nonelectrolytes*, 3rd ed., Van Nostrand Reinhold, New York, 1950.
89. Gardon, J.L., *J. Phys. Chem.*, **67**, 1935 (1963).

90. Bikerman, J.J., *The Science of Adhesive Joints*, Academic Press, New York and London, 1961.
91. Zisman, W.A. in *Contact Angle, Wettability and Adhesion*, Advances in Chemistry Series 43, Chapt.1, American Chemical Society, Washington, D.C., 1964.
92. Stobbe, B.D., "Corona Treatment 101", *Narrow Web Industry Magazine*, June 1996.
93. Liston, E.M., "Plasma Treatment for Improved Bonding: A Review", *J. Adhesion* **30**, 199 (1989).
94. Shut, J.H., "Plasma Treatment", *Plastic Technol.*, **64** (Oct. 1992).
95. Kaplan, S.L. and Rose, P.W., "Plasma Treatment of Plastics to Enhance Adhesion: An Overview", Chapter 2, Technical Paper, Plasma Science, Inc., Lorton, VA.
96. Shield, J, *Adhesives Handbook*, Butterworth Publishers, Ltd., London, 1970.
97. Briggs, D, Brevis, D.M., and Konieczo, M.B., *J. Mater. Sci.*, **11**, 1270 (1976).
98. Schulz, J., Carré, A., and Mazeau, C., *Inst. J. Adhesion Adhesives*, **4**, 163 (1984).
99. Tavakoli, S.M. and Riches, S.T., "Laser Modifications of Polymers to Enhance Adhesion, Part 1", SPE-ANTEC, 1984, New Otleans, LA, Conference Proceedings.
100. The Loctite Design Guide for Bonding Plastics (Publication LT-2197), Loctite Corporation, Rocky Hill, CT.
101. Wegman, R.F., *Surface Preparation Techniques for Adhesive Bonding*, Noyes Publications, Park Ridge, NJ, 1989.
102. Ebnesajjad, S. and Ebnesajjad, C.F., *Surface Treatment of Materials for Adhesive Bonding*, William Andrew Publishing, Norwich, NY, 2006.
103. Wu, S., *Polymer Interfaces and Adhesion*, 1st ed., Marcel Dekker, New York, 1982.
104. Schollenberger, C.S., in *Handbook of Adhesives*, 3rd ed., Chap. 20 (Skeist, I., Ed.), Van Nostrand Reinhold, New York, 1990.
105. Coover, H.W., Dreifus, D.W., and O'Connor, J.T., in *Handbook of Adhesives*, 3rd ed., Chap. 27 (Skeist, I., Ed.), Van Nostrand Reinhold, New York, 1990.
106. *Handbook of Adhesives*, 3rd ed., Chap. 27 (Skeist, I., Ed.), Van Nostrand Reinhold, New York, 1990.
107. Tod, D.A., in *Handbook of Adhesion* (Peckham, D.E., Ed.), Longman Scientific Technical Ltd, Essex, UK, 1992, p. 470.
108. ASTM D1002: *Test Method for Strength Properties of Adhesives in Shear by Tension Loading*.

109. ASTM D3528: *Test Method for Strength Properties of Double Lap Shear Adhesive Joints by Tension Loading.*
110. Tod, D.A., in *Handbook of Adhesion* (Peckham, D.E., Ed.), Longman Scientific Technical Ltd, Essex, UK, 1992, p. 471.
111. ASTM D897: *Test Method for Tensile Properties of Adhesive Joints.*
112. ASTM D1876: *Test Method for Peel Resistance of Adhesives (T-Peel Test).*
113. ASTM D903: *Test Method for Peel or Stripping Strength of Adhesive Joints.*

CHAPTER 3

POLYMERS AS ELECTRICAL INSULATORS

To describe electrical properties of polymers means to thoroughly examine their responses to the electric field applied to them. Unlike metals, which respond readily to an electric field by electronic conduction, the response of polymers is more complex. The phenomena observed in polymers include polarization resulting from distortion and alignment of their molecules, which are under the influence of an applied electric field. In general, the electric conduction in polymers is very low, which makes them useful as very effective electric insulators. If there is any conduction in a polymeric substance, it is often caused by impurities that provide minute charge carriers in the form of electrons and/or ions. At sufficiently high fields, a variety of processes will cause a complete breakdown of the polymeric material serving as insulation. Depending on the mobility of existing charges and their ability to diffuse through the material, they would support a direct current (dc) conductance. If the diffusion of mobile charges will be spatially limited, the polymeric material will be polarized by the field, but the dc conductance will not occur, [1].

Polymers for Electricity and Electronics: Materials, Properties, and Applications, First Edition. Jiri George Drobny.

Thus the response of polymers to the imposed electric field can be classified as:

1. dielectric properties and
2. bulk conductive properties.

Dielectric properties are characterized by the *dielectric constant*, representing polarization and by the *tangent of the dielectric loss angle* representing relaxation phenomena. Bulk conductive properties are characterized by two parameters, namely, *dielectric strength*, representing breakdown phenomena and *conductivity*, representing electrical conduction.

Besides these two fundamental properties, there are two other special groups of electrical properties: (1) The piezoelectric, pyroelectric, ferroelectric, triboelectric, photovoltaic, and photo-conducting properties and (2) electrostatic charges. This chapter will examine predominantly the function of polymers as insulators. Electronic conduction and specialty polymers that are used for other applications are the main topic of Chapter 4.

Typical requirements for an electrical insulator are: it must have a high enough dielectric strength to withstand an electrical field between the conductors; it must have a high insulation resistance to prevent leakage of current across the conductors. It must posses a good arc resistance to prevent damage in case of arcing. Insulating materials must be mechanically strong to resist shocks, vibrations, and other mechanical forces. It must maintain integrity under a wide variety of environmental hazards (e.g., humidity, temperature, and radiation).

There are several significant differences between insulation materials used for alternating current (ac) and dc, especially for high-voltage direct current (dc). Insulations for that service are often formulated to include a small amount of conductivity [2].

In summary, the key electrical properties that help meeting the above requirements are dielectric constant, dielectric strength, dissipation factor, and volume and surface resistivity.

3.1 DIELECTRIC CONSTANTS OF POLYMERS

When thinking about the dielectric behavior of polymers, it is convenient not to consider the long-chain molecules, but rather the component repeat units [3]. For most polymeric materials, the degree of polymerization is >100, so that the effect of chain ends can be neglected for

TABLE 3.1 Dielectric Constants of Selected Polymeric Substances

Material	Dielectric Constant
PTFE [Poly(tetrafluoroethylene)]	2.1
PP (Polypropylene)	2.2
PE (Polyethylene)	2.3–2.65
PS (Polystyrene)	2.4
ABS (Acrylonitrile–butadiene–styrene)	2.6
Modified PPO (Polyphenylene oxide)	2.7
PC (Polycarbonate)	3.0
PET [Poly(ethylene terephtalate)]	3.0–3.3
Acrylic resins	3.0
PBT (Polybutylene terephtalate)	3.1
PPS [Poly(phenylene sulfide)]	3.2
PEEK [Poly(ether etherbetone)]	3.3
LCP (Liquid-crystal polymer)	3–4
Acetal copolymers	3.7
PVC [Poly(vinyli chloride)]	4.0
PA 66 (Polyamide 66)	4.6
PVF [Poly(vinyl fluoride)]	8.2–8.5
PVDF (Polyvinylidene fluoride)	8.2–9.5

most purposes. Each segment is constrained due to its chemical linkage to the rest of the molecular chain on each side. Typical values of dielectric constants of polymeric substances are shown in Table 3.1.

3.1.1 Nonpolar Polymers

The nonpolar polymers represent the simplest case, since permanent dipoles do not exist and the dielectric constant has an essential linear dependence on density. Branching and the degree of crystallization have some effects on the dielectric constant [3].

For PE or other similar assemblies of $-CH_2-$ units, the Clausius–Mossotti relation becomes

$$\frac{\varepsilon - 1}{\varepsilon + 2} = \mathrm{K} \cdot \rho \tag{3.1}$$

where $K = 0.327 \times 10^{-3}$ and ρ is density

From that $$\varepsilon = 2.276 + 2.01(\rho - 920) \times 10^{-3} \tag{3.2}$$

Conjugated polymers that contain multiple carbon–carbon bonds can be expected to have somewhat higher relative permittivities, since the bond polarizabilities of multiple bonds are higher than those of single C—C bonds [3].

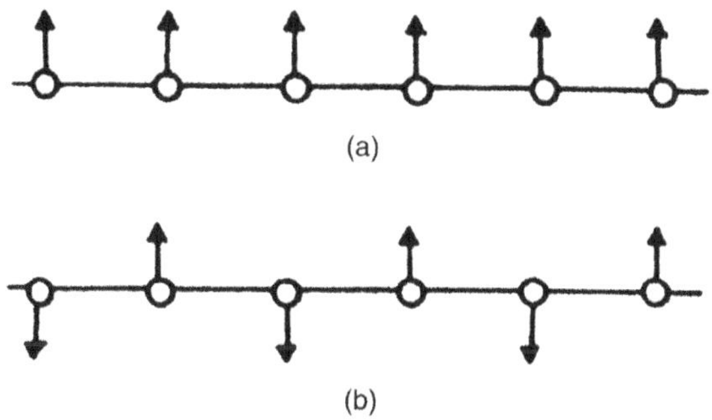

FIGURE 3.1 Schematic arrangement of polar groups in a polymer chain: (a) balanced dipoles and (b) additive dipoles. (Bueche, F., *Physical Properties of Polymers,* John Wiley & Sons, 1962, p. 275)

3.1.2 Polar Polymers

When permanent dipoles are present in a polymeric chain, two situations are possible [4]: (1) The whole polymer backbone, together with its polar group, is rigidly fixed in a single conformation and (2) the backbone is flexible with freely rotating pendant groups.

In Case 1, the resulting moment will depend on whether the individual segments reinforce or balance each other. For example, in PTFE, the high dipole moments of the alternating $-CF_2-$ groups balance each other exactly and its dielectric constant is low. The opposite situation is in PVC, where the C—Cl dipole moments are additive in a planar zigzag conformation (see Fig. 3.1) and its dielectric constant is high.

In Case 2, because of the free rotation, the observed dipole moment is an average from many different conformations. In theoretical considerations, the concept of mean-square moment is commonly used [5].

3.1.3 Polarization of Polymers

The difference between the dielectric constant of a polymer and that of free space is due to the restricted movements of charges within the polymer. Under the influence of the applied field, positive charges move with the electric field and an equal number of negative charges move against it. As a result there are no net charges anywhere within the polymer. However, there is a net positive charge at the surface where the positive direction of the field emerges and a negative charge where the field enters. Thus the field within the polymer is produced by a larger field outside it and the normal components have the ratio given by the dielectric constant. The charges are bound in the dielectric. This general process is called *polarization*. They cannot move through it; otherwise, they would produce conduction, not polarization. There are three fundamental components of molecular polarization [6]: electronic (Section 3.1.3.1), atomic (Section 3.1.3.2), and orientation (Section 3.1.3.3) (see Fig. 1.8). Electronic, atomic, and orientation polarizations are all due to charges that are locally bound in atoms molecules, or the structures of solids and liquids in addition, charge carriers usually exist that can migrate for some distances through the dielectrics. If a dielectric consists of several components of different properties, interfacial polarization occurs as a result of the accumulation of charges at the interface of the two adjacent media. This type of polarization is common in polymers and is closely related to the breakdown of polymers. Often polymers with additives, fillers, or blended polymers may be subjected to this type of polarization. Impurities may have the same effect (see Section 3.1.4).

3.1.3.1 Electronic Polarization An electric field will cause a slight displacement of the electron cloud of any atom in the polymer molecule relative to its positive nucleus. Electronic polarization occurs in all materials and is responsible for refraction of light. Typically, it occurs within 10^{-15} s and produces a value of dielectric constant of ~2.

3.1.3.2 Atomic Polarization An electric field also distorts the arrangement of atomic nuclei in a polymer molecule. It requires ~10^{-13} s and is less significant in ordinary polymers than in other dielectrics (e.g., ceramics).

3.1.3.3 Orientation Polarization If the polymer molecules already possess a permanent dipole moment, the moment will tend to be aligned by the applied field to give a net polarization in that direction. This polarization typically requires 10^{-9} s. A very high field, typically 10^5 V mm^{-1} is required to produce a significant alignment of the dipoles. Thus in most polymers orientation polarization is not significant [7].

3.1.4 Effect of Additives

In the formulation of polymer compositions, the use of polar plasticizers, antioxidants, stabilizers, and processing aids will increase the resulting dielectric constant. Conductive fillers (e.g., conductive carbon blacks) enhance dielectric properties beyond those of the base polymer. Rubber loaded with carbon black possesses a high conductivity, whereas unloaded rubber is an insulator. The conductive filler disperses in the polymer and usually forms domains that are separated by insulating layers of the polymer. In these systems, interfacial polarization becomes extremely large. The dielectric constant of such systems depends on the distribution of carbon black and is therefore useful in its evaluation by measuring the dielectric constant at different frequencies. With high loadings of conductive black, it is possible to prepare conductive vulcanizates [8] exhibiting a σ of 10^{-3}–10^{-1} $(\Omega \cdot \mathrm{cm})^{-1}$. Usually, the carbon black particles are separated by insulating polymer layers. In these systems, the interfacial polarization becomes extremely large. At carbon black concentrations of 30–60% by weight, the dielectric constant increases rapidly to reach values >100 [9–11].

Besides the effects of carbon blacks, other fillers with high dielectric constant (e.g., TiO_2, $SrTiO_3$, and $BaTiO_3$) are also effective. The inhomogeneity in the materials may cause interfacial polarization, and consequently increase of the value of ε [12].

3.1.5 Frequency Dependence of the Dielectric Constant

The dielectric constant of insulators varies with frequency. When evaluated at different frequencies, spectra are obtained that indicate relaxation processes and resonance.

Dipoles, permanent or induced, will rotate to align themselves with the outside field. If the field is alternating, the dipoles will rotate in sympathy with it. If the frequency of the field is equal to the natural frequency of the dipoles, a resonance will occur. If a dipole is in a media, which causes a heavy friction, the rotation will be slowed down and relaxation is observed. The rotation of the dipole will fall further and further in phase behind the applied field [12].

3.1.6 Temperature Dependence of the Dielectric Constant

In nonpolar polymers, where polarization is mainly electronic, the dielectric constant is independent of temperature. The Clausius–Mossotti equation (Eq. 1.6) can be roughly used to relate the polarizability to the dielectric constant. It seems that the dielectric constant is independent of temperature. However, since the density of the polymer is temperature dependent (decreases with increasing temperature) and ε is linearly dependent on density, the final effect is that the dielectric constant of nonpolar polymers decreases slightly with increasing temperature [13].

The Clausius–Mossotti equation no longer holds for polar polymers (see Section 1.6) and there is no satisfactory relation existing to accurately relate ε and T. Based on calculations using the Kirkwood equation [14, 15] the temperature dependence of the dielectric constant is also determined by the $1/T$ factor. Experimental studies indicated that the

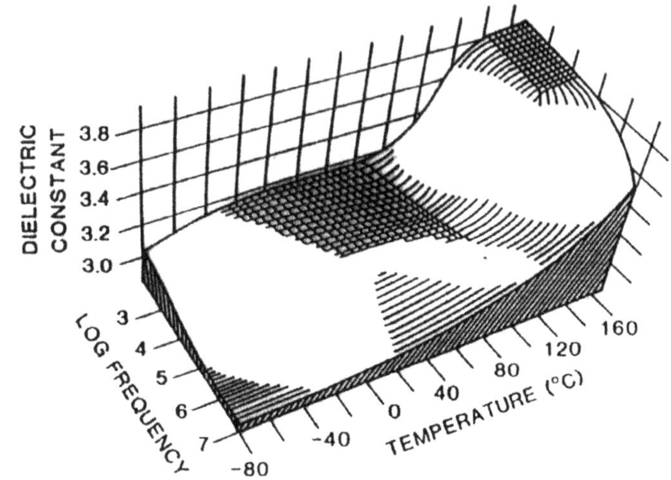

FIGURE 3.2 Dielectric constants as functions of frequency and temperature. (Ku, C.C. and Liepins, R., *Electrical Properties of Polymers: Chemical Principles,* Hanser Publishers, Munich, 1987, p. 52)

dielectric constant of practically all polar polymers increases with increasing temperature, indicating that this dependence is mainly governed by the change in the intra- and intermolecular interactions [14].

Polymeric materials with a dielectric constant independent of temperature can be prepared by copolymerization of monomers that are capable of cancelling the effects. An example is styrene (ε decreases slightly with temperature) and poly(methyl methacrylate) (PMMA) (a polar polymer whose dielectric constant increases with temperature). A copolymer of styrene and methyl methacrylate in proper proportion will have a dielectric constant independent of temperature within a certain temperature range [14].

Studies of the simultaneous effects of frequency and temperature on dielectric constants produce three-dimensional (3D) map surfaces (Fig. 3.2) [16]. This 3D map indicates that the dielectric constant increases with temperature and decreases with frequency. However, in certain frequency and temperature ranges, plateaus appear in the ε surface. The molecular relaxation processes in polymers that give rise to dielectric relaxation, in general, also give rise to mechanical relaxation (see Section 2.8.2.2).

3.2 DIELECTRIC RELAXATION

When an externally applied field is alternating, it causes an observable lag in attainment of equilibrium. This lag is commonly referred to as relaxation.

Dielectric relaxation, also known as dielectric dispersion or dielectric loss, is the exponential decay with time of the polarization in a dielectric when the externally applied field is removed. The relaxation time is defined as the time in which this polarization is reduced to 1/e times its original value, where e is the base of natural logarithm. Dielectric relaxation is the cause of anomalous dispersion in which the dielectric constant decreases with increasing frequency.

The basic theory of dielectric relaxation behavior, pioneered by Debye, begins with a macroscopic treatment of frequency dependence. This treatment rests on two essential premises: the exponential approach to equilibrium and the applicability of the superposition principle.

Orientation of molecular dipoles is a relatively slow process if comparison is made with electronic transitions or molecular vibrations, which have frequencies generally above 10^{12} Hz [17]. Only when sufficient time is allowed after the application of an electric field for the orientation to attain equilibrium will the maximum polarization, corresponding to the highest observable relative permittivity, be realized in a material. If ample time is allowed, then the observed relative permittivity is called the *static relative permittivity* (or static dielectric constant) ε_s. On the other hand, if the polarization is measured immediately after the field is applied, allowing no time after the field is applied, then the observed *instantaneous relative permittivity*, ε_∞ will be low and due to deformational effects alone. We can expect that somewhere in between these two extremes of time scale there will be dispersion from high to low relative permittivity [17].

We can start the examination of this rate effect by considering the application of an alternating electric field E, amplitude E_0, and angular frequency ω, across the dielectric material:

$$E = E_0 \cos \omega t \tag{3.3}$$

This will produce polarization, which alternates in direction, and, if the frequency is high enough, the orientation of any dipoles, which are present, will inevitably lag behind the applied field. The phase lag is denoted δ and this phase lag in the electric displacement D will be

$$D = D_0 \cos(\omega t - \delta) \tag{3.4}$$

which can be written as:

$$D = D_1 \cos \omega t + D_2 \sin \omega t \tag{3.5}$$

where

$$D_1 = D_0 \cos\delta \qquad \text{and} \qquad D_2 = D_0 \sin\delta$$

From that, we can define the two relative permittivities

$$\varepsilon' = \frac{D_1}{\varepsilon_0 E_0} \qquad \text{and} \qquad \varepsilon'' = \frac{D_2}{\varepsilon_0 E_0} \tag{3.6}$$

linked by the relation

$$\frac{\varepsilon''}{\varepsilon'} = \tan\delta \tag{3.7}$$

It is convenient to combine these two quantities into a *complex relative permittivity* (or complex relative constant):

$$\varepsilon^* = \varepsilon' - i\varepsilon'' \tag{3.8}$$

The meaning of the real and imaginary parts can be appreciated when considering a material in a capacitor, as shown in Fig. 3.3 (C_0 is the capacity when empty). The current I flows in the external circuit after application of an alternating voltage given by the real part of $V(V = V_0 e^{i\omega t})$ can be calculated from the complex relative permittivity [18]. The graphs of ε' and ε'' versus frequency of the applied field (on a logarithmic scale) referring to as debye dielectric dispersion curve are shown in Fig. 3.4.

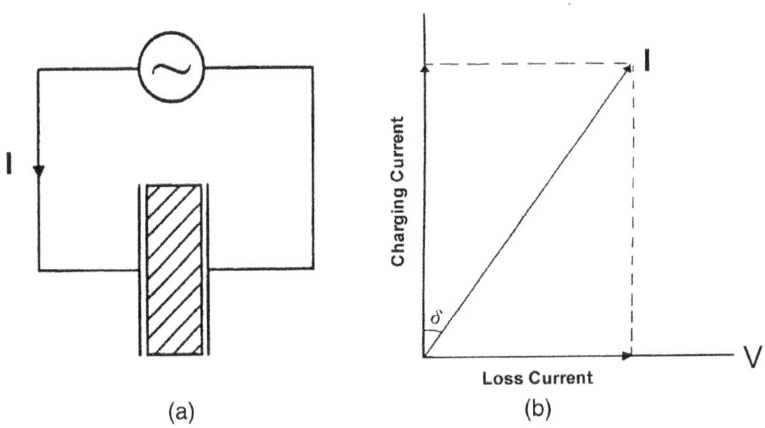

FIGURE 3.3 Alternating current losses in a dielectric: (a) diagram of an ac circuit with a dielectric, (b) diagram of a complex current-applied voltage relationship V applied ac voltage; I-current, δ-loss angle.

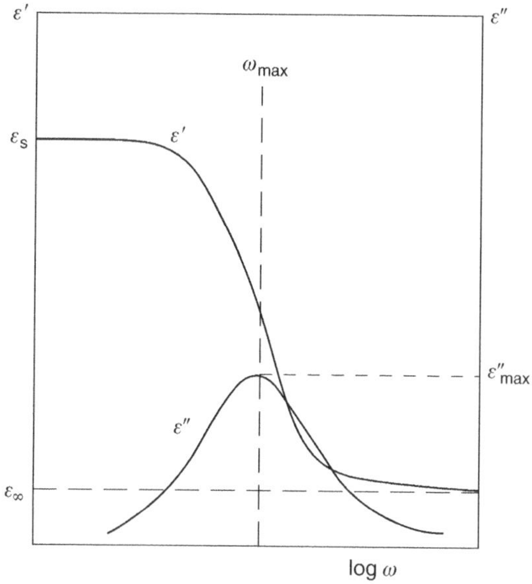

FIGURE 3.4 Debye dielectric dispersion curves. (Blythe, T. and Bloor, D., *Electrical Properties of Polymers*, Second edition, Cambridge University Press, Cambridge, UK, 2005, p. 63)

3.3 TANGENT OF THE DIELECTRIC LOSS ANGLE

The current I, which flows in the external circuit after application of an alternating voltage $V = V_0 e^{i\omega t}$ may be calculated as follows:

$$I = \varepsilon^* C_0 \frac{dV}{dt} = i\omega\varepsilon^* C_0 V = \omega C_0(\varepsilon'' + i\varepsilon')V \tag{3.9}$$

This implies that we have a capacitive component of the current

$$I_C = i\omega C_0 \varepsilon' V \tag{3.10}$$

which leads the voltage by 90°, and a resistive component,

$$I_R = \omega C_0 \varepsilon'' V \tag{3.11}$$

which is in phase with the voltage. Work can only be done by the latter component and the physical meaning of the useful quantity tan δ, previously defined by Eq. 3.7 becomes apparent:

$$\tan\delta = \frac{\varepsilon''}{\varepsilon'} \sim \frac{\text{energy dissipated per cycle}}{\text{energy per cycle}} \tag{3.12}$$

ε'' is called the dielectric loss factor or loss index, δ is the loss angle, and tan δ is the dielectric loss tangent or dissipation factor [19].

The ratio of the two, $\varepsilon''/\varepsilon'$ or the tangent of the dielectric loss angle, is commonly employed as a direct measure of the dielectric loss. Relaxation time, τ is defined as:

$$f(\tau) = \frac{1}{e} \tag{3.13}$$

that is, the time during which the polarization decayed to 1/e of its original value. It is a measurable quantity. For the maximum loss factor, the angular frequency, called the *critical frequency*, should be

$$\omega_{\mathrm{m}} = \frac{1}{\tau} \tag{3.14}$$

For examining the data, it is more convenient to use the critical frequency f_{m} instead of ω_{m}.

$$f_{\mathrm{m}} = \frac{\omega_{\mathrm{m}}}{2\pi} = \frac{1}{2\pi\tau} \tag{3.15}$$

The factors affecting the loss tangent are polymer morphology (branches, etc.), degree of crystallinity of the polymer, molecular weight, and plasticization.

Polymer Morphology

Branches rotate independently and will cause small-to-moderate dispersion.

Degree of Crystallinity

Most polymers exhibit more than one region of dielectric loss. Usually a minimum of two regions occurs in amorphous polymers, whereas in crystalline polymers loss regions can be expected to arise from both phases. The rubber-to-glass transition (at T_g) is definitely the source of a major dispersion. Most crystalline dispersion modes occur between T_g and the crystalline melting point (T_m). In studies of crystalline polymers, the peaks are labeled with Greek letters.

Molecular Weight

Generally, the tan δ_{max} decreases as the molecular weight increases and at very high molecular weights it levels off (Table 3.2). The degree of cross-linking has a similar effect, with tan δ decreasing with the increasing degree of cross-linking.

TABLE 3.2 Values of tan δ Values of Polystyrene and Polybutadiene with a High Dispersity of Molecular Weights at 60 Hz

Polymer	Frequency (Hz)	Temperature (°C)	Average Molecular Weight (10^3)	tan δ
Polystyrene	3×10^3	25	18	0.00220
			27	0.00144
			33	0.00127
			41	0.00124
			54	0.00078
			96	0.00051
			240	0.00043
Polyisobutylene	800	20	3	0.00420
			15	0.00260
			50	0.00220
			100	0.00155
			200	0.00150

Plasticization

The addition of plasticizers has a profound effect on tan δ. An example of PVC plasticized with dioctyl phalate (DOP) is shown in Fig. 3.5. As the amount of DOP is increased the peak value decreases and is shifted toward lower temperatures. The half-width of the curve becomes wider, which flattens the curve and decreases the value of tan δ. For practical applications, spectra in a proper temperature range are more important than a single value of tan δ. They should be generated at the operating frequency. Different plasticizers have different spectra because of their different chemical nature, polarity, compatibility with the polymer, viscosity, and so on. An example of the effect of different plasticizers on the value of tan δ of PVC is shown in Fig. 3.6.

The majority of applications of polymers in electrical and electronics technology require materials with low dielectric loss tangent values. The exception is the dielectric heating. The efficiency of dielectric heating is frequently compared by means of a comparison coefficient J [20], which is defined as:

$$J = \frac{1}{\varepsilon \tan \delta} \tag{3.16}$$

because the heat developed is proportional to the frequency and the product of ε and tan δ [20]. Some of the J values are listed in Table 3.3. These data can be used in actual practical applications [20].

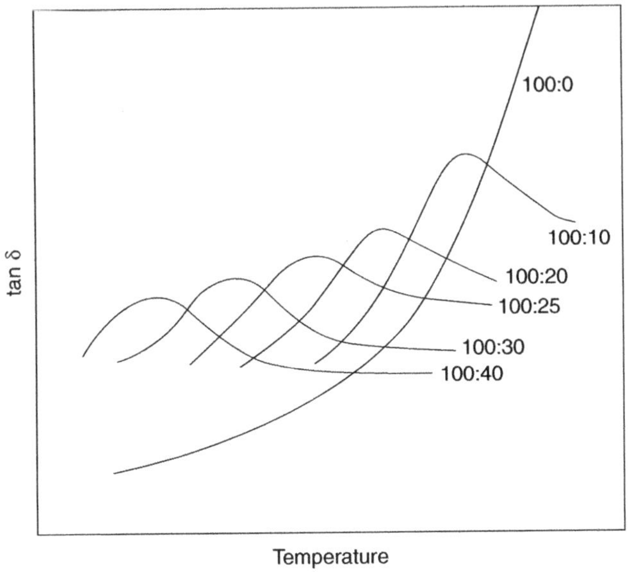

FIGURE 3.5 The effect of plasticizer (DOP) on the tan δ of PVC. (Ku, C.C. and Liepins, R., *Electrical Properties of Polymers: Chemical Principles,* Hanser Publishers, Munich, 1987, p. 87)

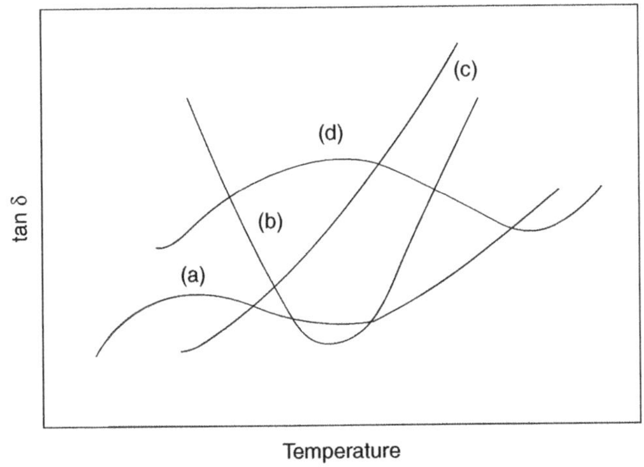

FIGURE 3.6 The effect of different kinds of plasticizers on the tan δ value of PVC: (a) DOC; (b) dibutyl phthalate, (c) dioctyl sebadente and (d) tricresyl phosphate. (Ku, C.C. and Liepins, R., *Electrical Properties of Polymers: Chemical Principles,* Hanser Publishers, Munich, 1987, p. 88)

An example of a J value use for the high-frequency dielectric heating of a phenol–formaldehyde resin ($J = 1.9$) and polystyrene ($J = 1330$) at the same duty cycle [20] follows: heating of phenol–formaldehyde

TABLE 3.3 Comparison Coefficients *J* of High-Frequency Dielectric Heating for Some Polymers (Ku, C.C. and Liepins, R., *Electrical Properties of Polymers: Chemical Principles*, Hanser Publishers, Munich, 1987, p. 93)

Polymer	*J*
Polystyrene	1330
Polyethylene	1100
ABS Resin	40
Poly(ethylene terephtalate)	20–35
Poly(vinyl chloride)	20
Polymethyl methacrylate	4–15
Polyamides	1.5–15
Urea–formaldehyde resin	3.8
Melamine resin	2.4
Phenol–formaldehyde resin	1–1.9

resin (30 s) and heating of polystyrene at the same duty cycle: $30 \times (1330/1.9) = 21{,}000$ s or 5.83 h.

The heat generated in the polymeric material is essentially based on the tan δ loss, although the final result will not only depend on the relaxation loss. The conductivity of the material may also contribute to the tan δ [21].

Additional details and a theoretical discussion of the dielectric relaxation can be found in chapter 3 of the following two publications: Ku and Liepins (1987) and Blythe and Bloor (2005).

3.4 DIELECTRIC BREAKDOWN OF POLYMERS

3.4.1 Introduction

When a piece of dielectric material is placed between electrodes and subjected to a steadily increasing voltage, at one point the insulating barrier will be exceeded and a sudden, catastrophic total breakdown will occur. With very high voltage, a great amount of electrical energy is released and the material will burn out in the breakdown area. Frequently, the breakdown occurs at a location with some imperfection or nonuniformity. In the material, it is rather difficult to define a property that represents an intrinsic resistance of the material to high voltage. Such resistance, referred to as dielectric breakdown strength (E_B), can be as high as 100 MV m^{-1} (25.4 kV mil^{-1}) for a homogeneous solid. Because it depends on unpredictable, statistic conditions of the material, the results are stochastic in nature [22].

Because of the difficulty in measuring a property that predicts the ability of the material to serve as an insulator, the industry adopted special tests, which simulate the conditions of the application.

Factors, such as aging and other deterioration of the material under service conditions (heat, mechanical stress, UV radiation, ozone, chemicals), are very important. They almost always introduce dielectric weakness, thus the tests will at least express the changes.

The fundamental breakdown processes can be roughly divided into electronic and purely thermal processes. In addition, a mechanical breakdown process exists, in which the mechanical deformation under the applied electric field dominates the breakdown phenomena. Many breakdown theories have been proposed, but it is still difficult to decide, which mechanism dominates the breakdown process. The principal mechanisms of dielectric breakdown in polymers can be listed as [23]: (1) electronic, (2) thermal, (3) electromechanical, and (4) gas discharge.

3.4.2 Electronic Breakdown

Electronic breakdown is initiated by the small number of electrons that are available for acceleration by the applied field. By the analogy with the mechanism of sparking gases, one can imagine that a Townsend-like avalanche may occur whenever the field is high enough for a conduction electron to gain sufficient energy to excite more electrons by collisions [24]. This is known as electron multiplication. In order to predict absolute dielectric strengths, we need to have more detailed information than is yet available about electronic states and mobilities in polymers. For the present, we can only conclude that there is a satisfactory agreement between the form of theoretical results, based on a rather general electronic model, and the best experimental results [25]. To the extent that the model is a very reasonable one, we can say that we can understand intrinsic breakdown behavior.

3.4.3 Thermal Breakdown

Whenever there is sufficient conductivity present in a dielectric to produce appreciable joule (ohmic or resistive) heating in an applied field, the possibility of thermal runaway exists. In alternating fields, there may be additional heat generated by relaxation processes, described in Section 3.2, and that will hasten the onset of any thermal run-away condition [25]. Whether breakdown will eventually develop in this way or not will depend on the rate at which heat is conducted away to surroundings. The heat balance equation is expressed by the following continuity equation:

$$\begin{matrix}\text{Electric power} & & \text{Rate of increase} & & \text{Rate at which}\\ \text{dissipated in material} & = & \text{in heat} & + & \text{heat is}\\ \text{per unit volume} & & \text{content} & & \text{conducted away}\end{matrix}$$

Thermal breakdown of polymers is most likely to occur either at high ambient temperatures, where electrical conductivity may become appreciable [e.g., PA 66 > 90°C (194°F)] or at high frequencies, where the heat dissipation by dipolar relaxation processes may be high [26].

3.4.4 Electromechanical Breakdown

The mechanism is based on the fact that the breakdown characteristics of polymers in the temperature region near the melting (softening) point are similar to their changes of mechanical characteristics. The breakdown is caused by the mechanical deformation resulting from Maxwell stress under the applied electric field [27].

Electrodes attached to the surface of a specimen during a dielectric breakdown test will exert a compressive force on the specimen by mutual coulombic attraction of the electrodes as the voltage V is imposed. If this is sufficient to cause appreciable deformation at fields below the intrinsic breakdown, the dielectric strength will be reduced. With the Young's modulus Y of the dielectric material being tested, the highest apparent dielectric strength E_a of the material placed between electrodes (original distance d_0 and compressed distance d) will be [28]

$$E_a \simeq 0.6\left(\frac{Y}{\varepsilon_0\,\varepsilon'}\right)^{\frac{1}{2}} \tag{3.17}$$

where ε_0 is the dielectric constant of vacuum and ε', the dielectric constant of the material being tested.

Apparently, low values of breakdown strength of many rubbery materials are described quantitatively by Eq. 3.17 and most plastics fail by the electromechanical mechanism at high temperatures [28, 29].

3.4.5 Gas-Discharge Breakdown

The dielectric strength of a gas is much less than that of a solid insulator, which is on the order of 3 MV m^{-1} (76.2 V mil^{-1}) [26]. Consequently, during the application of a high voltage to a solid specimen, discharges are likely to occur at an early stage in any gas that is at the edges of the electrodes or may be occluded as bubbles in the solid. Such external or internal discharges tend to damage the solid, and repeated discharges lead to dielectric failure. Organic polymers are especially prone to this type of failure, because bombardment of the polymer molecules by ions in the discharge is able to break chemical bonds. The reactive free radicals hereby produced, together with any available oxygen (often dissolved in the polymer), enter into a degradative chain reaction, causing damage to the dielectric. Each discharge in a cavity is extinguished as charges are deposited on the end walls, thereby reducing the field within the cavity. With an applied direct voltage, discharges will recur only after surface charges deposited by the previous discharge have had time to leak away. When an alternating voltage is used, discharges will be repeated with every half-cycle and the damage becomes more severe [30]. The high temperatures in the discharges often cause the polymer to degrade carbon and gaseous products. This process is often accelerated by reaction of ozone and other reactive products, such as nitrogen oxides formed in the gaseous discharge [31].

From the above, it can be seen that the complex nature of polymer macromolecules is often compounded by such factors as partial crystallinity and polar groups. However, since the chain length, order, orientation, and linkage have known effects on mechanical properties, it is at least logical to seek links between structural parameters and electrical properties.

3.4.6 Surface Discharge and Tracking

In many practical situations, an insulator is most likely to fail through deterioration of its surface [32]. Deposition of dirt and moisture inevitably allow some conduction over the surface, although good insulators will recover quickly as the heating effect of the leakage current tends to clean the surface [33]. Under these conditions some polymers tend to suffer permanent damage, called *tracking*, that leads to complete electrical breakdown. Tracking starts as the surface dries out and narrow, dry bands form. Most of the voltage is then dropped across the higher resistance of the dry band which can cause sparks to pass through the surface layer. If these sparks char the polymer, a conductive track may develop across the surface of the insulator and will ultimately result in flash-over. The polymer sometimes bursts into flames at this stage [33].

Much research has been done to examine all factors affecting the electric breakdown. The following variables are found to have an effect on the dielectric breakdown of organic polymers: polymer structure and morphology, type and amount of plasticizers, and type and amount of fillers.

3.4.7 Treeing in Polymers

Electrical breakdown starts frequently with prebreakdown that is called treeing. Treeing has been recognized as a cause of long-term electrical failures in solid dielectrics. Continuous dc, as well as ac and pulsed dc excitation, may lead to treeing. Direct current treeing is of great importance in practical applications, such as ultrahigh voltage dc transmission [34].

Direct current treeing breakdown is presumably associated with space-charge formation in polymers [35]. Space charge is formed during the prebreakdown conduction phase, so the prebreakdown conduction mechanism influences the field distribution that will be present at the onset of breakdown condition [36].

In general, two kinds of trees exist in polymers [34]: (1) electrical and (2) water trees. The former are generated by an electrical field alone, whereas the latter are generated by an electrical field together with water or various chemicals. In addition, trees can form under a variety of initiation and growth circumstances. Thus a relatively large number of descriptive terms are used [37–39]. There are two generic forms of electrical tree: vented and bow-tie. Vented trees start at the boundary of the insulation next to on electrode and grow in tree- or bushlike fashion toward the opposite electrode. Bow-tie trees grow simultaneously toward electrodes from an interior discharge initiation points. Other electrical trees are called dendrites, branch- or bush-type trees, spikes, strings, bow-ties; water trees are generally referred to as plumes, streamers, deltas, broccoli, or bow-ties. Examples of electrical trees are shown in Fig. 3.7.

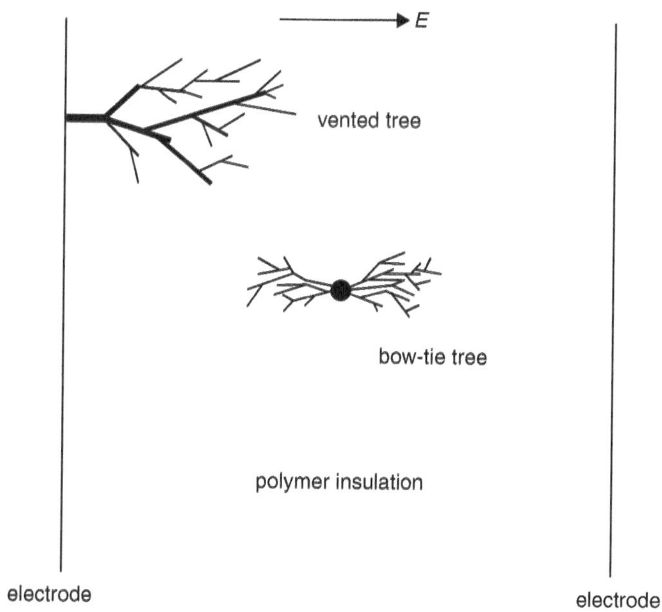

FIGURE 3.7 Examples of tree patterns. (Ku, C.C. and Liepins, R., *Electrical Properties of Polymers: Chemical Principles*, Hanser Publishers, Munich, 1987, p. 119)

Various staining techniques have been developed to render the trees permanently visible [40, 41]. Electrical trees that consist of hollow channels resulting from the decomposition of material are clearly visible in translucent or transparent solid dielectrics when examined under an optical microscope and transmitted light. On the other hand, fresh, unstained water trees appear diffuse and temporary [42]. The distinctions between electrical and water trees that have been emphasized are that water trees do not consist of permanent hollow channels as seen in electrical trees; water trees do not exhibit branching, although overlapping growth sometimes makes this hard to recognize; and growing water trees consist of strings of water-filled cavities [42]. An electric field and a stress concentration are required to initiate tree growth. The kind of tree that grows depends on the magnitude of the stress and the ambient conditions. In a dry environment, high stress is required and the trees are electrical. They grow by a partial discharge, causing decomposition of the material, and produce small hollow channels. In a wet environment, the voltage stress will determine whether electrical or water trees are produced. The electrical trees may grow slowly (although not as slowly as water trees) or rapidly as the result of an impulse, such as could be produced by a surge or lightning strike [43]. Water trees will be discussed in more detail in Section 3.4.7.2.

3.4.7.1 Electrical Trees There are many approaches to the problem of reducing or even eliminating electrical treeing in insulation systems. In the original paper–oil cables, treeing was overcome by careful drying and degassing of both paper and oil as well as the development of self-contained and pressurized oil-filled constructions [44].

The whole process of treeing can be divided into two distinct stages [45]: (1) the inception and (2) the propagation stage. The relative duration of each stage is different for different cases. The shape of the tree formed in each case is determined by the relative time difference.

3.4.7.1.1 Initiation of Electrical Trees In general, internal discharges are a common feature of polymeric insulators, because voids are very easily left in a material in the molding stage, or are associates with irregular, particulate contaminants. Since the relative permittivity of the gas filling the void is lower than that of the polymer, the electric field in the void will be greater than the field in the surrounding medium, depending to some extent on the shape of the void [30].

Electrical treeing of ac excitation is due to the internal partial discharge within the polymer under a high electrical field. The internal discharge leads to chemical decomposition of the polymer. A number of mechanisms for this process have been proposed [46]:

1. Electron injection.
2. Emission and action of *hot* electrons.
3. Electromechanical compression from the field, which leads to mechanical damage.
4. Presence of microvoids or occluded air or impurities around any incidental asperity resulting from imperfect wetting.
5. Fatigue cracking from repeated polarity reversals.
6. Local superheating and thermal decomposition.

3.4.7.1.2 Growth of Electrical Trees To achieve long service in an insulator, it is clear that gas discharges must be completely absent at the working voltage. For this reason, the voltage at which discharges start, commonly referred to as the discharge inception voltage (DIV), is an important characteristic of a product for high-voltage applications. Very sensitive DIV detectors incorporating visual displays are commercially available for testing purposes.

When the applied field is sufficient to initiate a discharge in a pre-existing cavity, the resulting damage produces a tubular channel in the polymer with a diameter of 1–5 μm, generally progressing in the field direction. The process develops more readily under ac conditions, because discharges continue to be triggered at each field reversal. The extending tubules of destruction progress in a way that seeks out pathways of predisposed weakness, (e.g., interfacial boundaries in nonhomogeneous materials). There is also a strong tendency for branching to occur.

The gas-discharge process will continue to drive the growth of the tree until the tree structure joints the electrodes. Large destructive currents can then flow, and complete breakdown of the insulation will ensue [47]. When voids are substantially absent in an insulating material, electrical trees still may occur, though generally at higher fields. Based on a great number of experiments, the following mechanisms are the main factors in growth of electrical trees [48]:

1. Internal gas discharge in the existing tree channel.
2. The gas discharge is pulsative and lasts <0.1 μs.
3. The velocity of extension of tree channel during a pulse discharge is $\sim 10^5$ cm s^{-1}.

4. The dielectric strength of the polymer films is nearly the same when two metal electrodes are used as when a plasma column discharge is the electrode.
5. The diameter of a tree channel is ~1 μm and is expanded gradually by repeated gas discharges.

Thus the extension of tree channels is caused by a partial dielectric breakdown of a polymer resulting from the strongly divergent electric field induced by the conductive plasma in the gas discharge. The partial dielectric breakdown might be due to electron avalanche, mechanical breakdown, thermal decomposition, or impulse discharge depending on the physical and chemical nature of the polymer [49].

3.4.7.1.3 Inhibition of Electrical Treeing There are many ways of reducing or even eliminating electrical treeing in insulation systems. The original treeing problems, which were very serious in paper–oil cables, were overcome by careful drying and degassing of both paper and oil, as mentioned previously or by development of self-contained and pressurized oil-filled constructions [50].

The following methods have been developed:

1. *Cable Construction:* Replacement of conductive tapes by extruded layers of semiconductive polymers [51, 52] applied by a so-called cross-head die. An example for the design of a typical power cable is shown in Fig, 3.8. In spite of all the process changes to minimize void formation, there still will be some imperfections generated by differential thermal expansion effects. These voids may form at the interfaces between the insulation and both the inner-metal (usually copper or aluminum) core and any outer-metal sheath. An incidental void next to the inner core would cause a gas discharge in the

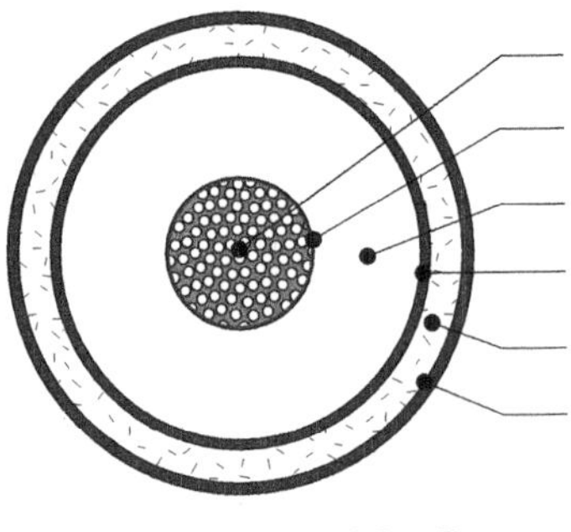

FIGURE 3.8 Cross-section of a power cable.

high-field region next to the core and the cable would have a lower discharge inception voltage. It is very difficult to prevent voids altogether, but the associated electrical breakdown problems may be solved by interposing a layer of conductive shielding between the metal and the insulation. Such shielding is frequently a sheet of a carbon black filled polymer (e.g., PE, ethylene–vinyl acetate copolymer, or EPDM rubber) [53].

2. *Process Changes:* (1) modifications of continuous vulcanization (CV) by a combination of steam heating with an ultrasonic generator, (2) radiation curing process (RCP) where the cable passes through an infrared heat source, a hot tube, in a pressurized inert gas atmosphere, (3) oil bath heating where steam and water have been replaced by the use of oil for both heating and cooling [54, 55], (4) a long extrusion die method, where a continuous vulcanization process [56] in which a very long extrusion die (up to 50 ft or 15.24 m) is used. In this process extrusion, heating and cross-linking are accomplished in separate steps, but all are contained within this die, (5) electron beam (EB) curing where irradiation of wire and cable insulation by an EB has been established and is widely used. In this process, voids can be generated, but they can also by avoided by controlling the dose rate [54].
3. *Tree- Retardant Additives*

There are many ways to prevent electrical trees by using a variety of special additives. Examples follow:

- Filling cavities with oil under pressure [57, 58], electronegative gas (e.g., SF_6 [59, 60], and even water to prevent partial discharges.
- Addition of small amounts of slightly incompatible liquids (e.g., diphenyl-*p*-phenylenediamine, phenyl-α-naphtylamine, and pentachlorophenol) [61].
- Use of voltage stabilizers (e.g., *o*-nitrotoluene, 2,4-dinitrotoluene, diphenyl amine, anthranilonitrile, etc.) [62].
- Inorganic fillers: Finely divided inorganic fillers were found to reduce treeing [63–68].

Variables affecting the tree formation are

- Polymer structure [69].
- Temperature, voltage, and strain [70].

3.4.7.2 Water Treeing As pointed out earlier, water trees develop in insulation exposed to water while also under a high electric field. It has been established [71] that they are rather different from the usual electrical trees. They appear to consist of strings of microcavities (1–5 μm in diameter) filled with electrolyte, possibly with some chemically modified inner surfaces, and in semicrystalline polymers, with some disordering of lamellae [71]. There is no direct connectivity between the cavities, although an electrolyte appears to be able to percolate among them through amorphous materials. Like an electrical tree, they grow in vented and bow-tie forms. Water trees can sometimes bridge the entire insulation between the electrodes without producing the final breakdown. It is usually an electrical tree that is responsible for the final breakdown, often starting from within a water tree.

Initiation and propagation of water trees depend on a multitude of factors: electric field strength, mechanical stress, type and concentration of ions in the water, constitution of the polymer, and the additives used in the polymeric compound.

3.4.7.2.1 Inhibition of Water Treeing Since water trees, just as electrical trees, are initiated by the stress concentration, contaminant particles and voids, as well as by presence of water, many of the methods used for the electrical trees also apply to water trees, particularly cable construction and chemical additives [72].

In cable design the power cables, which are completely enclosed in a continuous metal sheath, are watertight and do not exhibit treeing. This design is considered expensive and complicated to make [73]. Another complex patented design consists of different elements, such as a strand-blocking compound, an extruded semiconductive shield, and an extruded emission shield, which has a dielectric constant several times higher than that of the primary insulation [74].

Chemical additives used for the inhibition of water trees include commercial products, developed for this purpose, non-ionic surfactants [75], and electrolytic additives (e.g., NaCl, Na_2SO_4, Mg SO_4, $CaCl_2$) [76]. Here also voltage stabilizing additives are used and are similar to those used for suppressing electrical trees [77].

3.4.7.3 Effects of Space Charges Under the influence of a high electric field, charges may enter a dielectric by injection from electrodes and become trapped there as space charges. Space charges can also arise from ionic impurities within the material. The fields from these charges add to, or subtract from, the externally applied electric field, and can generate high-field distortion in their vicinity. Consequently, space charges will likely have bearing on the breakdown behavior [78]. Under a dc condition, space charges build up gradually over time. Charge separation originated in the body of the dielectric would have the effect of increasing the field near the electrodes. This may be even more damaging to the life of the material in service [78]. In developing materials for high-voltage insulation, it is important to check the susceptibility of the material to build up space charges, and if necessary, to take steps to reduce it. Attention to cleanliness will avoid the presence of ionizable impurities in the material and at its interface with electrodes [78].

In the case of ac fields, the build up of space charges is expected to be much less than under steady dc conditions, because the field reversals continuously cancel the effects of the previous half-cycles. However, some measurements show that there can be some space charges developed even under ac conditions. This result is possibly explained by variations in composition across the layer and over the interface layer at the electrodes [79].

3.4.8 Long-Term Endurance Under Electric Stress

The discussions up to this point (except of treeing) were concerned mainly with short-term effects. But for practical purposes and for the goal of manufacturing reliable electrical equipment, it is important to understand long-term behavior of dielectrics under an electric stress and determining the average service life of an insulator.

The breakdown process even under short-term electrical stress is rather complex and multistaged with elements of several of the mechanisms discussed. For long-term applications, the very distinctive aging process of polymers has to be considered.

In aging studies, the ultimate objective is to determine a relationship between effective working stress and service life. In evaluation of results, the intrinsically stochastic nature of the breakdown process has to be taken into account. The statistical method of dealing with times to failure developed by Weibull [80] can be applied here. In this procedure, the distribution function of breakdown probability Ψ is in the form of the *Weibull distribution*:

$$\Psi(t, G) = 1 - \exp(-Ct^a G^b) \tag{3.18}$$

where G is the effective value of field strength from the applied voltage, t is time, and a, b, and C are constants of polymers depending on temperature and other environmental conditions [81].

An example of a Weibull plot for high-grade PE with an applied field of 100 MV m^{-1} is shown in Fig. 3.9. Figure 3.10 shows the endurance or

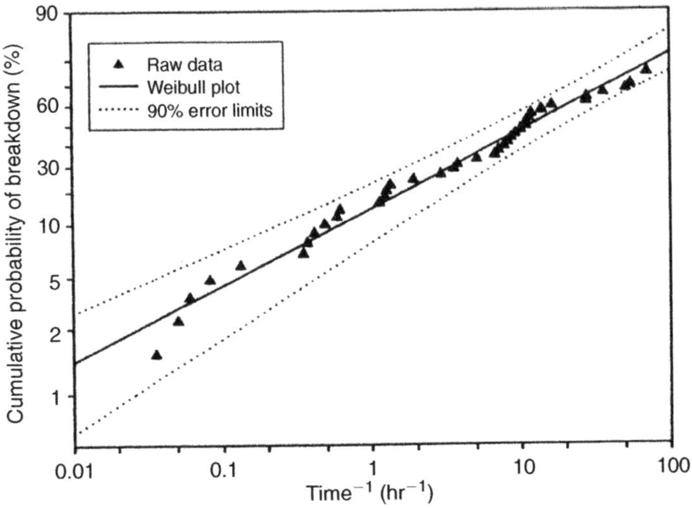

FIGURE 3.9 Weibull plot of the cumulative probability of breakdown for a set of PE samples as a function of time under an applied field 100 MV m^{-1} at room temperature. (Blythe, T. and Bloor, D., *Electrical Properties of Polymers*, Second edition, Cambridge University Press, Cambridge, UK, 2005, p. 202)

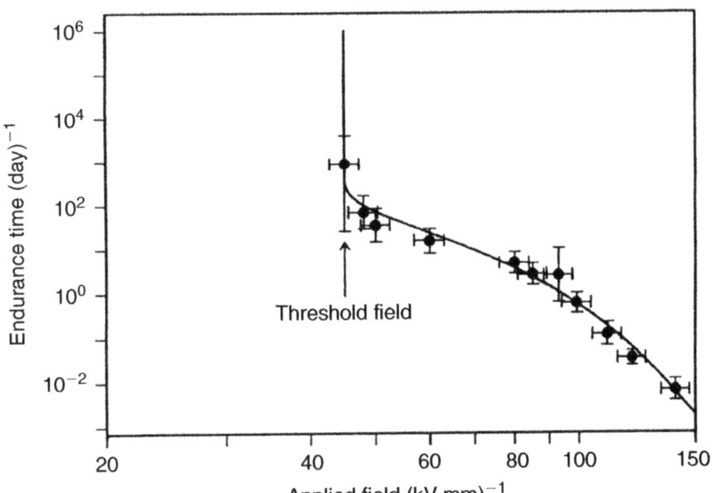

FIGURE 3.10 Electrical endurance curve for a polyethylene film. (Blythe, T. and Bloor, D., *Electrical Properties of Polymers*, Second edition, Cambridge University Press, Cambridge, UK, 2005, p. 203)

life curve obtained by plotting representative or characteristics failure times (corresponding to 63.2% cumulative probability values from Weibull plots) as a function of applied stress at room temperature. These refined endurance data can be used as an empirical basis for predicting the life of the material at any particular applied field for engineering design purposes [82]

The ac conditions are much more conducive to aging and treeing than comparable dc fields. Under the application of an alternative field ($E = E_0 \sin \omega t$) there will be a cyclic mechanical stress. The movement of polymeric chains under cyclic stress will cause an oscillating elastic energy storage and release in each half-cycle accompanied by some energy dissipation. This provides extra heat energy in just the right region to facilitate the molecular motion. Other fatigue processes, like buckling of the chains under compression, will likely enhance this further. The energy of dissipation depends on E_0^4 and will therefore be particularly strong where the field is locally enhanced, (e.g., at electrodes and where space charges are present) [83].

3.5 STATIC CHARGES

3.5.1 Introduction

A locally high concentration of electric charge q inside any material will decay exponentially with time t, current flowing away under the influence of the flow of the charge (initial concentration q_0) [84]:

$$q = q_0 \exp\left(\frac{-t}{\tau}\right) \tag{3.19}$$

where the time constant τ, called the relaxation time for decay of free charge, depends on the product of the inherent charge-storage capacity of the material, as expressed by its permittivity, $\varepsilon_0 \varepsilon'$, and its bulk resistivity, ρ:

$$\tau = \varepsilon_0 \varepsilon' \rho \tag{3.20}$$

For metals, τ is extremely short (hardly measurable) (e.g., the value for Cu is on the order of 10^{-18}s), but for common insulating polymers (e.g., PE) it can be very long. This means that polymers can retain charges for very long times, sometimes as long as several years, which presents an important aspect of their insulating character [84].

Although charges may become trapped under certain conditions within the bulk of a polymer specimen, it is the surface of the polymer that is the most vulnerable toward charging effects. As with conduction in polymers, the charging process can be either electronic or ionic in origin. Out of the many different ways in which charges can be imparted, the most common is simple contact charging: Whenever any two different materials merely come into contact with each other, there is always some redistribution of electrons or ions across the interface [84]. Our understanding of even these elementary processes has been slow to develop, partly as a consequence of contact charging being essentially a surface effect, suffering from the usual complications of high concentrations of impurities and structural defects that tend to be more prevalent at surfaces.

Controlled charging is most easily accomplished by spraying a surface with ions and this lies at the heart of many industrial electrostatic processes (e.g., xerography) [85]. Figure 3.11 shows the way the surface of an insulating film can be charged by a corona. A high voltage (e.g., 10 kV) applied to a needle electrode produces a very high field just at the vicinity of the needle tip, causing the air to ionize smoothly at that region [85]. The air ions of the same sign as the applied voltage are repelled and move along the divergent fields toward earth. They are intercepted by the insulator surface and the surface charge density builds up until the field at the needle tip is so reduced that the corona is extinguished. The final charge voltage depends on the voltage applied to the needle [85]. In order to charge large areas of film, the needle may be replaced by a fine wire stretching across the full width of the film, which may also be moving [85].

Static charges on polymers prove to be a distinct nuisance in many situations. Unless special precautions are taken, plastics are always liable

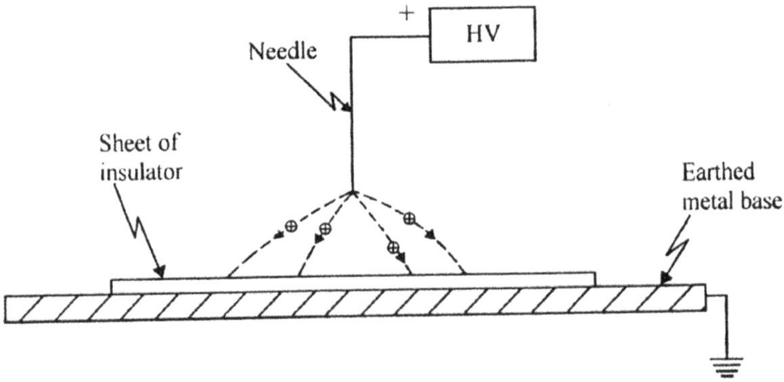

FIGURE 3.11 Charge of a film by a corona (HV = high voltage). (Blythe, T. and Bloor, D., *Electrical Properties of Polymers*, Second edition, Cambridge University Press, Cambridge, UK, 2005, p. 218)

to accumulate charges in use and processing, which attract dirt and dust. These spoil the appearance of molded parts and make handling of fibers and film products difficult by causing them to cling to metal work, and cause discomfort and ignition hazards due to sparking.

On the other hand, this ability of polymeric materials to hold charges is an advantage that can be utilized in technologies, such as electrostatic powder coating, xerography, and electret transducers.

3.5.2 Contact Charging of Polymers

When two materials are touched or rubbed together, electric charge is usually transferred from one to the other. If one of these materials is a good insulator, so that the charge persists on its surface for a long time, the effects of the charge transfer are readily apparent: Sparks may be produced and light objects may be so strongly polarized by the field that they fly toward the charged insulator. Although such phenomena have been known for a long time, the phenomenon of contact charging is still not well understood. The charge acquired by an insulator from a metal that touches it may depend not only on the nature of the insulator, but also on the circumstances, on the type of metal, and on the type and duration of the contact. So it is not really possible to define a "typical" charge density, even for a given insulator [86]. Nevertheless, it is useful to have some idea of the range of the charge density commonly encountered. The charge density is usually in the range from $\sim 10^{-5}$ to $\sim 10^{-3}$ C m^{-3}. A charge density of 10^{-4} C m^{-2} corresponds to one elementary charge for about every 10^4 surface atoms [86]. Some of the available data of the charge density for organic materials are shown in Table 3.4 [87–98].

The total charge transfer from a metal to a polymer was found to be dependent on the time of contact [89]. Additional observations indicated that the increase in charge with time of contact is due to viscoelastic deformation of the polymeric material that causes an increase of the area of contact [99]. Repeated contacts increase the charge onto an insulator, it increases with the number of contacts, but the rate of increase may become very slow [91]. There is enough evidence [100] that there are three main ways in which charge can transfer from one body to another: by transfer of electrons, ions, and materials carrying charge. Results from numerous experiments indicate that charge transfer is most often due to electrons [101].

3.5.2.1 Charge Transfer by Electrons The spontaneous redistribution of electrons across an interface between two materials is most completely documented and understood in the case of metal–metal

TABLE 3.4 Charge Density Observed on Various Materials after Contact with Metals

Material	Charge Density (Cm^{-2})	Conditions
Polyethylene	5×10^{-6}–10^{-4}	In air, mercury contact
	5×10^{-5}	In air, sliding contact
	2×10^{-5}	In vacuum
Poly(tetrafluroethylene)	3×10^{-5}	In vacuum
	7×10^{-5}	In air, sliding contact
	2×10^{-4}	In vacuum, sliding contact
	2×10^{-3}	In vacuum, prolonged rolling contact
Polyamide	10^{-3}	In air, mercury contact
	10^{-3}	In vacuum, prolonged rolling contact
Polycarbonate	10^{-3}	In vacuum, prolonged rolling contact
Polyimide	3×10^{-3}	In vacuum, prolonged rolling contact
Silicone rubber	10^{-4}	In vacuum, intimate contact
Solid paraffins	10^{-5}or less	In air, mercury contact
Anthracene	10^{-3}	In vacuum, elastic contact assumed

contacts. In polymers with chemically saturated structures, that is, σ-bonded backbones, it is difficult to observe any electronic conductivity at all, and what conductivity there is, usually depends on adventitious ions. Consequently, any improvements in the quality of insulation are generally won by careful preparation and purification, so as to avoid as much as possible the presence of ionic impurities, including catalyst residues, products of oxidation, and dissociable end groups. Conversely, useful levels of conductivity can be achieved by making composite materials in which a conductive component (most frequently a powder) is dispersed in the insulating polymeric matrix. The conductive additive is dominant in the conduction process where the insulating polymer is limiting the overall conductivity by forming barriers between the particles.

3.5.2.2 Charge Transfer by Ions As pointed out in Section 3.5.2.1, it is generally assumed that the charge transfer processes are essentially based on electrons, but there is also a possibility of contact charging by movement of ions across the interface. This applies particularly to polymers, that have ionic structures or contain ionic ingredients (e.g., antistatic agents that bloom out on the surface). In that case, ionic charge transfer will dominate [102]. The factors affecting the charging are

- The degree of dissociation of the ionic component. Often the presence of water with its high permittivity promotes ionization, making more ions available for transfer.

- The mobility of the ions, which depends on their size and solvation (most frequently by water). Greater salvations will reduce mobility.
- The relative stability of the positive and negative ions on the two surfaces. This factor determines the final equilibrium.

Examples of contact charges by ion transfer are ionomers, (e.g., poly(styrene-*co*-*N*-methyl-4-vinylpyridinium toluene sulfonate) or partially sulfonated polystyrene [102]. The ionic contact charging is utilized in industrial electrophotography. Ionic ingredients are used there to ensure effective charging of the developer systems, consisting of charged toner powder particles carried by beads charged with the opposite sign [103].

3.5.3 Electrets

Electret is a dielectric material that has a quasipermanent electric charge or dipole polarization. The name comes from ***electr**ostatic* and *magn**et***; drawing analogy to the formation of a magnet by alignment of magnetic domains in a piece of iron. An electret generates internal and external electric fields, and is essentially the electrostatic equivalent of a permanent magnet.

There is also a similarity between electrets and the dielectric layer used in capacitors. The difference is that dielectrics in capacitors possess an induced polarization that is only transient, dependent on the potential applied on the dielectric, while dielectrics with electret properties exhibit quasipermanent charge storage or dipole polarization as well. Some materials also display ferroelectricity; (i.e., they react to the external fields with a hysteresis of the polarization). Ferroelectrics can retain the polarization permanently because they are in thermodynamic equilibrium, and are used in ferroelectric capacitors (see also Section 1.15). Although electrets are only in a metastable state, those fashioned from very low leakage materials can retain excess charge or polarization for many years.

There are two types of electrets:

1. Real-charge electrets that contain excess charge of one or both polarities, either on the dielectric's surfaces (a surface charge) or within the dielectric's volume (a space charge).
2. Oriented-dipole electrets contain oriented (aligned) dipoles. Ferroelectric materials are one variant of these.

Cellular space charge electrets with internal bipolar charges at the voids provide a new class of electret materials that mimic ferroelectrics. Hence, they are known as ferroelectret. Ferroelectrets display strong piezoelectricity, comparable to ceramic piezoelectric materials.

Electret materials are quite common in nature. Quartz and other forms of silicon dioxide, for example, are naturally occurring electrets. Today, most electrets are made from synthetic polymers (fluoropolymers, polypropylene, and polyethylene terephthalate, etc.). An example is a thin (12-μm) FEP film, metallized on one side charged by depositing electrons directly in them by exposure to a low-energy (10–40 keV) electron beam [104]. The quasipermanent internal or external electric fields created by electrets can be exploited in various applications. For example, they are used in electret microphones, in copy machines, in some types of air filters, for electrostatic collection of dust particles, and in electret ion chambers for measuring ionizing radiation or radon. Additional applications are shown in Table 3.5.

An electret microphone is still the most widely used application; for that purpose electrets with charge densities as high as 1 mCm^{-2} and time constant for decay in excess of 20 years are used [126]. A schematic of an

TABLE 3.5 Examples of Electret Applications

Application	Reference
Accelerometer	105
Adaptive optics	106
Antifouling coatings	107
Audio speakers	108
Blood flow monitor	108
Bone healing implant	109
Cartridge for disk records	110
Coin sensors	111
Deformable mirror	106
Displays	112
Flexure-mode transducers	113
Fuse and detonator for explosives	114
Headphones	110
Heat exchanger membrane	115
Heat scanner	116
Impact detector	117
Medical ultrasonic imaging	110
Optical scanners	118
Pacemaker	109
Personal verification devices	119
Photocopying process	120
Shutters	121
Telephone	122
Ultrasonic transmitters	123
Vibrational fan	124
Vibration damper	125

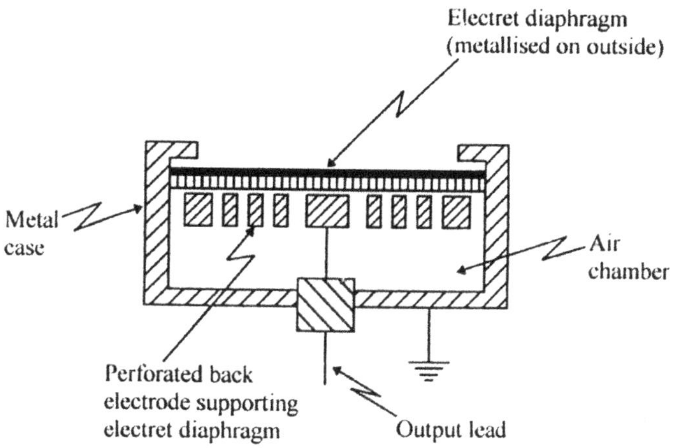

FIGURE 3.12 Basic design of an electret microphone. (Blythe, T. and Bloor, D., *Electrical Properties of Polymers*, Second edition, Cambridge University Press, Cambridge, UK, 2005, p 289)

electret microphone is shown in Fig. 3.12. The electret film forms the diaphragm on which the sound waves impinge. As the film moves, the electric field from the electret to the counterelectrode varies and induces the signal in the circuit connected to it. A preamplifier is incorporated in the microphone unit. The frequency response of the electret microphone is comparable to that of a condenser microphone without the requirement of any high bias potential [127]. This type of microphone can be easily miniaturized and is widely used in telephones and tape recorders [127].

FURTHER READINGS

Blythe, T. and Bloor, D., *Electrical Properties of Polymers*, 2nd ed., Cambridge University Press, Cambridge, UK, 2005.

Riande, E. and Diaz-Calleja, R., *Electrical Properties of Polymers*, Marcel-Dekker, New York, 2004.

Ku, C.C. and Liepins, R., *Electrical Properties of Polymers-Chemical Principles*, Hanser Publishers, Munich, 1987.

REFERENCES

1. Ku, C.C. and Liepins, R., *Electrical Properties of Polymers: Chemical Principles*, Hanser Publishers, Munich, 1987, p. 4.

2. Ku, C.C. and Liepins, R., *Electrical Properties of Polymers: Chemical Principles*, Hanser Publishers, Munich, 1987, p. 11.
3. Blythe, T. and Bloor, D., *Electrical Properties of Polymers*, 2nd ed., Cambridge University Press, Cambridge, UK, 2005, p. 48.
4. Blythe, T. and Bloor, D., *Electrical Properties of Polymers*, 2nd ed., Cambridge University Press, Cambridge, UK, 2005, p. 49.
5. Blythe, T. and Bloor, D., *Electrical Properties of Polymers*, 2nd ed., Cambridge University Press, Cambridge, UK, 2005, p. 50.
6. Ku, C.C. and Liepins, R., *Electrical Properties of Polymers: Chemical Principles*, Hanser Publishers, Munich, 1987, p. 25.
7. Ku, C.C. and Liepins, R., *Electrical Properties of Polymers: Chemical Principles*, Hanser Publishers, Munich, 1987, p. 27.
8. Goul, V.E. et al. *Electrically Conductive Polymeric Materials*, Izv. Khimia, Moscow, 1968 (in Russian).
9. Carter, W.C., Nagat, M., Schmieder, W.G., and Smyth, C.P., *Trans. Faraday Soc.* **42A**, 213 (1946).
10. Thirion, P. and Chasset, P., *Trans. Inst. Rubber Ind.* **27**, 364 (1951).
11. Lukomskaya, A.I. and Dogadkin, B.A., *Kolloidnyi Zhurnal*, **22**, 576 (1960) (in Russian).
12. Ku, C.C. and Liepins, R., *Electrical Properties of Polymers: Chemical Principles*, Hanser Publishers, Munich, 1987, p. 51.
13. Ku, C.C. and Liepins, R., *Electrical Properties of Polymers: Chemical Principles*, Hanser Publishers, Munich, 1987, p. 54.
14. Ku, C.C. and Liepins, R., *Electrical Properties of Polymers: Chemical Principles*, Hanser Publishers, Munich, 1987, p. 55.
15. Hedwig, P., *Dielectric Spectroscopy of Polymers*, John Wiley & Sons, Inc., New York, 1977.
16. Redish, W., *Trans. Faraday Soc.*, **46**, 459 (1950).
17. Blythe, T. and Bloor, D., *Electrical Properties of Polymers*, 2nd ed., Cambridge University Press, Cambridge, UK, 2005, p. 58.
18. Blythe, T. and Bloor, D., *Electrical Properties of Polymers*, 2nd ed., Cambridge University Press, Cambridge, UK, 2005, p. 59.
19. Blythe, T. and Bloor, D., *Electrical Properties of Polymers*, 2nd ed., Cambridge University Press, Cambridge, UK, 2005, p. 60.
20. Ku, C.C. and Liepins, R., *Electrical Properties of Polymers: Chemical Principles*, Hanser Publishers, Munich, 1987, p. 92.
21. Ku, C.C. and Liepins, R., *Electrical Properties of Polymers: Chemical Principles*, Hanser Publishers, Munich, 1987, p. 93.
22. Blythe, T. and Bloor, D., *Electrical Properties of Polymers*, 2nd ed., Cambridge University Press, Cambridge, UK, 2005, p. 186.
23. Ku, C.C. and Liepins, R., *Electrical Properties of Polymers: Chemical Principles*, Hanser Publishers, Munich, 1987, p. 142.

24. Blythe, T. and Bloor, D., *Electrical Properties of Polymers*, 2nd ed., Cambridge University Press, Cambridge, UK, 2005, p. 187.
25. Blythe, T. and Bloor, D., *Electrical Properties of Polymers*, 2nd ed., Cambridge University Press, Cambridge, UK, 2005, p. 192.
26. Blythe, T. and Bloor, D., *Electrical Properties of Polymers*, 2nd ed., Cambridge University Press, Cambridge, UK, 2005, p. 195.
27. Stark, K.H. and Garton. C.G., *Nature* (*London*), **176**, 1225 (1955).
28. Blythe, T. and Bloor, D., *Electrical Properties of Polymers*, 2nd ed., Cambridge University Press, Cambridge, UK, 2005, p. 194.
29. Ku, C.C. and Liepins, R., *Electrical Properties of Polymers: Chemical Principles*, Hanser Publishers, Munich, 1987, p. 153.
30. Blythe, T. and Bloor, D., *Electrical Properties of Polymers*, 2nd ed., Cambridge University Press, Cambridge, UK, 2005, p. 196.
31. Kouno, T., I.E.E.E. *Trans. Elect. Insul.*, **EI-15**, 153 (1980).
32. Blythe, T. and Bloor, D., *Electrical Properties of Polymers*, 2nd ed., Cambridge University Press, Cambridge, UK, 2005, p. 200.
33. Ku, C.C. and Liepins, R., *Electrical Properties of Polymers: Chemical Principles*, Hanser Publishers, Munich, 1987, p. 93.
34. Ku, C.C. and Liepins, R., *Electrical Properties of Polymers: Chemical Principles*, Hanser Publishers, Munich, 1987, p. 107.
35. Ieda, M. and Nawata, M., I.E.E.E. *Trans. EI-12*, **19** (1977).
36. Watson, D.B., *J. Phys.*, **D5**, **410** (1972).
37. Bahder, G. and Katz, C. in "*1972 Annual Report, Conference on Electrical Insulation and Dielectric Phenomena*," National Academy of Sciences-National Research Council, Washington, DC, (1973) p. 190.
38. Bahder, G., Katz, C., Lawson, J.H., and Vahlstrom, W., *I.E.E.E, Trans. Power Apparatus Systems*, **PAS-93**, 977 (1974).
39. Inoue, S. et al., *Dainichi Nippon Cables Rev.*, **57**, 40 (April 1974).
40. Matsubara, M. and Yamanouchi, S. in "*1974 Annual Report, Conference on Electrical Insulation and Dielectric Phenomena*", National Academy of Sciences-National Research Council, Washington DC, (1975), p. 270.
41. Ashcraft, A.C. and Eichhorn, R.M., *I.E.E.E. Trans. Elect. Insul.*, **EI-13**, 198 (1978).
42. Ku, C.C. and Liepins, R., *Electrical Properties of Polymers: Chemical Principles*, Hanser Publishers, Munich, 1987, p. 108.
43. Lanctoe, T.P., Lawson, J.H., and McVey, *I.E.E.E. Trans. Power Apparatus Systems*, **PAS-98**, 912 (1979).
44. Ku, C.C. and Liepins, R., *Electrical Properties of Polymers: Chemical Principles*, Hanser Publishers, Munich, 1987, p. 117.
45. Ku, C.C. and Liepins, R., *Electrical Properties of Polymers: Chemical Principles*, Hanser Publishers, Munich, 1987, p. 110.

46. Bahder, G., Dakin, T.W., and Lawson, J.H., *Conference Internationale des Grandes Reseaux Electriques*, Paris, Paper No. 15–05 (1974).
47. Blythe, T. and Bloor, D., *Electrical Properties of Polymers*, 2nd ed., Cambridge University Press, Cambridge, UK, 2005, p. 199.
48. Ieda, M. and Nawata, M., in "*1972 Annual Report, Conference on Electrical Insulation and Dielectric Phenomena*", National Academy of Sciences-National Research Council, Washington DC, (1973), p. 143.
49. Ku, C.C. and Liepins, R., *Electrical Properties of Polymers: Chemical Principles*, Hanser Publishers, Munich, 1987, p. 115.
50. Ku, C.C. and Liepins, R., *Electrical Properties of Polymers: Chemical Principles*, Hanser Publishers, Munich, 1987, p. 117.
51. McKean, A.L., Oliver, F.S., and Trill, S.W., *I.E.E.E. Trans. Power Apparatus Systems*, **PAS-86**, 1 (1967).
52. Fujisawa, Y., Yasui, T., Kawasaki, Y., and Matsamura, H., *I.E.E.E. Trans. Power Apparatus Systems*, **PAS-87**, 1899 (1968).
53. Ku, C.C. and Liepins, R., *Electrical Properties of Polymers: Chemical Principles*, Hanser Publishers, Munich, 1987, p. 119.
54. Takaota, M., Seki, M., Oshima, H., and Hasegawa, M, *Wire J.* 64, (June 1980).
55. U.S. Patent 3,997,288 (December 14, 1976) to Furukawa Electric Co.
56. Otani, K., *Jpn. Plastics*, 15 (November–December, 1974).
57. Hayami, T., *I.E.E.E. Trans. Power Apparatus Systems*, **PAS-88**, 897 (1969).
58. Hayami, T. and Yamada, Y., in "*1972 Annual Report, Conference on Electrical Insulation and Dielectric Phenomena*", National Academy of Sciences-National Research Council, Washington, DC, (1973), p. 239.
59. Morita, M., Hanai, M., and Shimanuki, H., in "*1972 Annual Report, Conference on Electrical Insulation and Dielectric Phenomena*", National Academy of Sciences-National Research Council, Washington, DC, (1973), p. 299.
60. Mears, W.H., *Minutes of the I.E.E.E. Insulated Conductors Committee, Denver, CO, 1973*.
61. Fujiki, S., Furusawa, H., and Matsuba, H., *I.E.E.E. Trans. Power Apparatus Systems*, **PAS-90**, 2703 (November–December 1971).
62. Hunt, G.H., U.S. Patent 3,629,110 (December 21, 1971); to Simplex Wire and Cable Co.
63. Olyphant, Jr., M. Insulation (February–April 1963).
64. Yoda, B. and Sekii, Y., *I.E.E.E. Trans. Power Apparatus Systems*, **PAS-90**, 2682 (November–December 1971).
65. MacKenzie, B.T., Prober, M., and Lever, R.C, in "*1972 Annual Report, Conference on Electrical Insulation and Dielectric Phenomena*", National Academy of Sciences-National Research Council, Washington, DC, 1973, p. 299.

66. Reynolds, E.H., in "*1972 Annual Report, Conference on Electrical Insulation and Dielectric Phenomena*", National Academy of Sciences-National Research Council, Washington, DC, (1973), p. 463.

67. Kind, D. and Schiweck, L., in "*1969Annual Report, Conference on Electrical Insulation and Dielectric Phenomena*", National Academy of Sciences-National Research Council, Washington DC, (1970), p. 128.

68. Singh, N., Bruhin, A.C., and Lever, R.C., in "*1972 Annual Report, Conference on Electrical Insulation and Dielectric Phenomena*", National Academy of Sciences-National Research Council, Washington DC, (1973), p. 247.

69. Ku, C.C. and Liepins, R., *Electrical Properties of Polymers: Chemical Principles*, Hanser Publishers, Munich, 1987, p. 126.

70. Ku, C.C. and Liepins, R., *Electrical Properties of Polymers: Chemical Principles*, Hanser Publishers, Munich, 1987, p. 127.

71. Blythe, T. and Bloor, D., *Electrical Properties of Polymers*, 2nd ed., Cambridge University Press, Cambridge, UK, 2005, p. 209.

72. Ku, C.C. and Liepins, R., *Electrical Properties of Polymers: Chemical Principles*, Hanser Publishers, Munich, 1987, p. 138.

73. Ku, C.C. and Liepins, R., *Electrical Properties of Polymers: Chemical Principles*, Hanser Publishers, Munich, 1987, p. 139.

74. Bahder, G., Eager, G.S., and Silver, D.A., U.S. Patent 3,885,085 (May 20, 1975), to General Cable Corporation.

75. Soma, K. and Kuma, S., *1980 I.E.E.E. International Symposium on Electrical Insulation*, Conference Record 80 CH1496–9-EI, Boston June 12–14, 1980, p. 212.

76. Ku, C.C. and Liepins, R., *Electrical Properties of Polymers: Chemical Principles*, Hanser Publishers, Munich, 1987, p. 140.

77. Matsuba, H., Kawai, E., and Sato, A., *Proceedings 8th Symposium on Electrically Insulating Materials* (Japanese), III-3, (1975), p. 133.

78. Blythe, T. and Bloor, D., *Electrical Properties of Polymers*, 2nd ed., Cambridge University Press, Cambridge, UK, 2005, p. 210.

79. Blythe, T. and Bloor, D., *Electrical Properties of Polymers*, 2nd ed., Cambridge University Press, Cambridge, UK, 2005, p. 211.

80. Weibull, W., *J.Appl. Mech.* **18**, 293 (1951).

81. Blythe, T. and Bloor, D., *Electrical Properties of Polymers*, 2nd ed., Cambridge University Press, Cambridge, UK, 2005, p. 203.

82. Ku, C.C. and Liepins, R., *Electrical Properties of Polymers: Chemical Principles*, Hanser Publishers, Munich, 1987, p. 183.

83. Blythe, T. and Bloor, D., *Electrical Properties of Polymers*, 2nd ed., Cambridge University Press, Cambridge, UK, 2005, p. 208.

84. Blythe, T. and Bloor, D., *Electrical Properties of Polymers*, 2nd ed., Cambridge University Press, Cambridge, UK, 2005, p. 217.

85. Blythe, T. and Bloor, D., *Electrical Properties of Polymers*, 2nd ed., Cambridge University Press, Cambridge, UK, 2005, p. 218.
86. Ku, C.C. and Liepins, R., *Electrical Properties of Polymers: Chemical Principles*, Hanser Publishers, Munich, 1987, p. 224.
87. Donald, D.K., *J. Electrochem. Soc.*, **115**, 270 (1968).
88. Wagner, P.A., *J. Appl. Phys.*, **27**, 1300 (1956).
89. Davies, D.K., Static Electrification, American Institute of Physics Conference Series, No. 4, p. 29 (1964).
90. Northage, F. and Backstrom, G., *Static Electrification*, American Institute of Physics Conference Series, No. 27, p. 84 (1975).
91. Lowell, J., *J. Phys.*, **D9**, 1571 (1976).
92. Haenen, H.J.M., *J. Elestrostat*, **2**, 151 (1976).
93. Wahlin, A. and Backstrom, G., *J. Appl. Phys.*, **45**, 2058 (1974).
94. Davies, D.K., in *1973 Annual Rep. Conf. Elec. Ins. Diel. Phenomena*, National Academy of Sciences–National Research Council, Washington, DC, p. 1.
95. Medley, J.A., *Br. J. Appl. Phys.*, **4** (Suppl. 2), 528 (1953).
96. Davies, D.K., *J. Phys.*, **D2**, 1533 (1962).
97. Davies, D.K., *Adv. Stat. Electrification I.*, **10** (1970).
98. Cotrell, G.A. *J. Phys.*, **D11**, 1533 (1978).
99. Homewood, K.P. and Rose-Innes, A.C., *Static Electrification*, American Institute of Physics Conference Series, No. 48, p. 233, (1979).
100. Harper, W.R., *Contact and Frictional Electrification*, University Press, Oxford, UK (1967).
101. Ku, C.C. and Liepins, R., *Electrical Properties of Polymers: Chemical Principles*, Hanser Publishers, Munich, 1987, p. 228.
102. Blythe, T. and Bloor, D., *Electrical Properties of Polymers*, 2nd ed., Cambridge University Press, Cambridge, UK, 2005, p. 243.
103. Blythe, T. and Bloor, D., *Electrical Properties of Polymers*, 2nd ed., Cambridge University Press, Cambridge, UK, 2005, p. 245.
104. Sessler, G.M. and West, J., *J. Appl. Phys. Lett.*, **17**, 507 (1970).
105. Edelman, S., Roth, S.C., and Mayo-Wells, J.F., U.S. Patent *3,970,862* (July 20, 1976).
106. Sato, T., Ishida, H. and Ikeda, O., *Appl.Opt.*, **19**, p. (1980).
107. Latour, M. et al., *Studies Elect. Electro. Eng.*, **2**, 175, (1979).
108. Hunklinger, S., Sussner, H., and Dransfeld, K., *Advances in Solid State Physics* (Treusch, J., Ed.) Vieweg, Braunschweig (1976), Vol. **XVI**, p. 267.
109. Hausler, E., Lang, H., and Shreiner, F.J., *IEEE Frontiers Eng. Health Care*, **4**, 333 (1980).
110. Tamura, M., Yamaguchi, T., Ogaba, T., and Yoshimi, T., *J. Audio Eng. Soc.* **23**, 21 (1975).

111. Crane, G.R., *IEEE Trans. Sonics Ultrason.*, **SU-25**, 393 (1978).
112. Gallantree, H.R. and Quilliam, R.M., *Marconi Rev.* **39**, 189 (1976).
113. Toda, M. and Osaka, S., U.S. Patent 4,234,245 (November 18, 1980).
114. Bloomfield, P.E., *National Bureau of Standards, Interagency Report* No. 75–724 (R) (1975).
115. Edelman, S. and Ballard, L.D.,U.S. Patent Application Ser. No. 30,966, April 17,1979.
116. Cohen, J., Edelman, S., and Veretti, C., *Nat. Bur. Stds. Tech News Bull.*, **56**, 52 (1972).
117. De Reggi, A.S., Edelman, S., Roth, S.C., *Nat. Bur. Stds. Interagency Rep.*, **76**, 1078 (1976).
118. Lee. J.K. and Marcus, M.A., *Ferroelectrics* **32**, 93 (1981).
119. Radice, P.R., UK Patent GB-2,032,669A.
120. Bergman, J.G., Crane, G.R., Balman, A.A., and O'Bryan, Jr., H.M., *Appl. Pys. Lett.*, **21**, 497 (1972).
121. Toda, M., *Trans. IEEE Jpn.*, **61**, 507 (1978).
122. Wagers, R.S., *J. Appl. Phys.*, **51**, 5797 (1980).
123. Ohigashi, H., *J. Appl. Phys.*, **47**, 949 (1976).
124. Sussner H. et al., *Phys. Lett.*, **A45**, 475 (1973).
125. Edelman, S., Roth, S.C, and Grisham, L.R., *Shock Vib. Bull.* **39**, 1 (1969).
126. Blythe, T. and Bloor, D., *Electrical Properties of Polymers*, 2nd ed., Cambridge University Press, Cambridge, UK, 2005, p. 248.
127. Blythe, T. and Bloor, D., *Electrical Properties of Polymers*, 2nd ed., Cambridge University Press, Cambridge, UK, 2005, p. 249.

CHAPTER 4

SPECIALTY POLYMERS

4.1 INTRINSICALLY CONDUCTIVE POLYMERS

The essential difference between insulating and intrinsically conductive polymers is their structure. Typically, the latter contain an unsaturated conjugated structure, while the insulators do not. Few conjugated polymers have structures approaching the ideal of fully extended, defect-free backbones. Most are either partially crystalline or amorphous. Thus the polymer backbone is distorted and its conjugation is disrupted. Consequently, the structure is defined by a distribution of somewhat extended, but separated, conjugated regions. Polymers with conjugated backbones have an inherently higher density of conjugated regions with a higher degree of connectivity than polymers with saturated backbones containing only localized conjugated regions. A consequence of this is that conjugated polymers can be made metallic while the latter are at best semiconductors [1]. Note that inorganic polymers (e.g., poly (sulfur nitride), SN_x) are conductive. Poly(sulfur nitride) is a head-to-head polymer with a cis-configuration prepared by solid-state polymerization of S_2N_2 and forms crystals with a metallic luster and exhibit electrical conductivity parallel to the polymer chains of $\sim 4 \times 10^5\ S\ m^{-1}$ and a value ~ 50 times smaller perpendicular to the chains at room temperature. The

Polymers for Electricity and Electronics: Materials, Properties, and Applications,
First Edition. Jiri George Drobny.

polymer becomes superconducting at 0.26 K (−459°F) [2]. However, because the field of intrinsically conductive organic polymers is much wider and has many useful applications, this chapter will concentrate on this class of polymers.

4.1.1 Conjugated Polymers

Conjugated polymers are those having their main chains containing uniformly distributed (conjugated) double bonds in the following way:

$$-CH{=}CH-CH{=}CH-CH{=}CH-CH{=}CH-$$

Such double bonds have been known for a long time to be the source of color in organic molecules including polymers, referred to as conjugated polymers.

All conjugated polymers possess conjugated π-electrons, that is, electrons that are delocalized rather than being part of one valence bond. Excitation energies of conjugated π-electrons are usually in the visible range and they are, therefore addressed as being optically active [3].

The electrooptical properties and other features, such as the operation voltage and the charge injection energy barrier, depend on the work function of the electrode materials and the energy difference between the highest occupied molecular orbital HOMO (valence band) and the lowest unoccupied molecular orbital LUMO (conduction band) of the conjugated polymers. This is defined as the energy or band gap ($E_{g.}$).

The highest energy π orbital and the lowest energy π^* orbital are designated as the HOMO and LUMO, respectively. The energy gap (see above) decreases with the increase in conjugation.

One of the conjugated polymers is polyacetylene (PAc) known for the semiconductivity of its films, since the latter part of the 19th century [2]. It was only in the 1970s, when a group of scientists discovered that *trans*-polyacetylene films exposed to a vapor of arsenic pentafluoride and iodine, resulted in a dramatic increase of conductivity. The increase of conductivity by this method was by a factor $>10^7$ from its basic semi-conductivity, bringing its conductivity to values comparable to metals [4]. The work was considered to be a major contribution to science and lead to the Nobel Prize for Chemistry being awarded to Mac Diarmid, Heeger, and Shirakawa in 2000 for the discovery of conductive polymers.

In general, conjugated polymers are derived from monomers containing triple bonds and aromatic groups. This group of polymers includes polyacetylene, polypyrrole, polythiophene, polyaniline, poly(*p*-phenylene vinylene), poly(*p*-phenylene), polyfluorene, and poly(thienylene vinylene).

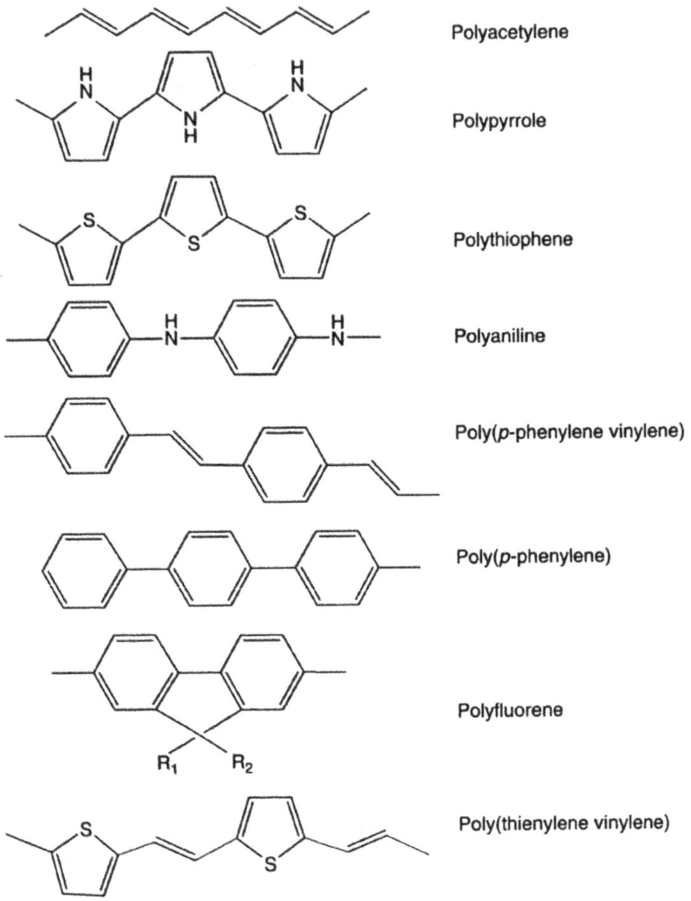

FIGURE 4.1 Molecular structures of the major classes of conjugated polymers. (*Introduction to Organic Electronic and Optoelectronic Materials and Devices* (Sun, S.-S. and Dalton, L.R., Eds.), 2008, p. 175)

Molecular structures of the major classes of conjugated polymers are shown in Fig. 4.1. The conjugated backbones are inflexible and the polymers are insoluble in solvents normally used for polymerization reactions, which limits the length of the polymer chains that can be attained. This problem can be overcome by either inclusion of a long flexible pendent group, which render the polymer soluble. Or by the preparation of a nonconjugated, soluble precursor polymer [5].

4.1.1.1 Synthesis of Conjugated Polymers The major methods used to synthesize conjugated polymers are [6] (1) pyrolysis, (2) direct synthesis by catalyzed addition polymerization, (3) condensation polymerization, and (4) electrochemical synthesis. By far, the most important techniques are *addition reactions* like the Ziegler–Natta catalysis, in which all the atoms present in the monomer are also incorporated into the polymer in condensation reactions, where a small molecule (e.g., water) is removed in the polymerization process. This method is still widely used. It yields free-standing films as well as foams, partially oriented stretched films, composites, and powders, depending on the experimental procedure used; all the generated materials, however, have been insoluble and infusible.

4.1.1.1.1 Pyrolysis Pyrolysis is the oldest way to produce conducting organic materials. The principle here is to eliminate heteroatoms (typically halogen, oxygen, nitrogen) from the polymer by heating it to form an extended aromatic structure, eventually approaching that of graphite. Graphite is a quintessential synthetic metal when treated with certain dopants. The product of polymer pyrolysis can be powder, film, or fiber, depending on the form and nature of the starting polymer, as well as the conditions of the pyrolysis. For example, polyacrylonitrile [7] and Kapton™, poly(44′-oxydiphenylene)pyromellitimide [8, 9], are both pyrolyzed in three steps. They yield conductive materials with conductivities under 100 S cm^{-1} (10^4 S m^{-1}).

4.1.1.1.2 Direct Synthesis

- Polyacetylene (PAc) can be synthesized by using the Ziegler–Natta catalyst system [10, 11]. The reaction is typically carried out in the temperature range −100 to 180°C (−148 to 356°F) with toluene as the solvent <80°C (176°F) and *n*-hexadecane >80°C. The chain configuration and color of the product depends on the temperature [11, 12]. The choice of catalyst and its concentration are both critical [13]. Only soluble catalyst systems form uniform films [12]. An alternative, a two-step process for the production of polyacetylene, involves starting from a nonconjugated polymer and converting it into PAc [14]. The precursor polymer is prepared by ring-opening metathesis polymerization of a monomer obtained from the reaction of *cyclo*-octatetraene and a substituted alkyne [14].

  In general, the different syntheses of PAc are quite involved and have variable results. Yet, there have been major improvements in the quality of PAc samples and consequent enhancement of physical properties. Nevertheless, the instability of PAc under a normal atmosphere prompted the focus of synthetic work to switch to other, more stable polymers, as the emphasis changed from fundamental studies to application at the turn of the 19th century [15].

- Polythiophenes (PTHs) offer simpler synthesis and the products have an improved stability when compared to polyacetylene [16, 17]. The most common method is to react dihalogenated thiophene with either magnesium or zinc to form a metal-substituted thiophene, a Grignard reagent, which can be polymerized either *in situ* or after intermediate purification by a number of different nickel catalysts. Oxidizing agents (e.g., $FeCl_3$ and AsF_3) can also be used to polymerize thiophene. Products from these methods are insoluble and have relatively low molecular weights. The electrical conductivity on doping is $\sim 10^3$ S m^{-1} [15]. Polymers with improved properties, including electrical conductivity, are obtained from polymerization of substituted thiophenes attaining values of conductivity as high as 10^5 S m^{-1} [17].
- Poly(*p*-phenylene vinylenes) (PPVs) can be synthesized by several methods. One of them is the polymerization of phenyldisulfonium salt to produce a soluble polymer, which can be converted to PPV by heating at 473–573 K (392–572°F) (in vacuum or treated with methyl alcohol and heating under HCl at 293 K (68°F). A similar route involves coupling of halogenated *p*-xylenes to make a brominated precursor with subsequent debromination [18]. Other methods lead to the implementation of commercial PPV manufacture in Germany, which is suitable for use in commercial light-emitting diodes (LEDs) [19].
- Polyaniline (PAni) is commonly synthesized by the oxidative coupling of aniline with ammonium persulfate in aqueous HCl. This produces partially protonated emeraldine salt that can be deprotonated with ammonia to form an emeraldine base. When the reaction is performed at 248 K (–13°F) the polymer is produced in nearly 100% yield, with a high molecular weight and greatly reduced defect content [20]. Other oxidizing agents can be employed in the preparation of PAni [21]. The use of a modified system and polymer synthesis below room temperature has facilitated the spinning of fibers with oriented polymer chains and conductivity along the fiber as high as 2×10^5 S m^{-1} [22].

4.1.1.1.3 Condensation Polymerization

Poly(*p*-phenylene)

Poly(*p*-phenylene) (PPP) and other polyarylenes are prepared by a variety of methods including oxidative polymerization of aromatic compounds, coupling of aromatic halogen compounds, condensation of aromatic nuclei via organometallic reagents, cycloaliphatic materials as precursors, and decomposition of diazonium salts.

Oxidative polymerization of benzene is one of the most effective techniques using benzene–aluminum chloride–cupric chloride systems at 35°C (95°F) [23, 24]. The polymer yield attains an almost quantitative value at an $AlCl_3/CuCl_2$ ratio of 2:1 [24].

Coupling of dihalobenzenes can be performed using two well-known coupling reactions commonly used in organic chemistry. One of them is the *Ullman reaction*, which uses using copper as a catalyst [25]:

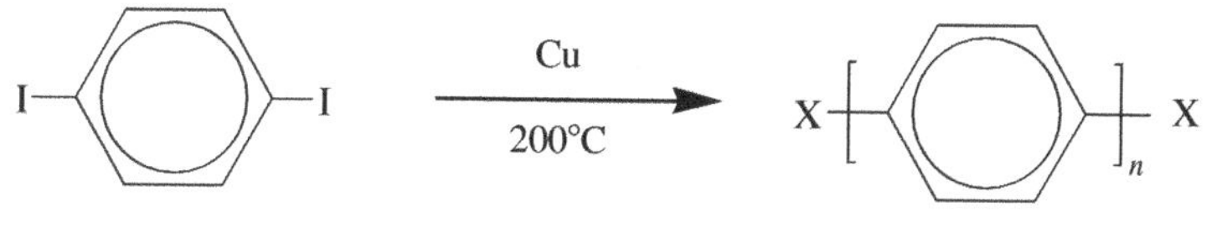

X = 1 or H

The other is the *Wurtz–Fittig reaction*, in which the catalyst is sodium or potassium [26]:

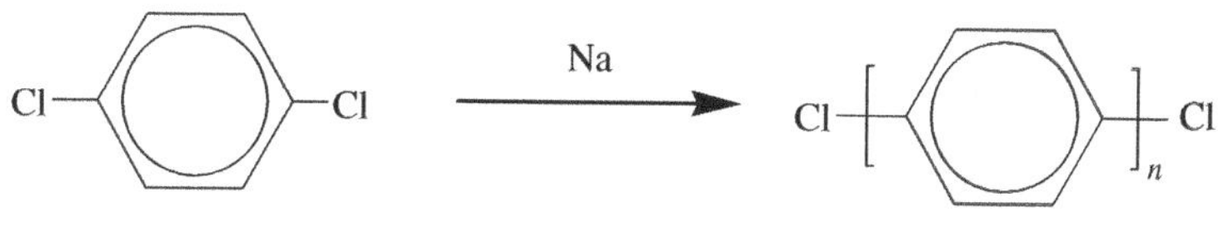

The temperature for the Ullman reaction is commonly ~200°C (392°F), although, when it depends on the reactivity of the aryl halide, it can be as low as 100°C (212°F) or as high as 300°C (572°F). The Wurtz–Fittig reaction has been used for the synthesis of polyphenylene rather extensively [27].

Higher poly(*p*-phenylenes) may be produced by condensation of aromatic nuclei via organometallic reagents. In this synthesis, a mono-Grignard derivative of a dihalobenzene is decomposed in the presence of catalytic amounts of a polyvalent metal chloride [28, 29]:

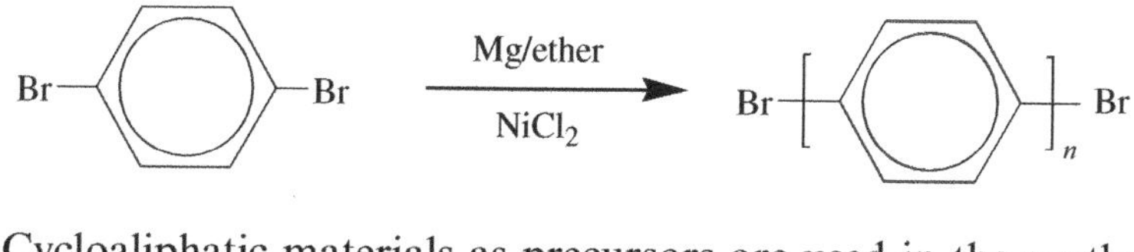

Cycloaliphatic materials as precursors are used in the synthesis of PPPs; in the early synthesis, which involved the dehydrogenation of poly(1,3-cyclohexadiene) with a Ziegler–Natta catalyst (tri-isobutyl aluminum–titanium tetrachloride) under a variety of conditions. The resulting polymer had a regular 1,4-structure with a molecular weight of ~10,000 [30]. Another method involves aromatization of the polymer by either bromination or chlorination, followed by pyrolysis or by treatment with chloranil in boiling xylene for 48 h provided excellent yields of PPP [31–34].

Decomposition of diazonium salts is the basis of the *Gatterman reaction*. In this synthesis, diazonium salts decompose in the presence of powdered copper. When this reaction is applied to the deamination of benzidine-3,3′-dicarboxylic acid via its diazonium salt, it generates poly(phenylenes) having carboxylic acid groups attached to the benzene rings. Thermal treatment of these alkali-soluble polymers produces PPPs [35–37].

Poly(*p*-phenylene sulfide)

One of the most thoroughly studied and used methods for the synthesis of poly(*p*-phenylene sulfide) (PPS) is the *Macallum reaction* this reaction involves treating halobenzenes with sulfur and sodium in a sealed container at 275–360°C (527–680°F) [38, 39]:

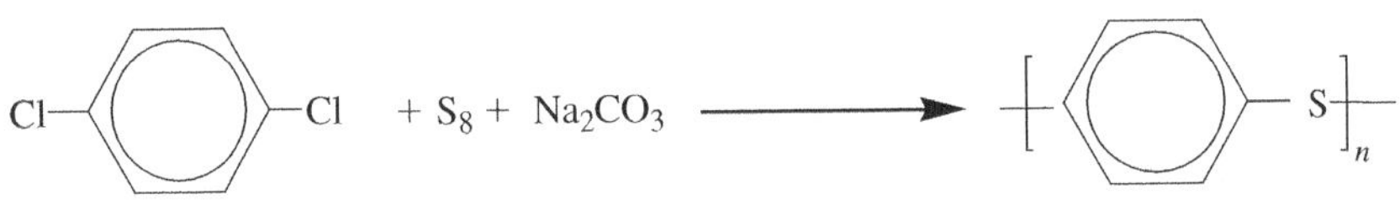

The product of this reaction is determined by the aryl/sulfur ratio.

Polyquinolines

High molecular weight soluble poly-2,6-(4-phenyl)quinoline is prepared using an acid-catalyzed Friedlander synthesis [40].

4.1.1.1.4 Electrochemical Synthesis The electrochemical oxidation of a wide range of conjugated monomers results in the deposition of an electrically deposited conductive film on the working electrode [41]. This method mostly has been used for the polymerization of heteroatom-containing monomers such as (pyrrole, thiophene, aniline, indole, etc.) and polycyclic monomers, (azulene, fluorine, pyrene, etc.) [42]. The most prominent example of a conducting polymer prepared by this technique is polypyrrole with conductivities in the range 40–100 S cm^{-1} (4×10^3–10^4 S m^{-1}) [43, 44].

4.1.2 Doping

Section 2.1 mentions the effect of doping conjugated polymers. For example, PAc is a semiconductor as are all conjugated polymers, but the doping process turns it into a system that finally possesses an electric conductivity comparable to metals with values in the range 10^4–10^5 S cm^{-1} (10^6–10^7 S m^{-1}). Doping of the semiconductor means that an oxidizing or reducing agent is introduced into the material.

In the case of conjugated polymers, this is usually carried out either by an exposure of the sample to a corresponding gas or placement in a corresponding liquid. [45]. Oxidizing atoms or molecules take up electrons from their surroundings. In a conjugated polymer, these are first π electrons from the HOMO level. Thus they act as electron acceptors (referred to as p-type dopants). As a result, holes, (i.e., positive charge carriers) are created. Reducing agents supply additional electrons, thus functioning as electron donors (n-type dopants). These electrons occupy states at and near the LUMO level [45]. With such a supply of electrons or holes, the number of free charges greatly increases. For example, for polyacetylene the resulting rise in the conductivity can be up to 17 orders of magnitude [46]. Polyacetylene is not an exception; doping results for all conjugated polymers can yield similarly high results.

The concentration of dopant atoms or molecules in the highly conductive state can reach 10^{21} cm^{-3}, which is much higher (several orders higher than that in inorganic semiconductors). The degree of doping of a conducting polymer is the doping charge numbers per monomer unit in the conjugated main chains; for example, the doping degree of polypyrrole is 0.33 [47]. When becoming dissolved in the material, dopants may have preference for the amorphous regions, but they also enter the crystalline regions [48].

The doping of conductive polymers can be carried out by charge transfer of donors or acceptors (chemical doping) or by electrochemical oxidation or reduction (electrochemical doping). Chemical doping includes p- or n-doping (see above). In the p-doping, the oxidant (acceptor) is I_2, Br_2, and AsF_5. After the p-doping, the main chains of the conjugated polymer were oxidizes to give out electrons and the dopants get electrons to become counteranions. In n-doping, the reductant is Na vapor and Na^+ $(C_{10}H_8)^-$. After n-doping, the main chains of the conjugated polymer were oxidized to get electrons and the dopants lose electrons to become cations. Electrochemical doping is performed by the electrochemical oxidation or reduction reactions. For electrochemical p-doping, the conjugated main chains are oxidized and the solution anions will dope into the polymer as counteranions. For electrochemical

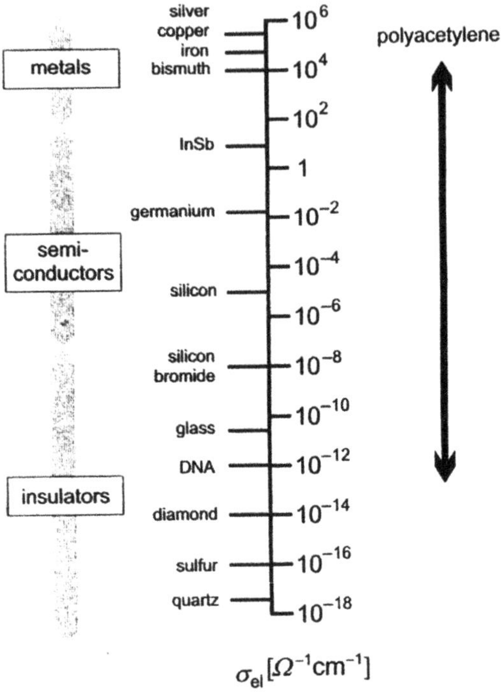

FIGURE 4.2 Ranges of conductivities of doped PAc compared to other materials. (Strobl, G., *The Physics of Polymers: Concepts for Understanding Their Structures and Behavior*, Springer-Verlag Berlin Heidelberg, 2007, p. 289)

n-doping, the conjugated main chains are reduced and the solution cations will dope into the polymers as countercations. Almost all the conjugated polymers can be doped electrochemically. But the stability of the doped state depends on the value of the doping potential [47].

Generally, the conductivity of the doped conducting polymers is in the range 10^{-3}–10^{3} S cm^{-1}. The range of connectivities of doped PAc in comparison with other materials is shown in Fig. 4.2. The highest conductivity of doped conductive polymers reached until now is 10^{7} S m^{-1} for the acetylene film after orientation by drawing [48]. Usually, the conductivity of the conductive polymers is dependent on temperature [49].

4.2 FERROELECTRIC POLYMERS

4.2.1 Introduction

All crystalline materials may be categorized into 23 crystallographic point groups. Of the 21 classes that lack a center of symmetry, 20 produce an electric dipole (i.e., polarization) when mechanically stressed. These materials are termed piezoelectric, the lack of the center of symmetry being a necessary condition to allow movement of the positive and negative ions to produce a dipole. Ten of these classes possess a permanent dipole and will respond to changes in temperature, as well as stress. These are defined as pyroelectric. These classes can be subdivided further into ferroelectric crystals, in which the dipole moments of the individual crystalline units can be reversed by application of an electric field [50].

Polymers consist of long chains built up by the repetition of small simple chemical units (monomers). With few exceptions, the main chains are formed by linking together carbon atoms. These types of links allow the formation of long flexible chains, and in general polymers as-made consist of these long chains intertwined in a completely disordered way. In contrast to ceramics, such random structures do not possess any regular features able to sustain a permanent polarization that could lead to active electric properties. The following sections describe a specialized class of polymers that can be partially ordered to give plastic materials that are piezo- and pyroelectric, light and tough, and can be processed using conventional plastics processing techniques.

4.2.2 Ferroelectric Polymeric Materials

While several polymers [(polystyrene, polypropylene, and poly(methyl methacrylate) (PMMA)] were reported to exhibit small, unstable piezoelectric effects, poly(vinylidene fluoride) (PVDF) when stretched and poled, was found to produce the largest known piezoelectric activity of all polymers investigated at that time [51]. Further research discovered that copolymers of vinylidene fluoride with trifluoroethylene (TrFE) and tetrafluoroethylene (TFE) crystallize into polar form without the stretching stage required for the PVDF homopolymer [52, 53]. This facilitates processing, and consequently opens a wider range of potential applications. Additional polymers, both semicrystalline and amorphous, were discovered later, namely, odd-numbered polyamides (semicrystalline), particularly PA11 [54], and a copolymer of vinylidene cyanide and vinyl acetate (amorphous) [55] exhibit a reasonable amount of piezoelectricity. Because PVDF homopolymer and the copolymer of vinylidene fluoride (VDF) and TrFE are the main polymers known to exhibit ferroelectricity, the following sections will concentrate on their synthesis, properties, and processing. Applications of these ferroelectric materials are discussed in Chapter 6.

4.2.2.1 Synthesis The two polymers are prepared from the respective monomers by free radical polymerization in suspension or emulsion, with water as the reaction medium at temperatures in the range of 20–60°C (68–140°F) and at pressures up to 60 MPa (8700 psi). Precise reaction conditions are required to produce polymers with the desired configurations that are necessary for their use as ferroelectric materials. Because of this, details of polymerization methods used to prepare commercially available electrically active materials are not generally accessible.

4.2.2.2 Properties Crystallization from the melt or solution produces the most common polymorph of PVDF, the *α-phase* (also known as Form II). The chain conformation is *trans gauche*$^+$ *trans gauche*$^-$ (tg$^+$ tg$^-$). Each chain conformation possesses a net dipole moment, originating from the CF_2 bond, with a component perpendicular to the polymer chain. In the case of the tg$^+$ tg$^-$ conformation, the dipole component normal to the chain axes are antiparallel and thus cancel each other. Therefore the α-phase is a nonpolar form [56]. Transformation from the α- to the polar *β-phase* (also known as form I) is achieved by mechanically stretching or rolling at elevated temperatures. This extends the polymeric chains to the all-trans conformation (as shown in Fig. 4.3). As the dipoles are normal to the chain axes, on a microscopic

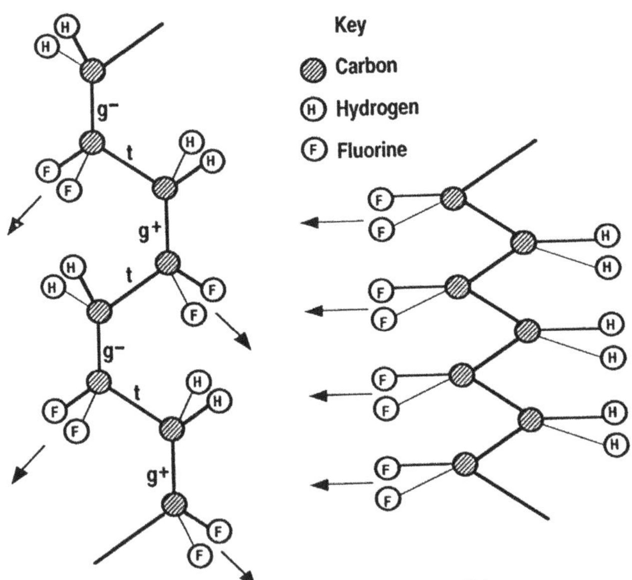

FIGURE 4.3 Schematic of the most common chain conformation of PVDF: (a) α-phase, (b) β-phase. Arrows indicate $-CF_2$ dipole directions or planes defined by a carbon backbone. (*Special Polymers for Electronics & Optoelectronics* (Chilton, J.A. and Goosey, M.T., Eds.), Chapman & Hall, London, 1995, p. 140)

scale, each crystallite has a net dipole and is piezoelectric. However, on the macroscopic scale there is zero polarization within the polymer film due to the random orientation of the crystallites. The final stage in rendering the PVDF film piezoelectric is known as *poling*, and involves the application of an electric field. This preferentially orients the crystallites in the direction of the poling field, thus producing a net polarization [57]. There are essentially two main poling techniques, conventional two-electrode poling (commonly known as "thermal" poling) and corona poling, each requiring optimization of poling time, temperature, and field to achieve maximum polarization [57].

In the copolymer of VDF and trifluoroethylene, the increased number or relatively large fluorine atoms prevents the formation of the $tg^+ tg^-$ conformation by replacing hydrogen atoms on adjacent carbon atoms. This forces the molecular chains to crystallize directly into the polar β-form. The copolymer film then undergoes a poling stage similar to PVDF to make it fully piezoelectric [57].

Ferroelectric materials have three main features:

1. Hysteresis loops formed by the electric displacement D and the electric field E.
2. Reversal of polarization by application of a step voltage.
3. A phase transition or Curie temperature, T_C, at which the material loses its ferroelectric properties and becomes paraelectric.

In linear dielectrics, D is directly proportional to E, increasing and decreasing linearly with E. In ferroelectric materials (e.g., PVDF and the VDF/TrFE copolymers) the surface charge density increases with the field in a nonlinear manner, and exhibits hysteresis as the field is decreased, with a significant amount of surface charge remaining when the field is completely removed. Subsequent applications and reversals of the electric field produce characteristic hysteresis loops.

The amount of surface charge density that remains when the field is completely removed is termed remnant polarization and is caused by a net alignment of dipoles within the crystalline phase of the material. When the direction of the electric field is reversed, the charge rapidly decreases and changes sign.

Unlike PVDF, the homopolymer, VDF/TrFE copolymers exhibit a distinct ferroelectric–paraelectric transition when heated between room temperature and the crystalline melting temperature. This transition can be seen on plots of dielectric constant as a function of temperature, when the dielectric constant shows a peak around the Curie temperature,

T_C, [58]. The T_C value increases with VDF content, ranging from 70°C (158°F) for the VDF/TrFE ratio of 56:44 to 121°C (250°F) for the 80:20 ratio.

Dielectric constant values of PVDF range from 10 to 12 at 1 kHz depending on the manufacturer. For VDF/TrFE copolymers it varies with molar composition, reaching its maximum at VDF/TrFE 55:45 with reported values as high as 15 [59]. These values are over two orders of magnitude lower than those of piezoelectric ceramics.

Dielectric loss (tan δ) values at room temperature range from 0.015 to 0.02 at 1 kHz for PVDF homopolymer and 0.015–0.025 for the VDF/TrFE copolymer, compared to 10^{-4} for some ferroelectric ceramics, such as lithium tantalite and lead germanate [60].

4.2.2.3 Processing Both the PVDF homopolymer and the VDF/TrFE copolymers are thermoplastics and therefore they can be fabricated into a variety of products by melt-processing methods.

The ease with which these thermoplastic materials can be shaped into large areas (e.g., films and sheets and complex shapes) is one of the significant advantages that polymeric ferroelectric materials have over ceramics.

4.2.2.3.1 Thin Films and Sheets Thin films (gauge ~25 μm) are suitable for the production of simple large area structures. Such films are made readily using conventional blown film extrusion or casting from solvents. In the next step, they are oriented (stretched) in one or two directions. In addition to that, very high fields (typically100 kV mm^{-1}) must be applied to the oriented films during the poling process to develop the ferroelectric properties. Practical considerations of voltage availability and safety therefore set an upper limit on the thickness of films that can be conventionally poled.

For some applications, sheets 1–2 mm thick are required. Such sheets are produced by extrusion using a flat die. The produced sheet is then cooled and wound into a roll. To fabricate a ferroelectric structure, the sheet has to be oriented and poled.

4.2.2.3.2 Injection Molding and Thermoforming To produce three-dimensional (3D) or shaped structures, injection molding or thermoforming are used. In these cases, the copolymer is suitable since it reverts to the oriented form on cooling, while PVDF homopolymer loses orientation when heated. The injection molded or thermoformed parts from the copolymer still must be poled with high fields to obtain active areas.

4.3 THERMOTROPIC LIQUID-CRYSTAL POLYMERS

General properties of liquid-crystal polymers (LCPs) are discussed at some detail in Section 2.10.2. Differences in chemistry and performance do exist among different materials. Generalizations concerning characteristics are based on commercially available information, although it is very likely that future commercial products will exhibit different properties.

The chemical structures and characteristic morphology of LCPs lead to a combination of attractive features: high strength, high modulus and impact strength, flame retardance, resistance to a wide range of aggressive chemicals, very low and readily adjustable coefficients of thermal expansion, excellent dimensional stability, and unique processability [61]. This balance of properties is the main reason electrical and electronic applications have been one of the major commercial end-uses for these thermoplastics.

Available commercial LCPs are based on hydroxybenzoic acid, therephtalic acid, and bisphenol, while others are based on hydroxybenzoic acid and naphthalene derivatives. Other products are classified generically as wholly aromatic polyesters with similar basic structures, but the final properties of different LCP polymeric systems vary significantly with their chemical compositions (i.e., mainly by monomer selection). Thermotropic LCPs are classified into three general polymer categories: Types I, II, and III, differentiated primarily by their thermal performance. The characteristics of these individual types are summarized in Table 4.1.

4.3.1 Processing

Thermotropic LCPs are easy to process because of the combination of their low-melt viscosity and high-melt strength. All current types of

TABLE 4.1 Classification of Thermotropic LCPs (*Special Polymers for Electronics & Optoelectronics* (Chilton, J.A. and Goosey, M.T., Eds.), Chapman & Hall, London, 1995, p. 260)

Property	Type I	Type II	Type III
Heat deflection temperature (°C)	~320	~220	~120
Tensile strength	High	High	Low
Impact properties	Low	Medium	High
Processability	Poor to good	Good to excellent	Good to excellent

TABLE 4.2 Anisotropy of Glass-Reinforced LCP

Specimen Thickness (mm)	Anisotropic Tensile Strength Ratio (TD/MD)[a] (%)	Anisotropic Flexural Strength Ratio (TD/MD)[a] (%)
1.6	35	34
3.2	55	54

[a] Here TD = transverse direction to flow and MD = machine direction to flow.

LCPs can be processed by injection molding and extrusion on conventional equipment capable of processing polymers up to 450°C (840°F). The strong anisotropic behavior of most LCP formulations offers opportunities and challenges for both part and mold designers. Table 4.2 illustrates the anisotropic nature of a Type II LCP filled with 30% glass fiber. Temperatures of the melt and mold, injection pressures, and time, as well as mold design, affect the part shrinkage. Although shrinkage can be anisotropic, most LCPs have exceptional dimensional stability [62]. Typical products are extruded rod, profile, film and sheet, pipe, tubing, fiber, and overcoating of glass fiber.

There are many different products with a wide selection of properties based on chemistry alone. Additional modified materials are prepared by coloring, reinforcing with fibers, with the addition of fillers, and so on. The following is a list of examples of typical systems [63]: unfilled (extrudable), glass fiber reinforced, carbon fiber reinforced, modified with fillers and fibers, modified with mineral filler only, filled by graphite, conductive, platable, and alloys.

4.3.2 Electrical Properties

Dielectric strength values of most LCPs range from 30 to 43 kV mm^{-1}, which is two to three times higher than most competitive materials, allowing thinner wall design and/or higher insulating capacity. An example of electrical properties and flammability resistance of glass-fiber reinforced nematic aromatic polyester (e.g., Vectra®, Registered Trademark of Ticona) is in Table 4.3.

4.4 IONOMERS

An ionomer is a polymer that contains both electrically neutral repeating units and a fraction of ionized units (usually no >15%). The combination of hydrophilic and hydrophobic groups in the same polymeric

TABLE 4.3 Electrical and Flammability Properties of Glass Fiber-Reinforced LCPS[a] (*Special Polymers for Electronics & Optoelectronics* (Chilton, J.A. and Goosey, M.T., Eds.), Chapman & Hall, London, 1995, p. 271)

			Grade		
Property	Unit	Test Method	15% Fiberglass	30% Fiberglass	50% Fiberglass
Volume resistivity	Ω · m	IEC 60093	10^{13}	10^{13}	10^{13}
Surface resistivity	Ω	IEC 60093	$>10^{15}$	$>10^{15}$	10^{14}
Dielectric strength	kV mm^{-1}	IEC 60243-1	34	31	28
Dielectric constant		IEC 60250			
1 MHz			3.0	3.7	4.7
10 MHz			2.9	3.2	
Dissipation factor		IEC 60250			
1 MHz			0.018	0.018	0.028
10 MHz			0.008	0.008	
Comparative tracking index		IEC 60112	200	175	
Flammability	Class	UL 94	V-0	V-0	V-0

[a] Vectra® Liquid Crystal Polymers (LCP), Product Information, Document VC-4 2/08 AM, Ticona Americas, 2008.

molecule results in unique properties and morphologies. Ionic groups may include a sulfonic or carboxylic group, as well as sulfonamides and sulfonimides [64]. An example of a commercially available ionomer is a copolymer of ethylene and methacrylic acid.

One of the widely used ionomer groups are the perfluorinated ionomers, which contain sulfonic and carboxylic groups. Within this group the emphasis has been placed on perfluoroionomers with a sulfonic group.

Perfluorinated ionomers are derived from melt-processable precursor polymers that are obtained by the copolymerization of TFE and perfluorinated vinyl ether containing a sulfonyl halide or carboxylic ester functional group. The ether linkage is required to give the vinyl group sufficient activity to allow the incorporation of a high percentage of the functional monomers. The different alkyl groups $-R_F$ used in the vinyl ethers $CF_2{=}CF{-}O{-}R_F$ are shown in Table 4.4. The important raw materials for the preparation of perfluorinated ionomers are TFE and hexafluoropropylene epoxide (HFPO). The steps involved in the large-scale manufacture of perfluorinated ionomer products (e.g., NAFION®, FLEMION® and ACIPLEX®), are shown in Fig. 4.4.

Most fluorinated ionomers are sold as flat sheets and films (e.g., extruded or solution cast films) or as composite membranes

TABLE 4.4 Some Alkyl Groups Used in Perfluorinated Vinyl Ethers

Molecular Weight	$-R_F$	Polymer Name	Company
446	$-CF_2-\underset{\displaystyle CF_3}{\underset{\vert}{CF_2}}-O-CF_2-CF_2-SO_2F$	Nafion®	DuPont
		Flemion®	Asahi Glass
		Aciplex	Asahi Kasei
422	$-CF_2-\underset{\displaystyle CF_3}{\underset{\vert}{CF_2}}-O-CF_2-CF_2-CO_2-CH_3$	Nafion®	DuPont
		Aciplex	Asahi Kasei
306	$-CF_2-CF_2-CF_2-CO_2-CH_3$	Flemion®	Asahi Glass

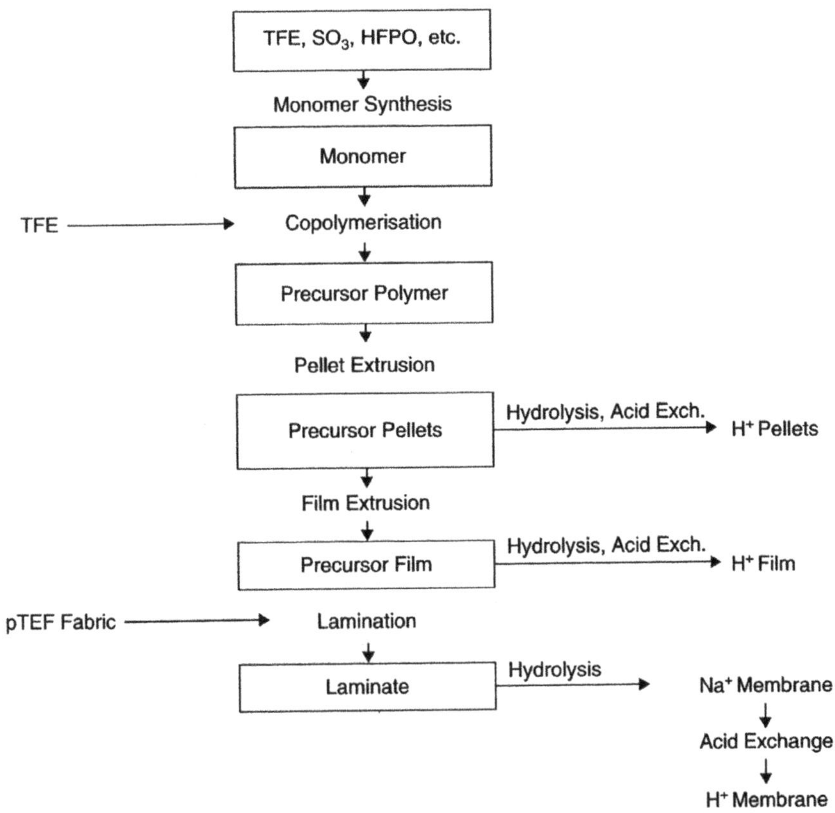

FIGURE 4.4 Steps in the manufacture of fluorinated ionomers.

containing fabric reinforcement added to one or more layers of the ionomer. Extruded capillary tubing is also available [65]. Some fluorinated ionomers dissolve in polar organic solvent or mixtures of such solvents with water under atmospheric conditions. Such polymers include partially fluorinated ionomers based on sulfonated polytrifluorostyrene, as well as perfluorinated ionomers of very low EW (equivalent weight, which is the weight in grams of an ionomer containing one equivalent of functional groups) [66].

Liquid compositions can be used to impregnate a porous matrix, particularly one made of expanded PTFE, (e.g., Goretex®). Battery separator based on porous polyethylene impregnated with liquid compositions of Nafion® is used in a vanadium redox battery [67].

Another important application for liquid compositions of fluorinated ionomers is as a binder in catalyst inks for fuel cells [68].

Perfluorinated ionomers are fairly expensive compared to other perfluorinated polymers. This is due to three factors: (1) the more expensive synthesis of the comonomer used; (2) the much larger amount of this expensive comonomer needed (typically 40–50% by weight), and (3) the relatively small scale of manufacture.

As a result, there have been numerous attempts to replace these expensive polymers with less expensive partially fluorinated ionomers. In the first approach, nonfluorinated monomers (such as styrene) are grafted on a commercially available PTFE or FEP film. An example is radiation grafting of styrene on FEP or PTFE films followed by sulfonation has been used to obtain partially fluorinated ionomer films. This approach takes advantage of the economics of scale film fabrication. In the second approach, a less expensive partially fluorinated monomer is used. In general, the chemical stability of the partially fluorinated ionomers is inferior to that of fully fluorinated ionomers [69].

A wide variety of inorganic oxides have been incorporated into fluorinated ionomers in an effort to increase the maximum use temperature, particularly under low humidity conditions. This is important for fuel cell applications, where a higher operating temperature increases the carbon monoxide tolerance and allows for easier dissipation of the waste heat. Ionomer composites also are useful for varied catalytic applications [70].

FURTHER READINGS

Conjugated Polymers: Theory, Synthesis, Properties and Characterization (Skotheim, T. and Reynolds, J.R., Eds.), CRC Press, Boca Raton, FL, 2007.

Grot, W., *Fluorinated Ionomers*, William Andrew Publishing, Norwich, NY, 2008.

Introduction to Organic and Optoelectronic Materials and Devices (Sun, S.-S. and Dalton, L.R., Eds.), CRC Press, Boca Raton, FL, 2008.

Blythe, T. and Bloor, D., *Electrical Properties of Polymers*, 2nd ed., Cambridge University Press, Cambridge, UK, 2005.

Special Polymers for Electronics and Optoelectronics (Chilton, J.A. and Goosey, M.T., Eds.), Chapman and Hall, London, 1995.

REFERENCES

1. Blythe, T. and Bloor, D., *Electrical Properties of Polymers*, 2nd ed., Cambridge University Press, Cambridge, UK, 2005, p. 306.
2. Blythe, T. and Bloor, D., *Electrical Properties of Polymers*, 2nd ed., Cambridge University Press, Cambridge, UK, 2005, p. 307.
3. Strobl, G., *The Physics of Polymers: Concepts for Understanding Their Structures and Behavior*, Springer-Verlag Berlin Heidelberg, 2007, p. 288.
4. Shirakawa, H., Louis, E.J., Mac Diarmid, A.G., Chiang, C.K. and Heger, A.J., *Chem. Commun.*, 578 (1977).
5. Blythe, T. and Bloor, D., *Electrical Properties of Polymers*, 2nd ed., Cambridge University Press, Cambridge, UK, 2005, p. 9.
6. Ku, C.C. and Liepins, R., *Electrical Properties of Polymers: Chemical Principles*, Hanser Publishers, Munich, 1987, p. 272.
7. Grassie, N. and McHeill, J.C., *J. Polym. Sci.* **17**, 707 (1958).
8. Rembaum, A., *J. Polym. Sci., Part C,* **29**, 157 (1970).
9. Bruck, S.D., *Polymer*, **6**, 319 (1965).
10. Shirakawa, H. and Ikeda, S., *Jpn. Pat.* No. **32**, 581 (1973).
11. Ito, T., Shirakawa, R., and Ikeda, S., *J. Polym. Sci.,Polym. Chem Ed.*, **12**, 11 (1974).
12. Ku, C.C. and Liepins, R., *Electrical Properties of Polymers: Chemical Principles*, Hanser Publishers, Munich, 1987, p. 275.
13. Shirakawa, H. and Ikeda, S., *Synthetic Metals*, **1**, 175 (1979/1980).
14. Blythe, T. and Bloor, D., *Electrical Properties of Polymers*, 2nd ed., Cambridge University Press, Cambridge, UK, 2005, p. 313.
15. Blythe, T. and Bloor, D., *Electrical Properties of Polymers*, 2nd ed., Cambridge University Press, Cambridge, UK, 2005, p. 315.
16. Roncali, J., *Chem. Rev.*, **92**, 711 (1992).
17. MacCullough, R.D., *Adv. Mater.*, **10**, 93 (1998).
18. Cho, B.R., *Progr. Polym. Sci.*, **27**, 307 (2002).
19. Becker, H. et al., *Synth. Metals*, **122**, 105 (2001).
20. Adams, P.M., Laughlin, P.J., Monkman, A.P., and Kenwright, A.M., *Polymer*, **37**, 3411 (1996).

21. Gospodinova, N. and Terlemezyan, L., *Progr. Polym. Sci.*, **23**, 1443 (1998).
22. Pomfret, S.J, Adams, P.N., Comfort, N.P., and Monkman, A.P., *Adv. Mater.*, **19**, 1351 (1998).
23. Kovacic, P. and Kyriakis, A., *J. Am. Chem. Soc.*, **85**, 945 (1963).
24. Kovacic, P. and Hsu, L.C., *J. Polym. Sci.*, **4**, 5 (1966).
25. Noren, G.K. and Stille, J.K., *Macromol. Revs.*, **5**, 385 (1971).
26. Edwards, G.K. and Goldfinger, G., *J. Polym. Sci.*, **16**, 589 (1955).
27. Ku, C.C. and Liepins, R., *Electrical Properties of Polymers: Chemical Principles*, Hanser Publishers, Munich, 1987, p. 280.
28. Yamamoto, T., Hayasaki, Y., and Yamamoto, A. *Bull. Chem. Soc. Jpn.* **51**, 2091 (1978).
29. Taylor, S.K., Bennet, S.G., Khoury, I., and Kovacic, P., *J. Polym. Sci. Polym. Lett. Eds.*, **19**, 85 (1981).
30. Ku, C.C. and Liepins, R., *Electrical Properties of Polymers: Chemical Principles*, Hanser Publishers, Munich, 1987, p. 281.
31. Marvel, C.S. and Hartzell, G.E., *J. Am. Chem. Soc.*, **41**, 448 (1959).
32. Le Febvre, G. and Dawans, F., *J. Polym. Sci.*, **2**, 3277 (1964).
33. Cassidy, P.E., Marvel, C.S., and Ray, S., *J. Polym. Sci.*, **3**, 1553 (1965).
34. Frey, D.A., Hasegawa, M., and Marvel, C.S., *J. Polym. Sci.*, **1**, 2057 (1963).
35. Berlin, A.A. and Parini, V.P., *Izv. Vyssh. Uchebn. Zaved. Khim., Khim. Technol.*, **1**, 122 (1958).
36. Berlin, A.A. and Parini, V.P., *Izv. Akad. Nauk. SSSR*, 1874 (1959).
37. Berlin, A.A., Liogonkii, V.I., and Parini, V.P., *J. Polym. Sci.*, **55**, 675 (1961).
38. Macallum, A.D., *J. Org. Chem.*, **13**, 154 (1948).
39. Macallum, A.D., U.S. Pat. 2, 513,188 (June 27, 1950) and 2,538,941 (January 23, 1951).
40. Stille, J.K., *Macromolecules* Vol. **14**, p. 870 (1981)
41. Waltman, R.J. and Bergon, J., *Can J. Chem.*, **64**, 76 (1986).
42. Blythe, T. and Bloor, D., *Electrical Properties of Polymers*, 2nd ed., Cambridge University Press, Cambridge, UK, 2005, p. 320.
43. Diaz, A.F., Kanazawa, K.K., and Gardini, G.P., *J. Chem. Soc.Chem. Commun.*, 635 (1979).
44. Kanazawa, K.K. et al., *Synthetic Metals*, **1**, 329 (1980).
45. Strobl, G., *The Physics of Polymers: Concepts for Understanding Their Structures and Behavior*, 3rd ed., Springer-Verlag, Freiburg, 2007, p. 302.
46. Strobl, G., *The Physics of Polymers: Concepts for Understanding Their Structures and Behavior*, 3rd ed., Springer-Verlag, Freiburg, 2007, p. 288.
47. *Introduction to Organic Electronic and Optoelectronic Materials and Devices* (Sun, S.-S. and Dalton, L.R., Eds.), CRC Press, Boca Raton, FL, 2008, p. 177.

48. Winokur, M.J., in *Handbook of Conducting Polymers* (Skotheim, T.A., Eisenbaumer, R.L. and Reynolds, J.R., Eds.), Marcel Dekker, NY, 1998, p. 714.
49. *Introduction to Organic Electronic and Optoelectronic Materials and Devices* (Sun, S.-S. and Dalton, L.R., Eds.), CRC Press, Boca Raton, FL, 2008, p. 178.
50. *Special Polymers for Electronics & Optoelectronics* (Chilton, J.A. and Goosey, M.T., Eds.), Chapman and Hall, London, 1995, p. 186.
51. Kawai, H., *Jpn. J. Appl. Phys.*, **8**, 975 (1969).
52. Yamada, T., Ueda, T., and Kitayama, T., *J. Appl. Phys.*, **52**, 948 (1981).
53. Lando, J.B. and Doll, W.W., *J. Macromol. Sci. (Phys)*, **B2**, 205 (1968).
54. Mather, S.C., Scheinbeim, J.I., and Newman, B.A., *J. Appl. Phys.*, **56**, 2419 (1985).
55. Miyata, S., Yoshikawa, M., Tasaka, S., and Ko, M., *Polym. J.*, **12**, 857 (1980).
56. *Special Polymers for Electronics & Optoelectronics* (Chilton, J.A. and Goosey, M.T., Eds.), Chapman and Hall, London, 1995, p. 190.
57. *Special Polymers for Electronics & Optoelectronics* (Chilton, J.A. and Goosey, M.T., Eds.), Chapman and Hall, London, 1995, p. 191.
58. *Special Polymers for Electronics & Optoelectronics* (Chilton, J.A. and Goosey, M.T., Eds.), Chapman and Hall, London, 1995, p. 199.
59. Highashihata, Y., Sako, J., and Yagi, T., *Ferroelectrics*, **32**, 85 (1981).
60. *Special Polymers for Electronics & Optoelectronics* (Chilton, J.A. and Goosey, M.T., Eds.), Chapman and Hall, London, 1995, p. 198.
61. Special *Polymers for Electronics & Optoelectronics* (Chilton, J.A. and Goosey, M.T., Eds.), Chapman and Hall, London, 1995, p. 258.
62. *Special Polymers for Electronics & Optoelectronics* (Chilton, J.A. and Goosey, M.T., Eds.), Chapman and Hall, London, 1995, p. 267.
63. *Vectra® Liquid Crystal Polymers (LCP), Product Information*, Document VC-4 2/08 AM, Ticona Americas, 2008.
64. Grot, W., *Fluorinated Ionomers*, William Andrew Publishing, Norwich, NY, 2008, p. 1.
65. Grot, W., *Fluorinated Ionomers*, William Andrew Publishing, Norwich, NY, 2008, p. 2.
66. Grot, W., *Fluorinated Ionomers*, William Andrew Publishing, Norwich, NY, 2008, p. 33.
67. Tian, B., Yan, C.W. and Wang, F.H., *J. Membr. Sci.*, **234**(1–2), 51–54, (May 1, 2004).
68. Grot, W., *Fluorinated Ionomers*, William Andrew Publishing, Norwich, NY 2008, p. 34.
69. Grot, W., *Fluorinated Ionomers*, William Andrew Publishing, Norwich, NY 2008, p. 36.
70. Tampani, T., Jalani, N., Choi, P., and Datta, R., *J. Electrochem. Soc.*, **152** (2), A316–A325 (2003).

CHAPTER 5

COMMERCIAL POLYMERS USED IN PRACTICAL ELECTRICAL AND ELECTRONIC APPLICATIONS

Current total worldwide annual consumption of commercial polymers (i.e., plastics, rubbers, and thermoplastic elastomers) is ~285 million metric tons with plastics amounting to ~89.5% of the total, rubbers 9%, and thermoplastic elastomers ~1.5% [1–3]. The sector of electrical and electronic applications is the fourth largest market for plastics, ~7% of the total amount [4], which is distributed as shown in Table 5.1.

This chapter deals with the use of standard commodity polymeric materials in the usual, well-established applications in both electrical and electronic fields including some newer applications (e.g., conductive adhesives, inks, and tapes). The use of and applications for specialty polymers (e.g., intrinsically conductive polymers, etc.) will be the subject of Chapter 6.

5.1 POLYMERIC MATERIALS USED AS INSULATORS

The largest volumes of industrial polymers are used in practical electrical and electronic applications where they serve mainly as insulators. The

Polymers for Electricity and Electronics: Materials, Properties, and Applications, First Edition. Jiri George Drobny.

TABLE 5.1 Plastics in Electrical and Electronic Applications

Thermoplastics	75%
Thermosets	15%
Composites	9%
Others[a]	1%

[a] Fibers, various coatings, adhesives, sealants, and potting compounds.

reason is that they are good dielectric materials with readily controllable electrical properties. However, this does not mean that they are inert in an electrical field; they may respond differently to different conditions (e.g., such as applied voltage, temperature, and different environmental conditions).

Most polymeric materials serving as electrical insulators may be required to operate in conditions of varying voltage with both direct current (dc) and alternating current (ac) at frequencies up to 10^5 MHz or even higher in some cases [5]. The voltages can vary from a few microvolts in communication equipment up to several million volts in power distribution systems. The currents may vary from several picoamps to thousands of amps. In addition, the parts made from polymeric materials are required to serve under wide ranges of temperatures, humidities, and other environmental conditions. Typical applications for polymers are listed in Table 5.2.

Some chemistry, properties, and established manufacturing processes are covered in detail in Sections 2.11–2.13. There are many ways to manufacture products from plastics, conventional elastomers, and thermoplastic elastomers. The following sections will concentrate mainly on the material used and only a brief mention will be made of the specific process. In general, practical applications including manufacture of finished products can be divided into the following broad categories: (1) thermoplastics, thermosets, composites, and thermoplastic elastomers and (2) rubber (conventional) processed by vulcanization.

5.1.1 Thermoplastic Resins and Elastomers

This group is defined as polymers that are hard and stiff; they can be softened by heating and returned to the initial state by cooling to the ambient temperature. Thermoplastics are processed by heating them to the extent that the formed melt can be shaped by the usual melt-processing methods (e.g., extrusion, injection molding, transfer and compression molding, blow molding, rotational molding and thermoforming) (see Chapter 2).

TABLE 5.2 Examples of Typical Applications for Polymers in Electrical and Electronic Industries

Application	Type of Material, Process Used
Wire and cable insulation and jackets	Thermoplastics, rubber, TPEs[a] (extrusion, coextrusion)
Connectors	Thermoplastics, TPEs[a] (injection molding, transfer, and compression molding, overmolding)
Molded electrical accessories	Thermoplastics, thermosets, rubber, TPEs (injection, compression, transfer molding, overmolding)
Transformers, coils, bushings	Thermoplastics, thermosets (transfer molding, injection molding, compression moldings), coatings, potting compounds, coil impregnation, wire insulation (extrusion coextrusion)
Integrated circuits, transistors, diodes, and other discrete devices	Thermoplastics, TPEs[a] (injection and transfer molding), coating, potting, dip and powder coating, specialty adhesives, junction coatings, photoresists (spin coating)
Printed circuit boards	Laminates, conformal coatings, solder masks, masking tapes, adhesives, vibration dampers, photoresists
Hybrids	Conductive and nonconductive adhesives, sealants, conformal coatings
Fuel cells	Polymer electrolyte membrane (PEM), perfluorosulfonate ionomers
Lithium ion polymer battery	Poly(ethylene oxide), poly(acrylonitrile), PVDF[b] gel
Organic light-emitting diode (OLED)	Highly conductive polymers

[a] Thermoplastic elastomers.
[b] Polyvinylidene fluoride, PVDF.

The detailed classification and properties of thermoplastic resins are covered in Section 2.12.1. Here they are grouped into only two groups with regard to their performance as materials in general and as insulators in particular. From this point, they are divided here as standard and high-performance thermoplastics.

Thermoplastics can be recycled and reused by being added to virgin material of the same kind or added as a suitable reprocessed material to other plastics or simply reprocessed alone into certain products. This, of course, is a great advantage for the environment, as well as for raw material cost.

5.1.1.1 Standard Thermoplastics Standard thermoplastics (widely referred to simply as "plastics") include ABS (acrylonitrile–butadiene–styrene), acrylic resins, polyamides, thermoplastic polyesters, polyethylene, polypropylene, poly(vinyl chloride) (PVC), polyvinylidene chloride (PVDC), polystyrene, styrene–acrylonitrile copolymer (SAN), and poly (methyl methacrylate) (PMMA). These materials are used in many electrical and electronic applications, including wire and cable insulations, connectors, cores, coils, insulators, insulating films and sheets, bushings, semiconductive and static-dissipative parts and films, covers and shields, housings, handles, cases, keyboards, tubing, profiles, where exceptional performance characteristics are not required. Most of them can be made flame retardant by proper compounding. An example of a PVC formulation for a wire and cable insulation is in Table 5.3.

TABLE 5.3 Example of a PVC Formulation for Wire and Cable

Ingredient	Amount (parts by weight)	Name/Function
PVC	100	Base polymer
Drapex 6.8	5	Epoxidized soybean oil, plasticizer
Mark 6734	5	Barium–zinc stabilizer
DIDP	40	Diisodecyl phthalate, plasticizer
Atomite whiting	75	Calcium carbonate, filler
Hydral 710	45	Aluminum trihydrate, flame retardant
Elvaloy HP 411	80	Ethylene terpolymer, nonmigratory plasticizer and/or flex modifier
Irganox 1010	0.8	Phenolic primary antioxidant
DLTDP	0.32	Thioester secondary antioxidant
TiPure R960	3	White pigment
Total	**354.12**	

The formulation shown in Table 5.3 produces a material with the following mechanical and electrical properties:

Mechanical Properties	
Tensile strength [MPa (psi)]	7.0 (1014)
Elongation at break (%)	362
Electrical Properties	
Dielectric strength in oil (V mil^{-1})	533
Dielectric constant @ 1 kHz	5.84
Dielectric constant @ 1MHz	4.03
Dissipation factor @ 1 kHz	0.060
Dissipation factor @ 1 MHz	0.072
Volume resistivity · Ω · cm	1.65×10^{12}

5.1.1.2 High-Performance Thermoplastics High-performance thermoplastics are polymeric materials typically exhibiting high resistance to elevated temperatures, chemicals, high mechanical strength and impact resistance, high dielectric strength, and varied degrees of flame resistance. This group includes polycarbonate (PC), polyimide (PI), poly(amide-imide) (PAI), poly(phenylene sulfide) (PPS), poly(aryl sulfone) (PAS), poly(ether imide) (PEI), poly(propylene oxide) (PPO), polyarylates (amorphous aromatic polyesters), poly(ether–ether ketone) (PEEK), poly (arylether ketone) (PAEK), poly(tetrafluoroethylene) (PTFE), copolymer of ethylene–tetrafluoroethylene (ETFE), perfluoroalkoxy resins copolymer of tetrafluoroethylene (TFE) and perfluoropropylvinyl ether (PFA) and copolymer of tetrafluoroethylene and perfluoromethylvinyl ether (MFA), fluorinated ethylene–propylene (FEP), poly(vinylidene fluoride) PVDF, poly(chlorotrifluoroethylene) (PCTFE), copolymer of ethylene–chlorotrifluoroethylene (ECTFE), and ionomers. These materials are used for parts, wire and cable insulations, films and sheets for extended service at elevated temperatures, typically 250°C (482°F) or higher, for aggressive chemical environment, radiation, for specialized devices (e.g., fuel cells), organic light emitted diodes, lithium ion rechargeable batteries, electret devices, and so on. Fluorinated polymers and copolymers are intrinsically flame retardant and most of the remaining high-performance thermoplastics can be made flame retardant by suitable compounding.

5.1.1.3 Thermoplastic Elastomers Thermoplastic elastomers include a group of materials with different properties and generally good processing behavior. They are widely used in electrical and electronic applications. The following list shows specific examples for individual TPEs:

- Styrenic TPEs (styrenic block copolymer, SBCs) are frequently used as a replacement for conventional rubber in wire and cable insulations, power cords (Fig. 5.1), plugs, handles of power tools, and so on.
- Thermoplastic vulcanizates (TPVs) and engineering thermoplastic vulcanizates (ETPVs) are widely used as primary insulators and as jacketing insulators or both. Plug insulators for electrical and electronic assemblies used in, computer hardware, telephones, electronic appliances, and office equipment are another use. Examples of ETPVs are automotive spark plug and ignition coil boots shown in Fig. 5.2.
- Thermoplastic polyolefin elastomers (TPOs) are used in flexible cords, battery booster cables, appliance wires, low-voltage wires, control cables, hook-up wire, submersible pump wire, and cable

FIGURE 5.1 Power cords. (Courtesy Teknor Apex).

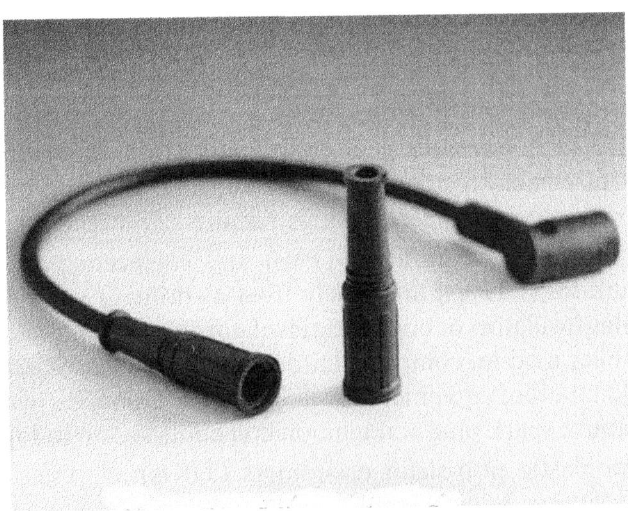

FIGURE 5.2 The ETPV spark plug and ignition coil boots. (Courtesy DuPont).

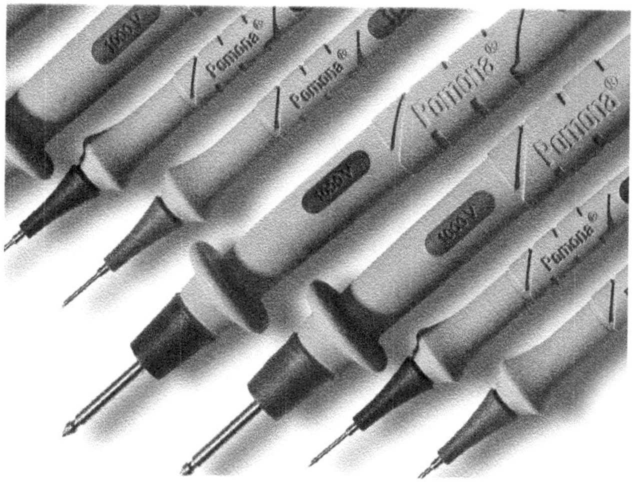

FIGURE 5.3 Electric probes. (Courtesy RTP Company).

jackets. Excellent electrical properties, ozone resistance, and water resistance are the main attributes in these applications [6]. Wire and cable products include both standard and fire-retardant grades. An example of electric probes made from TPO is in Fig. 5.3)

- Melt processable rubber (MPR) is used mostly for low-voltage (≤600 V) applications, where toughness, environmental endurance, and rubbery quality are required. Typical applications are flexible cords, welding cable with excellent spark burn-through resistance, and as jacketing in higher-voltage cable construction to provide durable protection to primary insulation [7].
- Thermoplastic polyurethanes (TPUs) due to their high-abrasion resistance, cut resistance, flexibility (at both ambient and low temperatures), and oil resistance, find a wide range of electrical and electronic applications: seismographic cable, audio wire, camera cables, computer cables, head-set cables, marine cable, welder cables, air gun control cable, communication wires, towing leader, abrasion resistant cable jacketing, and miscellaneous plugs (see Fig. 5.4) [7].
- Copolyester thermoplastic elastomers (COPEs) are tough materials exhibiting high tear resistance, flex-cut resistance, oil and solvent resistance, and environmental resistance. Because of that they find use as thin-wire insulation, hermetically sealed units, switches, and connectors, retractable coiled cables (Fig. 5.5), panel insulated covers [6, 8, 9]. In fiber optics technology, COPEs are used as the primary coating, buffer tube, and jacketing [6, 10].

FIGURE 5.4 Electrical and electronic plugs. (Courtesy Bayer MaterialScience, LLC).

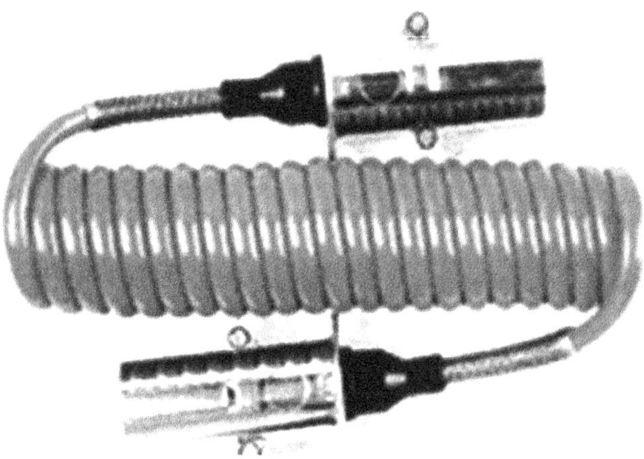

FIGURE 5.5 Retractable coiled electrical cable. (Courtesy DuPont).

- Polyamide thermoplastic elastomers (COPAs) are used mainly for high-temperature insulations, abrasion resistant cable jackets, and for nonkinking wires [7]. Another application is a computer keyboard cover (Fig. 5.6).

5.1.2 Thermosetting Resins

Thermosetting resins or thermosets are materials that react during processing to form cross-linked structures that cannot be remelted or reprocessed. Thermoset scrap must be either discarded or used as a

FIGURE 5.6 Computer keyboard cover. (Courtesy Arkema Inc.).

low-cost filler in other products. In some cases, it may be pyrolyzed to recover inorganic fillers (e.g., such as glass reinforcements and possibly monomers and/or other useful organic compounds). Thermosets can be supplied in liquid form or as a partially polymerized solid molding powder. In their uncured state, they can be formed to the finished product shape with or without pressure and polymerized by using chemicals, heat, or radiation.

Thermosetting resins generally provide one or more of the following advantages: (1) high thermal stability, (2) resistance to creep and deformation under load and high dimensional stability, and (3) high rigidity and hardness. These advantages are coupled with the light weight and excellent electrical insulating properties common to almost all plastics. The compression and transfer molding methods by which most of the materials are formed along with thermoset injection molding techniques, offer low-processing cost and highly mechanized production.

Thermosetting molding compounds consist of two major ingredients: (1) a *resin system*, which generally contains such components as curing agents, hardeners, inhibitors, and plasticizers and (2) *fillers and/or reinforcements,* which may consist of mineral or organic particles, inorganic or organic fibers, or chopped cloth or paper. The resin system usually exerts the dominant effect, determining to a great extent the cost, dimensional stability, electrical qualities, heat resistance, decorative possibilities, and flammability. Fillers and reinforcements affect all these properties to varying degrees, but their most dramatic effects are seen in strength, toughness, and sometimes electrical qualities [11]. The following list is of the most commonly used thermosetting resins in electrical and electronic applications: alkyds (typically dry polyester molding compound, cross-linked by diallyl phtalate (DAP) monomer, allylics [DAP and diallyl isophtalate (DAIP)], bismaleimides (BMI), epoxies, melamines, phenolics, polyesters (unsaturated polyesters, orthophtallic, isophtalic, vinyl esters), polyimides (thermoset), silicones, and ureas. Typical applications for thermosetting resins are housings, lighting fixtures, closures, wiring devices, buttons, circuit breaker insulation coil forms, capacitor and resistor encapsulations, cases, housings, switchgear components, adhesives, coatings, laminated resins, films for electric motors, insulations for wire and cable, chip carriers for integrated circuits, and printed wiring boards.

5.1.3 Fiber-Reinforced Polymeric Composites

As pointed out in Section 2.12.4, fiber-reinforced polymeric composites used in electrical and electronic applications consist essentially of a polymeric matrix combined with a reinforcing agent or agents, which most frequently are fibers. The reinforcement improves strength and stiffness properties, compared to neat resin. The *matrix* of polymeric fiber-reinforced composites is made from either a thermosetting or a thermoplastic resin.

Glass, in any form, including fibrous, is a very low conductor of electricity. In fact, the designation of conventional glass fibers as *E-glass* is derived from the early use of glass in electrical applications. Reinforced polyesters are widely used in applications designed to provide protection from electric shocks [12]. Typical applications for reinforced plastics in the electrical and electronic field are [13]:

- Avionic electronics
- Microelectronics
- Printed circuit boards
- Heat sinks
- Die attach
- Interlayer dielectrics
- Heat sinks

- Panels
- Lids and housings
- Electronic packaging
- Electrical contact pads

Plastics reinforced by carbon fibers will exhibit some degree of electrical conductivity, depending mainly on the type and contents of the fibers in the composite.

5.1.4 Rubbers (Conventional Elastomers)

Elastomers are normally good electrical insulators, particularly the nonpolar (not oil-resistant) types better than the polar (oil-resistant) ones (see Table 5.4). However, the compounding ingredients (e.g., fillers, plasticizers, and other additives) often have more influence on the electrical properties of the vulcanizate than the base elastomer. Compounds containing carbon black must be avoided if a very high resistance is required. For this purpose, silicone elastomers are the best choice [14].

It is possible to make rubber compounds antistatic and even conductive by incorporating sufficient amounts of special carbon black, graphite, and certain metal powders, but the conductivity of the vulcanizate is much lower than that of metals. It is difficult to make a clear-cut distinction between insulating, antistatic, and conducting rubber. In

TABLE 5.4 Dielectric Properties of Major Nonpolar and Polar Elastomers

Elastomer[a]	Dielectric Properties
Natural rubber (NP)	Good–Excellent
Styrene–butadiene copolymers (NP)	Good
Polychloroprene (P)	Poor–Fair
Butyl rubber (NP)	Good–Excellent
EPR, EPDM (NP)	Good–Excellent
Chlorinated polyethylene (P)	Fair–Good
Chlorosulfonated polyethylene (P)	Fair–Good
Ethylene–vinyl acetate copolymers (NP)	Good
Epichlorohydrin (P)	Poor
Butadiene–nitrile copolymers (P)	Poor
Silicone rubber (NP)	Excellent
Fluorocarbon elastomers (NP)	Good
Polyurethane rubber (NP)	Good
Acrylic (P)	Poor

[a] Nonpolar = NP and polar = P.

general, rubber materials having resistance $< 10^4$ Ω are considered conductive, between 10^4 and 10^8 Ω, antistatic, and above this value they are insulating [14]. Besides serving as an insulator or a conductor, the elastomer compound must exhibit a good retention of its electrical properties. For that reason, resistance to other factors, including heat, high humidity, and weathering must be considered in conjunction with electrical properties [14]. The effect of fillers on the resistivity of vulcanized natural rubber (NR) and styrene–butadiene rubber (SBR) is shown in Table 5.5.

Compounding conventional rubber materials is a rather complex task considering the usual number of components in the formulation. This is a major reason why conventional rubber is often replaced by thermoplastic elastomers. However, there are still certain applications, where conventional cured rubber is still preferred, such as in specialty cable jackets with exceptional heat resistance, ozone resistance, and oil resistance.

TABLE 5.5 Effect of Fillers on the Volume Resistivity of Vulcanized NR and SBR

Filler	Loading (% by volume)	Water Immersion (days at 70°C)	Volume Resistivity (Ω-m)	
			NR[a]	SBR[b]
None	–	0	4.4×10^{18}	3.7×10^{17}
		7	3.3×10^{18}	6.6×10^{17}
		14	1.4×10^{18}	6.1×10^{17}
Zinc oxide	50	0	8.2×10^{15}	3.9×10^{14}
		7	7.9×10^{15}	3.5×10^{14}
		14	5.3×10^{15}	1.2×10^{14}
Calcined clay	50	0	4.3×10^{17}	3.0×10^{17}
		7	2.1×10^{16}	2.9×10^{16}
		14	2.1×10^{16}	2.7×10^{16}
Whiting (water ground)	50	0	8.4×10^{17}	4.5×10^{17}
		7	1.9×10^{17}	2.3×10^{17}
		14	1.4×10^{17}	2.1×10^{17}
N990 Carbon black	25	0	1.5×10^{18}	4.5×10^{17}
		7	1.9×10^{18}	3.9×10^{17}
		14	1.1×10^{18}	3.5×10^{16}
N990 Carbon black	50	0	2.5×10^{12}	2.8×10^{17}
		7	3.7×10^{14}	1.3×10^{17}
		14	4.1×10^{14}	1.1×10^{17}
N765 Carbon black	25	0	9.1×10^{12}	1.8×10^{17}
		7	3.6×10^{15}	1.2×10^{17}
		14	9.5×10^{15}	1.0×10^{17}

[a] Press cured 15 min at 143°C (290°F).
[b] Press cured 30 min at 1603°C (3200°F).

5.2 WIRE AND CABLE TECHNOLOGY

5.2.1 Introduction

The wire and cable (W&C) industry produces a large variety of conductors. For the most part, the materials used for the conduction of electricity are copper and aluminum. The primary function of polymeric materials (e.g., plastics, rubber, thermoplastic elastomers, and specialized coatings) is to provide electrical insulation for these conductors. Other use, also very important, is as an outer sheath (jacket) that protects the W&C insulation from the effects of chemicals, moisture, and ozone, as well as other environmental factors, and from abrasion and cutting. Specialized polymers and compounds improve flame resistance of the coated wires and cables.

Materials used for insulation must exhibit superior electrical, sufficient physical, and aging properties to withstand different environments and still maintain their electrical integrity for a long time (typically, as much as 40 years). Rigorous requirements are imposed by Underwriters Laboratories (UL), Insulated Power Cable Engineers Association (ICEA), American Edison Illuminating Companies (AEIC) and various other national and international agencies to assure a satisfactory service of these materials.

5.2.2 Types of Wire and Cable Products

The conductors produced by the W&C industry can be roughly divided into two basic groups: wires and cables. The difference between them is that a *wire* is a single conductor and a *cable* is a group of two or more insulated conductors. If there was not any insulation on the two conductors, then it would not be a cable, it would still be a single conductor that would classify it as a wire. There are four basic categories of W&C: (1) single-conductor, (2) multiconductor, (3) twisted pairs, and (4) coaxial cable. The most widely used materials are copper and aluminum.

There are two types of wire: solid and stranded wire. Both types have different uses depending on the level of resistance and flexibility required. Solid wire is a single solid conductor that may be bare or insulated. This type of wire offers a little lower resistance. Solid conductors are the key to better performance at higher frequencies. Stranded wire is wire that is made up of several strands of bare conductor to form a single conductor when all strands are put together. This type of wire has more flexibility and a longer flex life before becoming unusable.

There are four basic cable types: twisted pairs, multiconductor, coaxial, and fiber optics cables. Twisted pairs cable consists of pairs of conductors that are twisted together. This cable is specifically intended for signal carrying. Twisting the pairs of conductors gives the cable some immunity to interference. Multiconductor cable is cable that is made up of many insulated conductors. This type of cable is common in control applications, but is almost never used in signal applications. Coaxial cable is the other popular configuration for cable. The signal on the two conductors in a coaxial cable is not the same because the shield carries the ground and the signal. Since the signal is not the same on both conductors, this configuration is an unbalanced line.

Fiber optic cables are divided into three kinds: plastic fiber, multimode fiber, and single-mode fiber. Plastic fiber is usually used in high-end audio signals. Multimode fiber is made of glass, ranges in diameters, and is used in the data world. Single-mode fiber is so fine that it can be seen only under a microscope.

5.2.3 Cable Manufacturing Processes

The primary process applied to the W&C manufacture is extrusion (Section 5.2.3.1), which is capable of applying the insulating and protecting materials in the desired diameter and length. Another important process is vulcanization that cross-links the conventional rubbers and makes them harder, stronger, more elastic, and insoluble. Clearly, thermoplastic materials do not require vulcanization (Section 5.2.3.2). This is their major advantage over conventional rubber as to productivity.

5.2.3.1 Extrusion The cross-head extrusion process is widely used to coat W&C with a polymer-based insulation. The basic procedure includes pulling of the W&C to be coated at a uniform rate by a cross-head die (see Fig. 5.7), where it is covered with the molten plastic or hot rubber compound. The usual cross-head die is at a 90° angle between the wire line and the extruder body axis. Cross-heads with a 30 and a 45° angle are also used; these prevent dead spots in the melt channel and provide a more streamlined interior.

The extrusion process for coating is used in most W&C that find usage in telecommunication and electrical applications, as well in the electronic industry. For more coatings, two extruders can also be used in tandem.

Wire coating is generally done until the use of single screw extruders, in which the cross-head extrusion process is carried out. The job of the extruder is to melt the resin and forward it to the die at an even and constant melt pressure and constant temperature. The cross-head

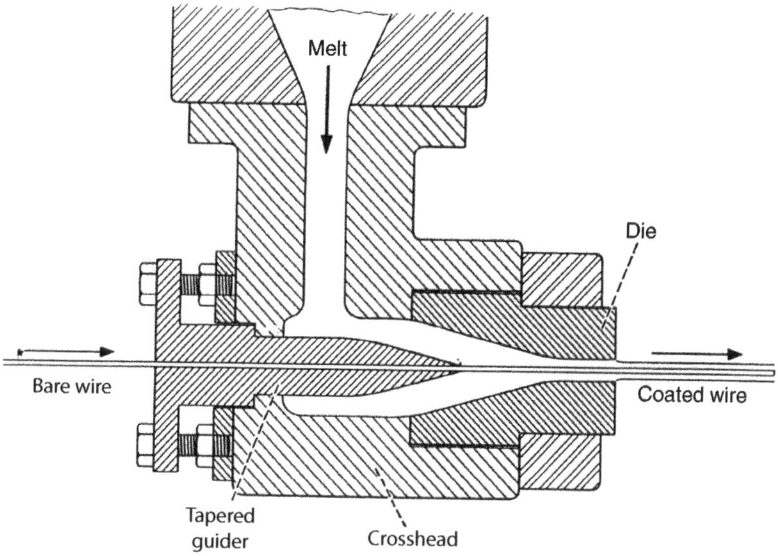

FIGURE 5.7 An example of a cross-head die (self-centered design).

extrusion process is carried out by using equipment in the line, which usually includes the following eight basic components:

1. Unwind station or some other W&C source for feeding the line.
2. Pretensioning station for setting the tension throughout the process.
3. Preheat station for preparing the wire for coating.
4. An extruder equipped with a cross-head die.
5. Cooling equipment for the coating.
6. Test stations for assuring that the wire is suitably coated.
7. Puller for providing constant tension throughout the process.
8. Winder for collecting the product.

Examples of modern W&C extrusion lines including control systems are shown in Figs. 5.8 and 5.9.

In this process, *primary insulation* is defined as the polymeric material applied directly onto the metal W&C to isolate the metal electrically. *Jacketing* (or *sheathing*) refers to the covering on a wire or a group of wires with an insulating coating or jacket for nonelectrical protection. Jackets are usually put onto primary wires.

After leaving the die, the coated wire is cooled in an air and/or water bath and then it is continuously tested for spark and concentricity

FIGURE 5.8 Detail of the wire and cable extrusion line. (Courtesy Davis-Standard, LLC).

FIGURE 5.9 Extruder system with a cross-head. (Courtesy Davis-Standard, LLC).

(roundness of the coating). Rubber compounds are extruded in a similar fashion, but the extruders have different design and also the downstream operations are somewhat different (see Section 2.13.3.2).

In some wire coating operations, polyethylene (PE) is used in its cross-linked version. The wire is first extrusion coated with PE and then at the end of the operation, the PE coating on the wire is cross-linked either chemically or by an electron beam (EB) irradiation.

There are essentially three chemical methods of cross-linking PE [15]: (1) the peroxide method, (2) the azo method, and (3) the silane method.

1. The peroxide method (often referred to as the Engel method) uses an organic peroxide (most frequently dicumyl peroxide) with a coagent. Usual curing temperatures for peroxide cross-linking range from 250 to 280°C (480 to 535°F). The process can be performed in a steam- or liquid-curing media.
2. The azo method is a two-stage method using azo compounds, such as 2,2′-azobis(2-ethoxypropane) and the articles can be cured at 240–270°C (465–520°F) in a steam or liquid-curing media.
3. The silane method uses mostly vinyl alkoxy silanes and requires a sufficient amount of moisture. There are essentially two silane cross-linking processes for PE, namely, the two-stage method (Sioplas®) and the one-stage method (Monosil®) [16]. The silane method has the advantage of low capital and equipment costs since it can use existing thermoplastic extrusion lines. It can take place at atmospheric pressure in a water bath at elevated temperatures. The moisture requirement is a drawback; its rate depends on the rate of diffusion of moisture in the polymer. This method is well established in W&C technology for low voltage ($<$1 kV) in power distribution, telecommunication, transportation and appliance cables, as well as heat-shrinkable tubing. [17].

An example of a PE formulation for chemical cross-linking by peroxide is shown in Table 5.6.

Polytetrafluoroethylene is used frequently in electrical and electronic applications. In fact, about one-half of the PTFE resin produced is used in these applications [18], with major use for insulation hookup wire for military and aerospace electronic equipment. It is also used as insulation for airframe and computer wires, as "spaghetti" tubing, and in electronic component. The wire insulation and tubing are produced by paste extrusion of PTFE fine powders and subsequent sintering at temperatures 365–380°C (689–716°F) [19].

TABLE 5.6 Formulation for a Chemically Cross-Linked Polyethylene Insulation Jacket

Ingredient	Amount (parts by weight)	Name/Function
Polyethylene, low density	100	Base polymer
Substituted quinoline	0.50	Antioxidant
N990 Black	60	MT (medium thermal carbon black)
Dicumyl peroxide (40%)	7	Cross-linking agent
Total	**167.50**	

The formulation shown in Table 5.6 produces a material with the following mechanical properties:

Mechanical Properties	
Tensile strength [MPa (psi)]	16.1 (2330)
Elongation at break (%)	310

Rubber-based insulations have been used in W&C manufacture for nearly a century. Rubber compounds (formulations) for W&C are designed primarily to possess the required electrical properties for their intended service, and the ability to perform satisfactorily under their anticipated conditions of use. While physical properties are of secondary consideration, there are definite requirements imposed by a variety of specifications. Vulcanizable (cross-linkable) elastomers finding applications in this area include NR, SBR, butyl and ethylene–propylene rubbers of both copolymer and terpolymer types [20]. Polyethylene originally used as a thermoplastic in communication W&C is now cross-linked and its uses extended to include high-voltage insulations, switchboard wire, and service dropwire. Chlorosulfonated polyethylene (hypalon), CR (neoprene), NBR (nitrile rubber), acrylic, epichlorohydrin, and silicone elastomers are also used in specialized applications as insulating or jacket materials. In addition, EPDM is widely used in a variety of applications as shown in Table 5.7. Examples of different rubber compounds used in wire and cable manufacture are in Tables 5.8–5.16. These examples show only basic concepts and may be used possibly as starting formulations for the given applications. It is recommended that additional information and guidance should be obtained from the pertinent raw material manufacturers.

Jacket or sheath rubber compounds are applied over insulations in cable structures and provide the wearing quality, resistance to elements, and overall strength required in service. Physical property requirements vary widely with the intended applications ranging from 1500 psi (10.3 MPa)

TABLE 5.7 The EPDM Wire and Cable Applications

Low-Voltage Applications	High-Voltage Applications
UL flexible cord	Industrial power cable
Submersible pump cables	Utility power cables
Appliance wire	Underground residential (URD) cable
Automotive ignition cable	
Insulation	
Track resistance insulators	
Welding cable insulation	

TABLE 5.8 The EPDM Low-Voltage Flexible Cord

Ingredient	Parts by Weight	Name/Function
EPDM (E/P = 75/25), Mooney 60	100	Base polymer
TMQ[a]	1	Antioxidant
Paraffin wax	5	Processing aid
Calcined clay	200	Filler
Paraffinic oil	70	Plasticizer
Vinyl silane[b]	2	Coupling agent, improves bond between elastomer and filler
Zinc oxide	5	Acid acceptor
TMPT[c]	2	Coagent, peroxide activator
Organic peroxide	7.5	Curing agent
Total	**392.5**	

[a] Polymerized 1,2-dihydro-2,2,4-trimethylquinoline.
[b] Improves interaction between polymer and filler.
[c] Trimethylolpropane triacrylate.

TABLE 5.9 Low-Cost EPDM Electrical Insulation

Ingredient	Amount (parts by weight)	Name/Function
EPDM, high ENB,[a] fast curing	100	Base elastomer
Silene 732 D	75	Precipitated silica filler
Nucap 200	75	Mercaptosilane treated clay, filler
Burgess KE	75	Clay, mineral filler
Paraffinic oil	90	Plasticizer
Carbowax 4000	1.5	Polyethylene glycol, dispersant
Stearic acid	1.0	Cure activator
TE 80	5.0	Processing aid
Zinc oxide	3.0	Cure activator
Tetramethylthiuram disulfide	3.0	Accelarator, sulfur donor

(*Continued*)

TABLE 5.9 (Continued)

Ingredient	Amount (parts by weight)	Name/Function
TMTM[b]	1.5	Accelerator
Zinc dibutyldithiocarbamate	2.25	Ultraaccelerator, antioxidant
4.4′-Dithiomorpholine	2.0	Accelerator
Sulfur	0.6	Curing agent
Total	**434.85**	

[a] Ethylenenobecne.
[b] Tetramethylthiuram monosulfide.

The formulation shown in Table 5.9 produces a material with the following mechanical properties:

Mechanical Properties	
Tensile strength [MPa (psi)]	8.28 (1200)
Elongation at break (%)	670

TABLE 5.10 Medium Voltage (15 kV) EPDM Wire and Cable Insulation

Ingredient	Amount (parts by weight)	Function
EPDM	100.00	Base elastomer
Zinc oxide	5.00	Activator
Agerite MA	0.55	Antioxidant
Translink 37	120.69	Mineral Filler
Lead oxide (red lead)	5.00	Ion scavenger, improves electrical properties
Vinyl silane	1.03	Coupling agent, improves bond of elastomer to filler and electrical properties
Paraffin wax	5.00	Process aid
Paraffinic oil	20.00	Plasticizer
DiCup 40 KE Peroxide	7.59	Curing agent (40% active)
Total	**265.78**	

[a] Measured after 14 days @ 75°C (167°F).

The formulation shown in Table 5.3 produces a material with the following mechanical and electrical properties:

RHC (Rubber hydrocarbon content) (%)	37.63 by weight, 58.9 by volume
Mechanical Properties (from Finished Cable)	
Tensile strength [MPa (psi)]	8.8 (1277)
Elongation at break (%)	337
Electrical Properties (from Finished Cable)	
Dielectric constant[a]	3.6
Power factor[a]	1.1
Dielectric strength (680 V mil^{-1})	

TABLE 5.11 Low-Voltage Cable Insulation (Chemically Curable Polyolefin Elastomer)

Ingredient	Amount (parts by weight)	Name/Function
Engage 8150	100	Base polyolefin elastomer
Calcined clay	125	Mineral filler
Calcium carbonate	125	Mineral filler
Paraffinic oil	80	Plasticizer, process aid
Agerite Resin D	1.50	Antioxidant
Vinyl silane	1.25	Coupling agent, improves bond of elastomer to filler and electrical properties
SR 350	3	Coagent, enhances peroxide cure
Vulcup 40 KE	10	Peroxide, 40%, curing agent
Total	**445.75**	

The formulation shown in Table 5.11 produces a material with the following mechanical properties:

Mechanical Properties	
Tensile strength [MPa (psi)]	7.93 (1150)
Elongation at break (%)	250
Dielectric constant @ 1 kHz	2.7

TABLE 5.12 Heat Resistant Appliance Wire and Ignition Cable

Ingredient	Amount (parts by weight)	Name/Function
EPDM, moderate curing rate	100	Base elastomer
Neoprene WRT	8	Polychloroprene, enhances cure
Hydrated alumina	100	Mineral filler
Zinc oxide	15	Cure activator, enhances heat resistance
Paraffinic oil	15	Plasticizer
Vinyl silane	1.5	Coupling agent for mineral filler, improves electrical properties
2-Mercaptotoluimidazole	2	Primary antioxidant
Vanox AM	1	Secondary antioxidant
Trimethylolpropane methacrylate	3	Coagent for peroxidic cure
Dicumyl peroxide	3.5	Curing agent
Total	**249.0**	

The formulation shown in Table 5.12 produces a material with the following mechanical properties:

Mechanical Properties	
Tensile strength [MPa (psi)]	11.8 (1700)
Elongation at break (%)	290

TABLE 5.13 The EPDM Motor Lead Wire Insulation

Ingredient	Amount (parts by weight)	Name/Function
EPDM high ENB, fast curing	50	Base elastomer
EPDM, medium ENB, low viscosity	50	Base elastomer
Nulok 321	100	Surface treated clay, filler
Paraffinic oil	30	Plasticizer
Polyethylene AC 617A	5.0	Low molecular weight PE, processing aid
Paraffin wax	5.0	Processing aid
Stearic acid	1.0	Activator of cure
Zinc oxide	3.0	Activator of cure
Sulfenamide	1.0	Accelerator
Mercaptobenzothiazole	1.0	Accelerator
Tetramethylthiuram monosulfide	1.0	Accelerator
Sulfur	1.5	Curing agent
Total	**248.5**	

The formulation shown in Table 5.13 produces a material with the following mechanical properties:

Mechanical Properties	
Tensile strength [MPa (psi)]	16.3 (2218)
Elongation at break (%)	650

TABLE 5.14 Heavy Duty Polychloroprene Jacket

Ingredient	Amount (parts by weight)	Name/Function
Neoprene W	100	Base elastomer
Paraffinic oil	12	Plasticizer
Vanfre AP-2	6	Processing aid
Agerite HP-S	1.5	Antioxidant
Agerite Stalite S	1.5	Antioxidant
N770 Black	25	SRF carbon black
N550 Black	25	FEF carbon black
Whiting (water ground)	35	Mineral filler
Vanax NP	1	Accelerator
Sulfads	1.5	Vulcanizing agent
DOTG[a]	0.5	Accelerator
Total	**218.0**	

[a] Diorthotolyl guanidine (rubber accelerator) = DOTG.

The formulation shown in Table 5.14 produces a material with the following mechanical and electrical properties:

Mechanical Properties	
Tensile strength [MPa (psi)]	16.1 (2330)
Elongation at break (%)	490

TABLE 5.15 Extra Heavy Duty Mining Cable Jacket

Ingredient	Amount (parts by weight)	Name/Function
Paracryl OZO	100	Base elastomer, NBR/PVC, ozone resistant
Stearic acid	1	Cure activator
Zinc oxide	3	Cure activator
Dioctyl sebacate	10	Plasticizer
Paraplex G-33	10	Low molecular weight polymeric plasticizer
Agerite Geltrol	2	Antioxidant
Titanium dioxide	5	White pigment
Hydrated silica	35	Mineral filler
Dixie clay	15	Mineral filler
Vanstay 3027	1.25	Heat stabilizer
MBTS[a]	2.50	Benzothiazyl disulfide, accelerator
TMTD[b]	3	Tetramethyl thiuram disulfide, accelerator
Sulfur	2	Curing agent
Total	**189.75**	

[a] 2-Mercaptobenzothiazole (rubber accelerator) = MPTS.
[b] Tetramethylthiuram disulfide (rubber accelerator) = TMTD.

The formulation shown in Table 5.15 produces a material with the following mechanical properties:

Mechanical Properties	
Tensile strength [MPa (psi)]	12.4 (1880)
Elongation at break (%)	620

tensile strength specified for flexible portable cords to a 3500-psi (24.1-MPa) minimum for heavy duty cables. Because of polychloroprene rubber its good weathering properties, oil resistance, flame resistance, and mechanical strength, is currently the most widely used jacketing material. Jacketing materials should be selected and designed to assure good aging of insulation over which they are applied. While electrical properties are not generally important, there are cases where required performance tests on finished cable may influence the selections of compounding materials for a given application. Several examples of different cable designs are depicted in Fig. 5.10 (2-kV power and control cable, with ethylene–propylene copolymer (EPR) insulation and a CSM cover), Fig. 5.11 (600-V power and control cable), Fig. 5.12 (Cross-section of a 225-kV power transmission cable), Fig. 5.13 (ethernet thin coaxial cable). Practically all wire and cable compounds are applied by extrusion and a large percentage of them are extruded and vulcanized in a continuous operation.

TABLE 5.16 High-Heat Acrylic Rubber Wire Insulation or Cable Jacket

Ingredient	Amount (parts by weight)	Name/Function
HyTemp 4054	100	Acrylic elastomer, easy processing
Stearic acid	1	Cure activator
Zinc oxide	9	Cure activator
Nulok 321	120	Silane treated clay, filler
Zeolex 23	60	Aluminum silicate, filler
Admex 760	15	Polymeric adipate, plasticizer
Agerite Stalite S	2	Antioxidant
Cumate	0.2	Ultraaccelerator
Polyethylene AC 617	6	Processing aid
Vanax 829	0.9	Cross-linking agent
Butyl Tuads	2.5	Sulfur donor and accelerator
Lead stearate	0.5	Stabilizer and external lubricant
Total	**314.1**	

The formulation shown in Table 5.16 produces a material with the following mechanical and electrical properties:

Mechanical Properties

Original

Tensile strength [MPa (psi)]	3.2 (460) original
Elongation at break (%)	350 original

Aged 70 h @ 175°C

Tensile strength [MPa (psi)]	6.9 (1000)	Change: +117%
Elongation at break (%)	80	Change: −73

Electrical Properties (at RT)[a]

Dielectric strength [V mil^{-1} (kV mm^{-1})]	283 (11.3)
Volume resistivity (Ω-cm)	7.2×10^{10}

[a] Room temperature = RT.

FIGURE 5.10 A 2-kV traction cable. (Insulation: EPR, jacket: CSM).

5.2.3.2 Vulcanization The Lead (Mold) Curing Process is one of the original ways to vulcanize large cables. In this method, the complete length of the cable is enclosed in an uncured rubber jacket with a

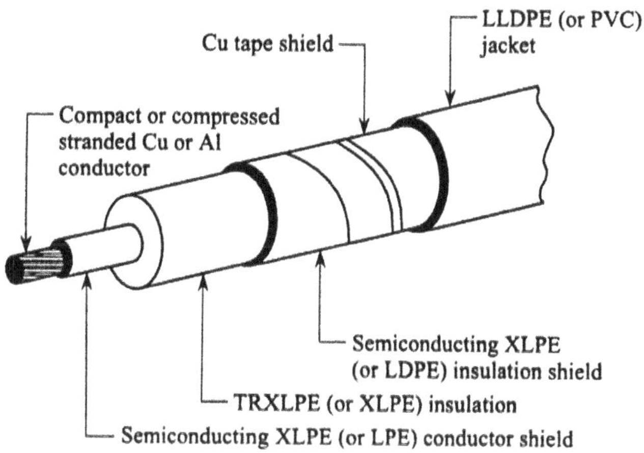

FIGURE 5.11 600 V Power and control cable. (Courtesy of The Okonite Company).

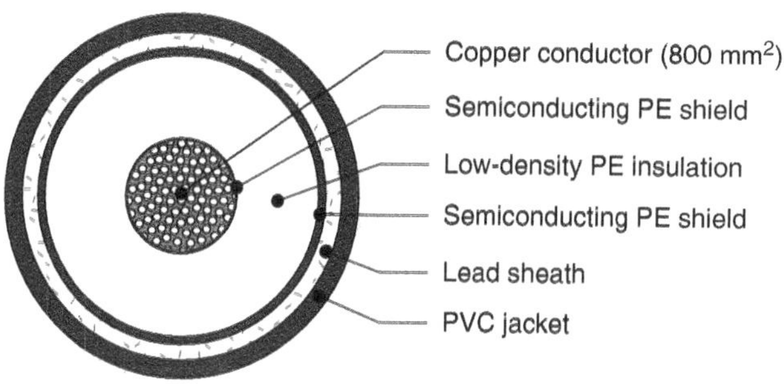

FIGURE 5.12 Cross-section of a 225 kV power transmission cable.

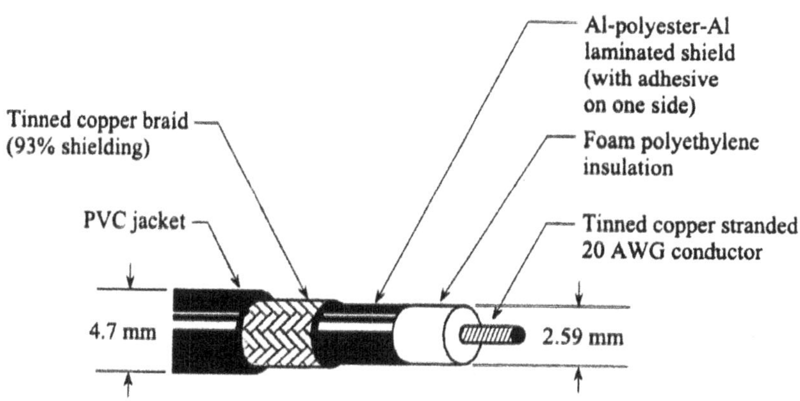

FIGURE 5.13 Ethernet thin coaxial bus cable. (Courtesy of Belden Wire and Cable Company).

temporary lead jacket extruded over it. It is then rolled on a reel and placed into an autoclave, where it is cured. Compared to current curing methods, the lead curing process is considerably less efficient and has several drawbacks. One of the most serious drawbacks is that the cables being cured on a reel during the autoclave process take a "set" and "memorize" the coil shape of the reel. This memorization of the reel eventually causes cable damage due to torque binding when the cable is reeled–unreeled or dragged in a particular application. The final result produces a twisting or cork-screwing effect and more quickly results in cable failure. This is especially disadvantageous for use in underground heavy mining equipment (long-wall miners, continuous miners, etc.) and shovels. Because of the above mentioned shortcomings, this method is all but obsolete and is used very seldom.

The most widely used process, particularly in North America, is the continuous vulcanization (CV operation). The production unit consists essentially of an extruder attached to a jacketed curing tube in which high-pressure steam is confined. The wire or cable with the compound applied by extrusion is discharged directly from the extruder head into the steam tube through which it is conveyed under tension. Steam pressures of 1.4 MPa (200 psi) and higher are generally used and tube lengths may be 60 m (200 ft) or longer. The heated tube is often divided into several separately controlled heat zones, which can be adjusted up to a temperature as high as 450°C (842°F). Discharge of the vulcanized product into the atmosphere may take place through mechanical seals or through a water seal in which pressure is maintained just under steam pressure in the adjacent curing tube. While originally developed and applied in the production of small-size, conductor insulated wire; mechanical refinements and new equipment design now permit the production of practically all types and sizes of wire and cable by this method [21]. Compounds for continuous vulcanization are designed to cure in seconds at elevated temperatures and their handling and processing is somewhat more critical than in conventional rubber manufacturing. The temperature control of these longer, more sophisticated lines allows for the precise control of the stocks and for the greater throughput under controlled conditions [22].

For the production of heavy-walled high-voltage cables (e.g., 15 kV power cables and especially those using EPDM and cross-linked PE insulations) some machines are tower-mounted and operated in a vertical configuration delivering the extruded cable downward into the curing chamber. For the thickest insulated cables, several extruders supply the appropriate extrudate for the overlapping layers. The melt is also extruded onto the conductor straight down to maintain concentricity. Because the cable is moving in the vertical direction, it never touches the tube.

An arrangement for vulcanizing both heavy- and intermediate-sized cables is referred to as catenary CV (CCV). The first 12–30 m (~40–100 ft) or more of the pipe assumes a catenary curve. The cable follows its natural curve when passing through the first part of the curing pipe, which minimizes cable drag along the bottom of the pipe [22]. The remainder of the pipe continues to slope at a slight angle and may be placed under the floor.

The recent developments in methods include the replacement of the high-pressure steam by hot nitrogen. This method, often called *dry cure*, reduces the number of microvoids in the insulation. Another alternative to steam cure is the pressurized liquid salt (PLCV) system. This process utilizes eutectic salt with a melting point of ~150°C (302°F) as the heat transfer medium. Because the heat transfer is rapid, relatively short pipes are used, and the pressure necessary is much lower than in the steam process [15]. An example of modern CV systems is shown in Fig. 5.14, which shows steam and gas heating systems. Figure 5.15 illustrates the typical Drycure™ configuration.

The Long land die process (sometime called MDCV). It uses a thin film of lubricant during extrusion that facilitates smooth movement inside the die (Fig. 5.16). As a result, the curing zone is relatively short, typically ranging from 12 to 18 m (40 to 60 ft). Another characteristic of this method is that the product travels in a horizontal line without undergoing excessive bending) [23]. Certain W&C rubber compounds can be vulcanized continuously by EB radiation (see Section 5.2.3.3).

5.2.3.3 Electron Beam Cross-Linking of Polymers in the W&C Process Electron beam radiation is an established industrial process with several advantages. Its *economic* advantage is based on the ability of the EB radiation processing systems to treat product lines at faster line speeds than conventional heat processing techniques. Radiation is essentially an ambient temperature process, there are fewer emission problems; this means that it offers an *ecological* advantage [24]. Moreover, radiation energy can be anywhere from 2 to 10 times more efficient in the conversion of energy required for cross-linking of thermoplastic polymeric systems or vulcanization of rubber compounds. This is partially due to the fact that electron radiation energy can be focused and directly aimed at the product [25]. With the advancements of

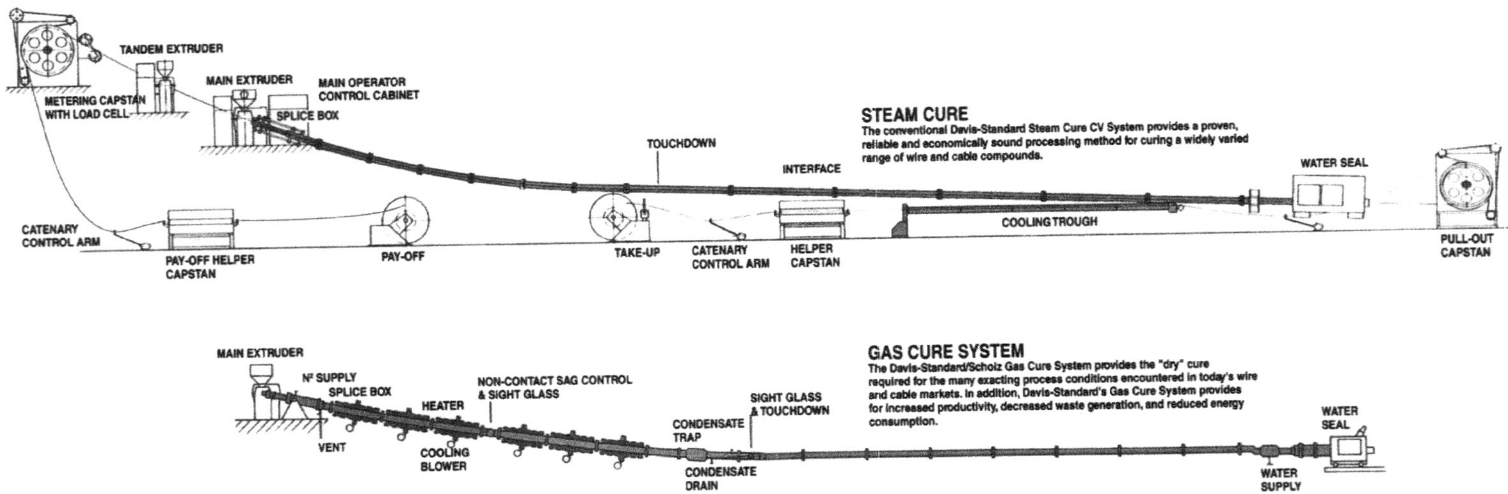

FIGURE 5.14 Wire and cable CV cure system. (Top: standard steam heating, bottom: gas heating). (Courtesy Davis-Standard, LLC).

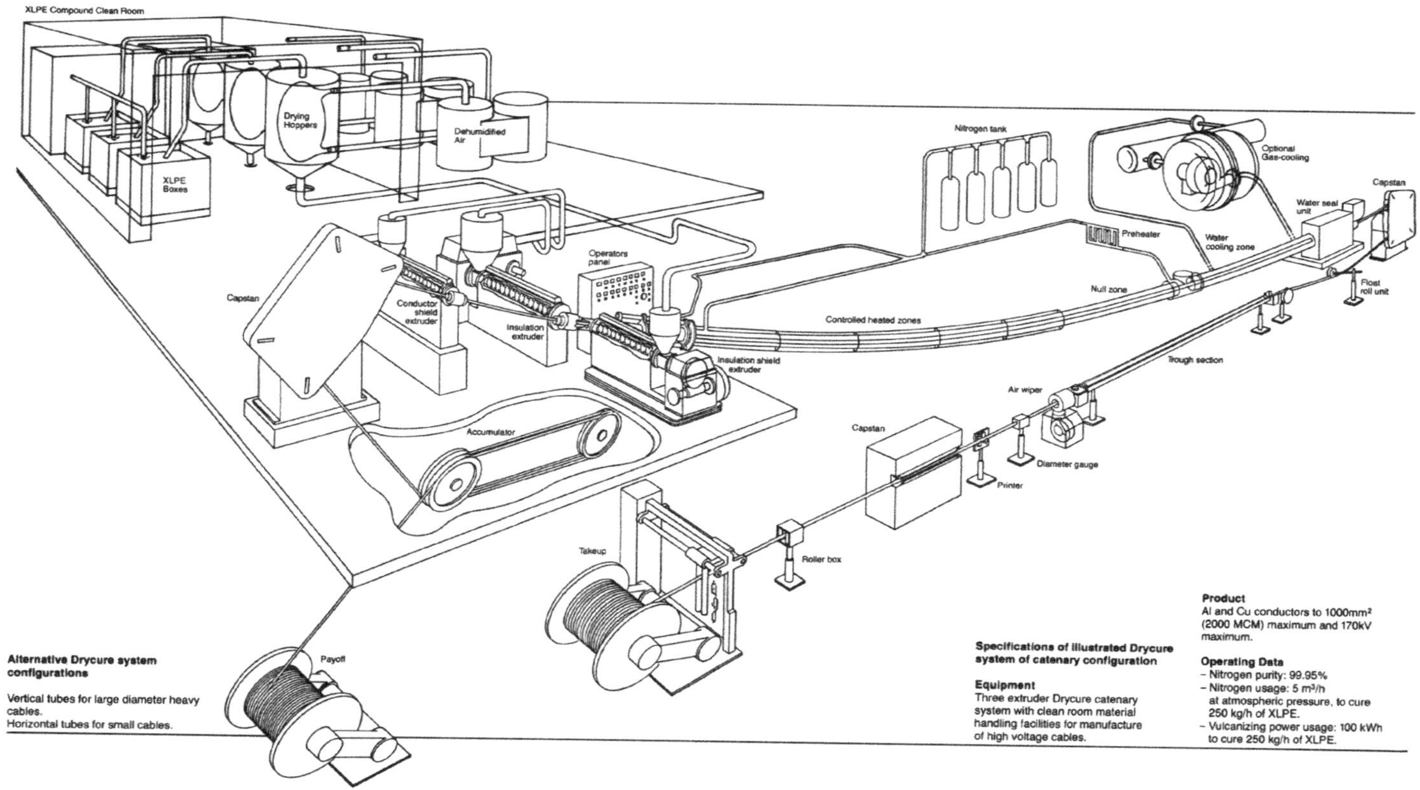

FIGURE 5.15 Typical configuration of the Drycure™ system. (Courtesy of Royle Systems Group).

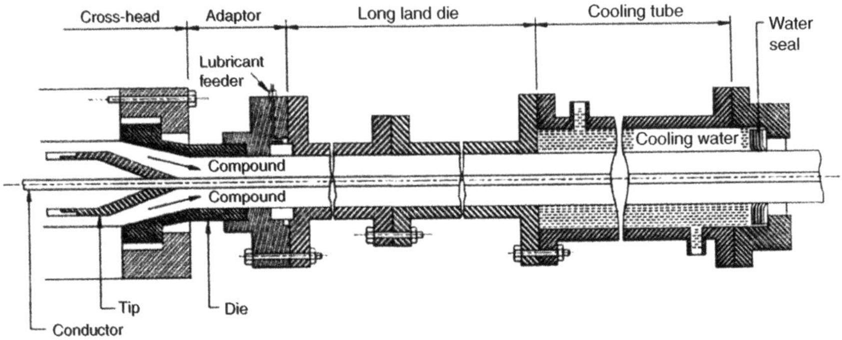

FIGURE 5.16 Long land die. (the MDCV Process). (*Power and Communication Cables: Theory and Practice* (Bartnikas, R. and Srivastava, K.D., Eds.), Institute of Electrical and Electronic Engineers, New York, 2000, p. 225)

polymer processing, today it is possible to carry out simple cross-linking of PE, PVC or other thermoplastics, or a more complex radiation curing of elastomers.

The principle of producing high-energy electrons is very simple. The electrons are emitted in a vacuum by a heated cathode and accelerated in the electrostatic field applied between cathode and anode. Acceleration takes place from the cathode, which is on a negative high-voltage potential to the grounded accelerator vessel as the anode. Usually an electron optical system is used to focus the accelerated electrons to the accelerator window plane.

The energy gain of the electrons is proportional to the accelerating voltage. It is expressed in electronvolts (eV), that is, the energy that a particle of unit charge gains by passing a potential difference of 1 V. The electrons leave the vacuum chamber and reach the process zone if their energy is high enough to penetrate the 5–25-μm thick titanium window foil.

Stopping the high-energy electrons by a material generates X-rays. For that reason, the electron accelerator and process zones have to be shielded to protect the operator. For electrons with energies up to 300 keV, a self-shielding with lead cladding up to ~1 in. thick is sufficient. For systems, where electrons with higher energy are generated, the equipment is shielded by a concrete or steel vault built around it. Examples of EB accelerators of different designs are shown in Figs. 5.17–5.19.

The basic electrical parameters of an EB processor are its acceleration voltage, the electron beam current, and the electron beam power. The

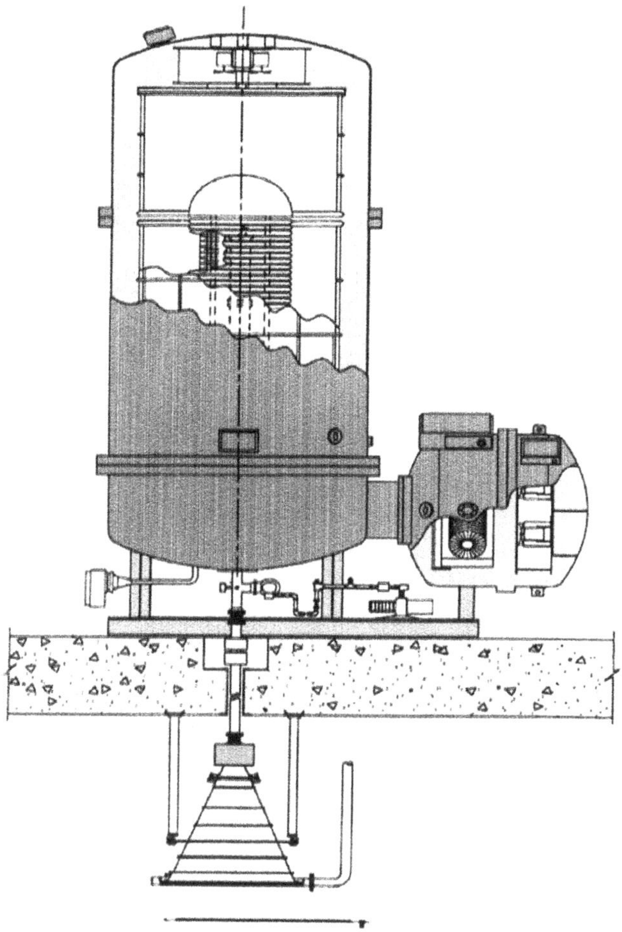

FIGURE 5.17 Dynamitron accelerator. (Courtesy of IBA Industrial).

ratio of EB power and of the input electrical power defines the efficiency of an electron accelerator [26]. The acceleration voltage determines the energy of the electrons. The absorption properties of the accelerated electrons in the processed materials are the absorbed dose, the depth dose profile, the penetration range, and the dose rate.

Process parameters involve line speed; if dose rate and line speed are combined, the dose delivered to the product to be cured can be calculated. A processor specific yield factor depends on the relationship between the beam current and line speed. The dose-speed capacity of a processor is given by the product of the line speed and the delivered dose at maximum EB power [27].

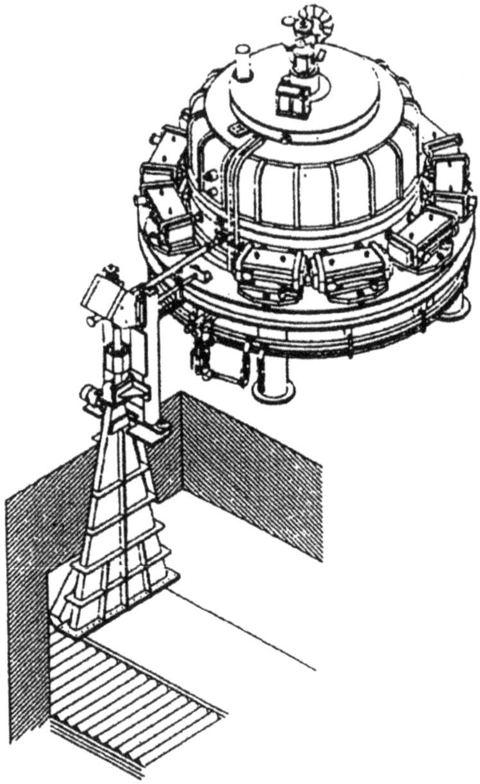

FIGURE 5.18 Rhodotron accelerator. (Courtesy of IBA Industrial).

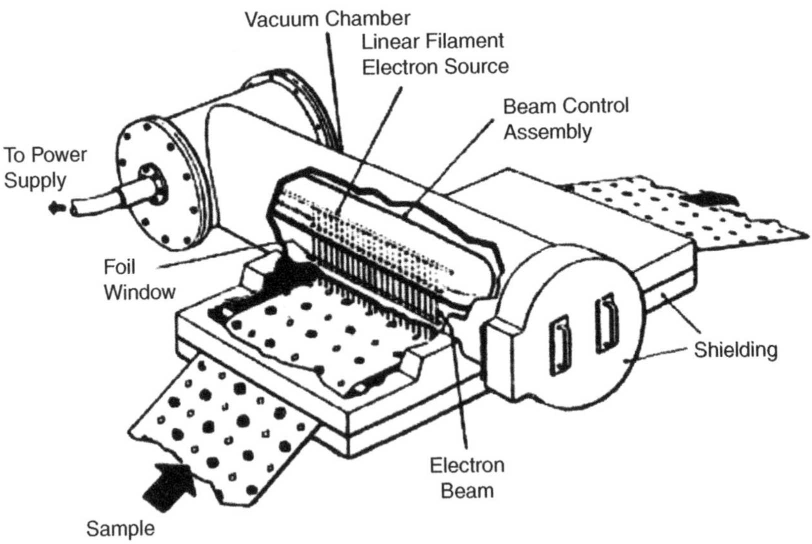

FIGURE 5.19 Example of a low-energy accelerator. (Courtesy of Energy Sciences, Inc.).

The above mentioned parameters are defined & listed below:

Acceleration Voltage: Potential difference between the cathode and the anode of the accelerator usually expressed in kilovolts and obtained using a voltage divider chain of the high-voltage unit.

Electron Beam Current: Number of electrons per second emitted from the cathode, measured in milliamps (mA) (1 mA = 6.25×10^{15} electrons s^{-1}) at the high-voltage unit.

Electron Beam Power: Product of the acceleration voltage and of the EB current, expressed in kilowatts (kW) (1 kW = 10 mA $\times$ 100 kV).

Absorbed Dose: Mean value of energy of the ionizing radiation absorbed by the unit of mass of the processed material. The unit of absorbed dose is 1 gray (Gy) = 1 J kg^{-1}. The older unit, used officially until 1986 and still used occasionally in the literature, is 1 megarad (Mrad) equivalent to 10 kGy.

Depth Dose Profile: The energy deposition produced by electrons of a given energy in an absorber can be expressed as a function of depth, the atomic number, and the atomic weight of the absorber. The energy loss dE/dx of the fast electrons (expressed in megaelection volts per centimeter (MeV cm^{-1}) is given as a function of the mass per unit of area (g cm^{-2}). It is directly proportional to the absorbed dose: $1/\rho\ dE/dx$ = dose x unit an area with dimensions joules per gramper squared centimeters (J $g^{-1}cm^2$). In actual EB processes, energy losses occur in the titanium window foil, in the gap between the foil, the coating, and in the substrate.

Dose Rate: Dose per unit time, expressed in Gy s^{-1} = J $kg^{-1}s^{-1}$. At a constant accelerating voltage, it is proportional to the EB current.

Line Speed: Speed of the material being irradiated, usually expressed in meters per minute (m min^{-1}) or feet per minute (ft min^{-1}). It determines the exposure time of the material.

Delivered Dose: The ratio of the dose rate and of the line speed.

$$\text{Delivered dose} = \text{Dose rate/Line speed}$$

At a given dose rate, the line speed can be adjusted to obtain the desired dose. On the other hand, the beam current can be controlled by the line speed to maintain a constant delivered dose at all line speeds.

Yield Factor: The yield factor is used to characterize the curing performance of an electron processor. It is a constant that relates the delivered dose to the beam current and line speed:

$$\text{Delivered dose} = \text{Yield factor} \times \text{Beam current/line speed}$$

The unit of yield factor is kGy × m min^{-1} mA^{-1}

Dose-Speed Capacity: Product of the delivered dose and of the line speed at the maximum beam power:

$$\text{Dose-speed capacity} = \text{Delivered dose} \times \text{Line speed.}$$

In usual EB operation, the dose speed capacity is measured in meters per minute (m min^{-1}) at 10 kGy. It is the most convenient unit to relate the curing performance of an EB processor to the desired process parameters (e.g., dose and line speed) [28].

When polymers are subjected to high-energy ionizing radiation, most of them undergo simultaneous cross-linking and degradation. These phenomena are described in terms of the G-value, the number of events occurring per 100 eV of absorbed energy [29, 30]. Because of that, this value can be useful in assessing the relative efficiency of cross-linking of different polymers. Examples of the G-values of different polymers are shown in Table 5.17.

As pointed out earlier, the EB penetration is proportional to the beam energy, but the dose is not uniform through the depth of its penetration. Typically, a dose increases to a maximum of about one-third through the entire range and is equal to the surface dose of about two-third through this range. Wire geometry also complicates the imparted dose at any specific region in the radiation process. These problems can be alleviated by the use of multiple EB units in tandem, multiple passes, or backscattering [31]. To meet the industry requirements, absolute uniformity of

TABLE 5.17 The G Values of Radiation Cross-Linkable Polymers

Polymer	G Value
Polyethylene	2
Polypropylene	0.6
Polyisoprene (natural)	1.3–1.8
cis-1,4-Polyisoprene (synthetic)	0.9
Polystyrene	0.04
Polyamide	0.3
Poly(vinyl chloride)	0.2
Styrene–butadiene rubber	3.8
Polychloroprene rubber	9.6
Ethylene–propylene rubber	1.8
Silicone rubber	2.2

cross-linking is not required as long as the gel fraction in the minimum region is above a designated critical value.

Polymer cross-linking usually requires doses of 50–150 kGy, depending on the chemical composition of the material. The dose within the material is not always uniform. With the electron energies of 500 keV the dose tends to increase with the depth of the material to about one-half of the maximum electron range and then decrease to nearly zero at a greater depth, where the electrons have dissipated most of their kinetic energy. A useful quantity is the depth, where the exit dose has diminished to 50% of the entrance dose. The electron energy should be enough to make this quantity, usually called R(50e) equal or greater than the thickness of the material [32].

In curing applications, electrons have to penetrate the reactive solids or liquids with typical masses per unit area of one to several hundred grams per square meter (g/m^2) ($1\ g\ m^{-2} = 1\ \mu m$ at unit density), as shown in the depth dose profiles (Fig. 5.20) [33]. The electron penetration range is related to the path length the electron travels during the energy degradation process. The electron penetration range is expressed in grams per square meters ($g\ m^{-2}$) as the function of electron energy is shown in Fig. 5.21.

Besides cross-linking and resulting in improvement of properties the EB radiation process may cause other changes, some of which are chemical in nature. These are

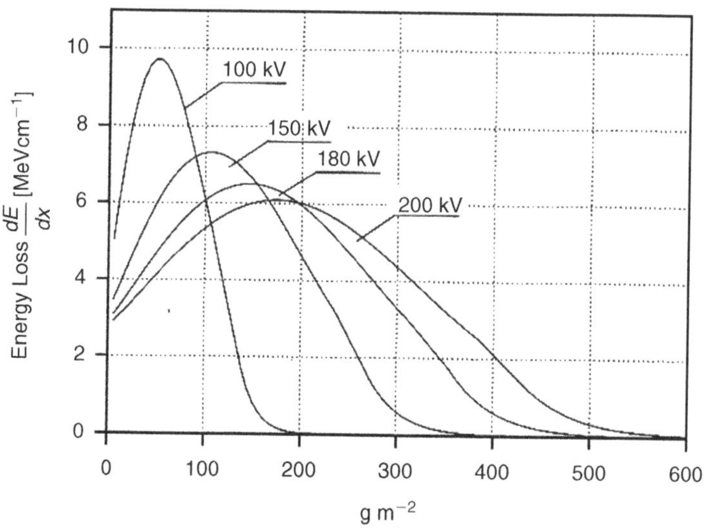

FIGURE 5.20 Example of a depth dose distribution for electron energies in the range 100–200 keV.

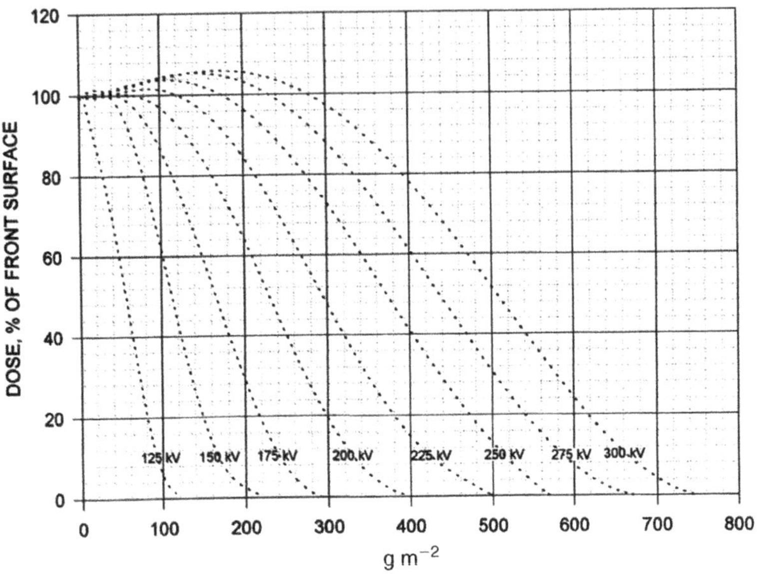

FIGURE 5.21 Electron penetration range in gram per square meter (g/m^2) as the function of electron energy. (Courtesy of Energy Sciences, Inc.).

- Simultaneous chain scission, which leads to the formation and accumulation of low molecular weight fractions in the polymer.
- Disappearance of any residual vinyl of vinylidene groups that may be present initially.
- The evolution of hydrogen and other gases with hydrogen amounting to 98–99%.

5.2.3.3.1 Electron Beam Cross-Linking of Thermoplastics Polyethylene cross-links readily when irradiated. In order to achieve its optimum properties (70–80% gel fraction) a dose of 200–300 kGy is required; however, the exact dose necessary for any specific grade PE is determined by the polymer molecular distribution, degree of branching, and so on. [34]. Most of the properties of PE are upgraded, tensile strength and modulus are increased, elongation increases at low doses, stress-crack resistance is improved, and the polymer no longer flows above its crystalline melt temperature [105–110°C (221–230°F) for low-density PE]. At this point, PE behaves as an elastomer. These are all desirable properties for a wire and cable insulation, and result from the generation of a three-dimensional (3D) cross-linked network. The cross-linked polymer will swell, but will not dissolve in hydrocarbon solvents. The magnitude of the insoluble gel fraction is an indication of the degree of property improvement that occurs. The tightness of the gel fraction, which is related to the gel swelling ratio, or hot modulus, is a quantitative indication of the degree to which the polymer was upgraded. Since the same property improvements can be achieved by peroxide-initiated cross-linking, some have been adopted as standard quality control test procedures by the wire and cable industry. Superficially at least, the two cross-linking techniques behave in a parallel fashion [30]. The comparison of properties of standard and irradiated LDPE is shown in Table 5.18 and typical applications for irradiated LDPEs are listed in Table 5.19. The comparison of standard and irradiated PVC is in Table 5.20.

TABLE 5.18 Comparison of Standard and Irradiated PE[a]

Property	Standard LDPE (not irradiated)	Irradiated LDPE
Tensile strength [psi (MPa)]	1800 (12.4)	2800 (19.3)
Elongation at break (%)	400	250
Oxygen index	16	29
Solder iron, seconds to cut-through	<1	>600
Chisel cut-through (°C)	102	125
Flexibility [psi (MPa)]	39 (0.27)	48 (0.33)
Swelling in ASTM Oil #3	Good	Excellent

[a] Test from insulated wire consisting of 37 strands of #29 AWG tinned conductor with 30 mil (0.75 mm) radial wall of insulation.

TABLE 5.19 Typical Applications for Irradiated LDPE

Application	Important Attributes
Locomotive power and control wires	Small size Overload capability Hot oil and grease resistance Flexibility
Nuclear plant control cables	Resistance to radiation Fluid resistance Good electrical properties Small size
Automobile (limited to antiskid cables)	Abrasion resistance Cut-through resistance Fluid resistance Thermal resistance
Aircraft wiring	Small size Light weight Thermal stability

TABLE 5.20 Comparison of Standard and Irradiated PVC[a]

Property	Standard PVC (not irradiated)	Irradiated PVC
Tensile strength [psi (MPa)]	2000 (13.8)	4200 (29)
Elongation at break (%)	200	170
Oxygen index	32	37
Solder iron, seconds to cut-through	<1	>600
Chisel cut-through (°C)	85	108
Flexibility [psi (MPa)]	23 (0.16)	80 (0.55)
Swelling in ASTM Oil #3	Good	Excellent

[a] Used in an insulated wire consisting of 19 strands of #32 AWG tinned copper with 10 mil (0.25 mm) wall of insulation.

5.2.3.3.2 Electron Beam Curing (Cross-Linking) of Elastomers Elastomers used in W&C manufacture have little or no crystallinity, and cross-linking is random at all temperatures. Generally, tensile strength values of radiation-cured elastomers are reported to be lower than those of sulfur-cured rubber [35]. It can be shown that main-chain scissions, prevalent particularly under certain irradiation conditions, have a distinct effect on the strength of cured elastomers.

There are certain polymers such as polyisobutylene, which by nature of their microstructure have a high yield of radiation induced scissions, regardless of how they were irradiated. The compound ingredients (antioxidants, oils, etc.) may retard the rate of cross-linking of an elastomeric compound. Thus, the presence of cure retarding impurities and exposure in an atmosphere containing oxygen may significantly contribute to the lower strength of radiation-cured elastomers. Another factor may be degradation of the polymer by ozone generated during irradiation.

Obviously, there are many subtle differences in the structure, morphology, or network topology between radiation- and sulfur-cured elastomers [36], but their physical properties may be nearly equal, provided that precautions are taken to avoid the occurrence of chain scissions. The comparison of radiation cross-linked and sulfur-cured NR (gum and carbon black reinforced compounds) as made in Table 5.21 illustrates this point. The following list covers the EB curing of individual elastomers.

Natural Rubber and *Cis*-Polyisoprene

The yield of cross-links in natural rubber and high *cis*-polyisoprenes was found to be about equal, however, polyisoprenes with high 1,2 and 3,4 configurations exhibited an exceptionally high degree of cross-linking. Oxygen has been found to increase the rate of scission and to decrease the degree of cross-linking [37, 38]. The most commonly reported physical properties of radiation cross-linked natural rubber and compounds made from it are modulus and tensile strength, obtained from stress–strain measurements. Figure 5.22 illustrates some of the results obtained from gum rubber and from a natural rubber compound reinforced by high-abrasion furnace carbon black (HAF) [39]. In Fig. 5.23, the tensile strength of radiation cured gum is compared to vulcanizates cured by sulfur and peroxide [37].

Clearly, large doses are required for a full cure. The dose levels for attaining a maximum tensile strength are within the 20–50 Mrad

TABLE 5.21 Comparison of Stress-Strain Properties of Natural Rubber Cross-Linked by Sulfur and Radiation

Gum (Compound without fillers)	Sulfur Cured	Radiation Cured
Tensile strength [MPa (psi)]	27.8 (4026)	18.6 (2700)
Elongation at break (%)	700	760
300% modulus [MPa (psi)]	2.3 (332)	1.7 (250)
Reinforced by 50 phr HAF[a] Carbon Black	**Sulfur Cured**	**Radiation Cured**
Tensile strength [MPa (psi)]	27.4 (3975)	22.8 (3300)
Elongation at break (%)	470	350
300% modulus [MPa (psi)]	14.8 (21.50)	17.9 (2600)

[a]HAF = High-abrasion furnace carbon black.

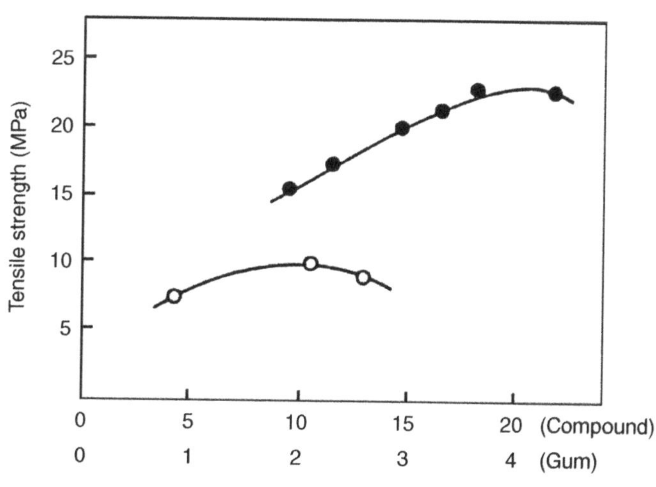

FIGURE 5.22 Tensile strength of radiation-cured purified natural rubber ○ gum; • compound (50 phr N330 carbon black). (Böhm, G.G.A. and Tveekrem, J.O, *Rubber Chem. Technol.* Vol. 55, p. 619 (1982))

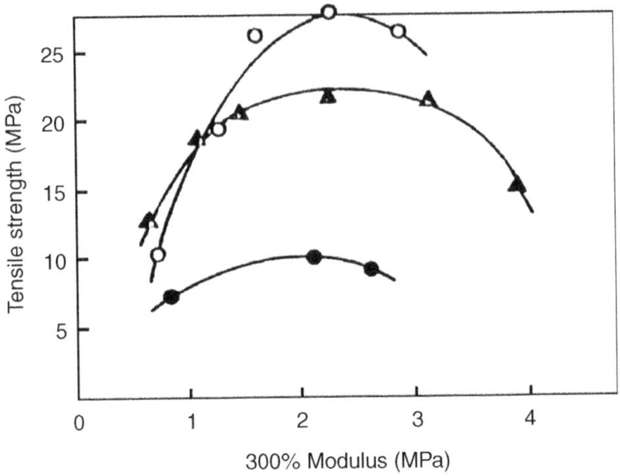

FIGURE 5.23 Tensile strength of radiation-cured purified natural rubber compared to the same cured with sulfur and organic peroxide ○ sulfur; ● irradiation in nitrogen; ▲ peroxide. (Böhm, G.G.A. and Tveekrem, J.O, *Rubber Chem. Technol.* Vol. 55, p. 620 (1982))

TABLE 5.22 Tensile Strength Data from EB Cured NR

Compound	Amount of Additive (phr)	Dose (kGy)	Tensile Strength [MPa (psi) reported]	Note
Gum rubber (SMR-5L) plus *o*-dichlorobenzene	3	140	9.0 (1305)	15 kGy s^{-1}
Gum rubber (smoked sheets)		600	4.8 (696)	Irradiated in air
Smoked sheets and N330 carbon black	50	500 140	19.6 (2842) 13.3 (1928)	Irradiated in air
Smoked sheets and N330 carbon black plus octyl acrylate	60 20	140	18.3 (2654)	Irradiated in air
Smoked sheets and N330 carbon black plus tetramethylene diacrylate	55 10	140	18 (2291)	Irradiated in air
Smoked sheets and N330 carbon black plus glyceryl triacrylate	50 5	130	16.6 (2407)	Irradiated in air
Pale crepe and N330 carbon black	50	160	15.8(2291)	Irradiated in air
Pale crepe and N330 carbon black plus octyl acrylate	50 20	160	16.6 (2407)	Irradiated in air

(200–500 kGy). When prorads (cross-link promoters, e.g., dichlorobenzene, maleimides, or acrylates) are used, the curing dose is reduced and higher tensile strength values are found. Tensile strength data from radiation-cured NR are shown in Table 5.22. The retention of the maximum tensile strength at elevated temperatures is greater for radiation cured than for chemically cured NR as reported in [40, 41].

Polybutadienes and Their Copolymers

Upon irradiation, 1,4-polybutadienes and poly(butadiene–styrene) form free radicals relatively readily and their concentration has been found to increase linearly proportional to dose up to ~100 Mrad (1000 kGy) [42].

The information on physical properties of radiation cross-linking of polybutadiene rubber and of butadiene copolymers was obtained in a similar fashion as for NR, namely, by stress–strain measurements. The addition of prorads allows further reduction of the cure dose with the actual value depending on the microstructure and macrostructure of the polymer and also on the type and concentration of the compounding ingredients (e.g., oils, processing aids, antioxidants) in the compound. Radiation-cured polybutadiene and its copolymers were reported to have superior tensile strength and ultimate elongation at high temperatures [43]. The abrasion resistance was found higher; most of the higher properties (e.g., hardness, resilience, and permanent set) were equal to those of chemically cured compounds of this category. Tensile strength values are nearly equal to those obtained by chemically cured acrylonitrile–butadiene rubber (nitrile rubber) (NBR) were reported [44–46]. Relatively low radiation doses were required for a complete cure of these elastomers. Radiation cured SBR was found to have a somewhat lower tensile strength and a lower maximum extensibility than sulfur-cured compounds, considerably longer fatigue flex life, and slower crack growth [47].

Polyisobutylene and Its Copolymers

Upon irradiation, polyisobutylene and its copolymers tend to degrade. There is a large number of studies determining the nature of the process and mechanisms and it was established that it involves formation and reaction of free radicals [48]. The radical concentration increases linearly with doses up to 100 Mrad (1000 kGy) [49].

Degradation is the predominant process in irradiation of polyisobutylene. Generally, the scission yield increases with temperature in accord with the temperature dependence of the formation of unsaturation [50].

It is possible to cross-link polyisobutylene and poly(isobutylene–isoprene) and chlorobutyl rubber by means of certain additives;

however, the physical properties of the vulcanizates prepared by conventional curing methods cannot be attained. The most promising results were obtained with chlorobutyl compounds containing thioether polythiols as prorads [51].

Ethylene–Propylene and EPDM

Commercial grades of EPRs contain 60–75 mol% of ethylene to minimize crystallization. The addition of a third monomer (e.g., 1,4-hexadiene, dicyclopentadiene, or 5-ethylidene-2-norbornene) generally produces amorphous faster-curing elastomers. A large number of such *terpolymers*, is available commercially. Their properties, performance, and also their response to radiation vary considerably depending on macrostructure, ethylene/propylene ratio, as well as on the type, amount, and distribution of the third monomer.

When EPDM is irradiated, the nature of the radical formed depends on the third monomer used [52, 53]. As in most cases, the radical concentration increases linearly with dose [54].

The cross-linking rate of EPR by radiation comes close to that of polypropylene. The EPDM terpolymers exhibit an enhanced cross-linking rate and it increases with the diene content. However, not only cross-linking rate, but also a greater yield of scissions results from the addition of the third monomer [54].

Radiation cured carbon black filled EPDM compounds exhibit tensile strength comparable to their analogues vulcanized by sulfur-accelerator systems [55]. Similar results were obtained from compounds containing up to 20 phr of acrylic additives (e.g., such as trimethylolpropane trimethacrylate, ethylene glycol dimethacrylate, triallyl phosphite, triallyl cyanurate, and others) [56]. Among the variety of EPDM grades, those containing ethylidene norbornene exhibit the fastest rate of radiation cross-linking [55–57].

Blends of PE and EPDM are used when greater flexibility is needed. Flame-retardant compositions of PE with EPM and EPDM are replacing PVC because of concerns over the release of toxic fumes when PVC is exposed to fire [58]. Example of a flame-retardant blend of PE and EPDM is shown in Table 5.23.

Polychloroprene

Irradiation of carbon black reinforced polychloroprene compounds produced maximum tensile strength of 20 MPa (2900 psi) at a dose of 20 Mrad (200 kGy), which is a value, obtained typically from chemically cured compounds.

TABLE 5.23 Example of an EB Curable Flame-Retardant Formulation based on a PE–EPDM Blend

Component	Function	Parts by Weight
PE–EPDM Blend	Base elastomer	100
Hydral® 710[a]	Aluminum trihydrate, flame retardant	250
Zinc oxide	Enhances aging resistance of the material	5
Mineral oil	Plasticizer, process aid	10
Silane A 1729[b]	Coupling agent, promotes bond between filler and the polymer blend	2
Antioxidant	Reduces aging effects	2
TMPTA[c] or TAC[d]	Prorad, enhances cross-linking by EB	5

[a] Almatis.
[b] Gelest Inc.
[c] Trimethylolpropane triacrylate - TMPTA.
[d] Triallyl cyanurate - TAC.

Silicones

Irradiation of silicones (polydimethylsiloxane) produces hydrogen, methane, and ethane. The gas yield at room temperature correlates with the concentration of cross-links formed [59]. This can be expected since double bonds cannot be formed. Cross-linking is a predominant process during irradiation of siloxane polymers. Chain scissions are negligible [60–62]. The cross-link density increases linearly with dose up to 160 Mrad (1600 kGy) [63].

The tensile strength of silicone elastomers irradiated by a dose of 60 kGy (6 Mrad) is 15% lower than that of PDMS cross-linked by peroxide, but when 0.14 mol% of vinyl unsaturation is substituted for the methyl groups, the tensile strength is 30% higher. Irradiation of PDMS containing 55 phr silica filler by a dose of 40 kGy exhibited an optimum tensile strength of 8 MPa (1160 psi) and excellent thermal stability at 523 K (482°F) [63].

Fluorocarbon Elastomers

Fluorocarbon elastomers represent the largest group of fluoroelastomers. They have carbon–carbon linkages in the polymer backbone and a varied amount of fluorine in the molecule. In general, they may consist of several types of monomers: polyvinylidene fluoride (VDF), hexafluoropropylene (HFP), chlorotrifluoroethylene (CTFE), tetrafluoroethylene (TFE), perfluoromethylvinyl ether (PMVE), ethylene or propylene [64]. Since the fluorocarbon elastomers being discussed here contain hydrogen in their molecules, they have the tendency to cross-link in addition to scission, which is common in fluoropolymers when they are exposed to radiation. The cross-linking predominates, but there is still a significant degree of chain scission [65].

5.2.3.4 Comparison of Irradiation and Chemical Cross-Linking of Elastomers Conventional methods of cross-linking *rubbe*r (vulcanization), such as CV, involve chemical reaction between the elastomer and cross-linking agent at very high temperatures. As pointed out earlier, irradiation by an EB is done at ambient temperatures. In general, irradiation curing offers the following advantages over the chemical-curing process [66]:

- Considerable energy savings.
- Considerable space savings in comparison with CV cure.
- For conductor sizes <#18 AWG or insulation wall thicknesses <20 mils (0.5 mm), manufacture by CV is impractical while irradiation cross-linking has no limits on physical size.
- Residual unreacted cross-linking agent in the chemical process will generally degrade electrical properties.
- Irradiation is generally more uniform than chemical process and yields better physical properties in many cases.
- Irradiation is more versatile and offers a more precise control over cross-linking density: temperature- and moisture-sensitive materials can be processed.
- The deleterious effects of steam are eliminated. The microvoids and bubbles in chemically cross-linked insulation are easily avoided by the use of irradiation. Likewise, a hole in the insulation in the CV equipment will result in water flowing for long distances longitudinally in the stranded conductor.
- Irradiation offers more design freedom. For example, with stranded conductors, it is necessary to use a tape separator between conductor and insulation with small wall thicknesses when curing with steam because, without it, the steam pressure forces the hot soft material deep into the strand making it impossible to strip. With irradiation the material configuration remains as extruded.

Radiation- and chemical-curing methods are compared concisely in Table 5.24.

Irradiation cross-linking in general provides several important improvements in *plastic* wire insulators [66]:

TABLE 5.24 Comparison of EB and Chemical Cures in W&C Manufacture

Electron Beam Cure	Chemical Cure
No special extrusion equipment needed	Special extrusion equipment required
Noncritical extrusion conditions	Critical extrusion conditions
No auxiliary requirement	Hi pressure steam
Instant start and stop	Slow start-up
Low scrap on starting and stopping	High scrap on start-up and stopping
High line speed	Low line speed
Wall thickness limitations	No wall thickness limitations
Compact, small space required	Large space requirements
Ozone, radiation are environmental safety factors	Heat and steam are environmental factors

- Higher tensile strength (at the expense of some loss of elongation), better abrasion resistance, and improved cut-through resistance. The latter property is remarkably improved above the melting temperature and is, in fact the single most important reason for the use of cross-linking of plastics.
- The surface is usually hard and slippery improving ability to pull the conduit or trays.
- Resistance to fluid attack is generally improved.
- Flame resistance is slightly improved as the cross-linking process also appears to render some of the volatile components less volatile.
- The surface of waxy materials (e.g., polyethylene) becomes easier to print on or color-code.

5.2.3.5 Radiation Curing of Coatings, Adhesives and Enamels

In general, most coatings subjected to EB irradiation are polymerized via radical chain reactions [67]. The high-energy electrons are absorbed directly in the coating itself where they create the initiating free radicals uniformly in depth. For a typical system [i.e., a coating 25 μm (0.01 mil)] the energy is deposited in a uniform manner throughout the coating even for electrons of relatively low energy (e.g., 110–140 keV) [68]. Because of the thin layer of the applied resin or compound, low-energy EB equipment as shown in Fig. 5.18, is sufficient for the polymerization and/or cross-linking reaction.

In the type of coatings used in W&C technology, the polymerization process propagates until the growing chain is terminated, usually by two radicals. Termination can also result from quenching of the radical, usually through its interaction with a scavenger (e.g., oxygen). For this reason, most industrial radiation curing processes are performed in an

TABLE 5.25 Typical Components of UV Curable Coatings

Component	Purpose	Concentration Range (%)
Oligomers	Control final cured film properties	30–80
Reactive diluents/ monomers	Reduce coating viscosity, contribute to coating performance and cure speed	20–70
Photoinitiators	Absorb UV light and initiate the polymerization reaction	2–10
Additives	Various performance enhancements	0–10

oxygen-free or oxygen-poor environment. The curing or polymerization takes place in a very short period of time, typically within a few milliseconds.

Other techniques of radiation, inducing free radical initiation [e.g., ultraviolet (UV)] energy sources can be used. This process requires the incorporation of sensitizers (photoinitiators) into the coating and has certain limitations; the chief one being its inability to provide uniform cures at depths in all but the most UV transparent of materials [69]. Typical components of a UV curable coating are shown in Table 5.25. Examples of applications using UV radiation are sensor switch encapsulants, coil terminators, strain relief for electrical connectors, printed circuit conformal coatings, printed circuit solder masks, conductive inks for rear window defroster, windshield wiper motor sealing adhesives, motor balancing compounds, component marking inks, tacking adhesives, and screen printed membrane switches.

The coating systems used for EB cure contain a high degree of unsaturation and are 100% solids. No solvents are used since the reactions occur rapidly at ambient temperatures. Typical compositions typically contain 70% of polymer or prepolymer and 30% cross-linking monomer [69]. In magnet and enameled wire coatings systems that can be modified readily for EB cure are acrylic coatings, polyurethane coatings, and unsaturated polyester coatings. To be suitable for the above applications, an additional cross-linking monomer must be included to achieve a rapid and complete cure in the compound. In some cases, adhesion promoters should be a part of the system [69]. An example of the use of high-strength enamel on leads for the winding of electric motors is illustrated by Fig. 5.24.

5.2.3.6 Process and Equipment for EB Curing of W&C Electron beam cross-linking of wire insulations and cable jackets is usually done with electron energies in the range of 500 keV–1.5 MeV, although some facilities use higher energies, up to 3.0 MeV. The required energy is

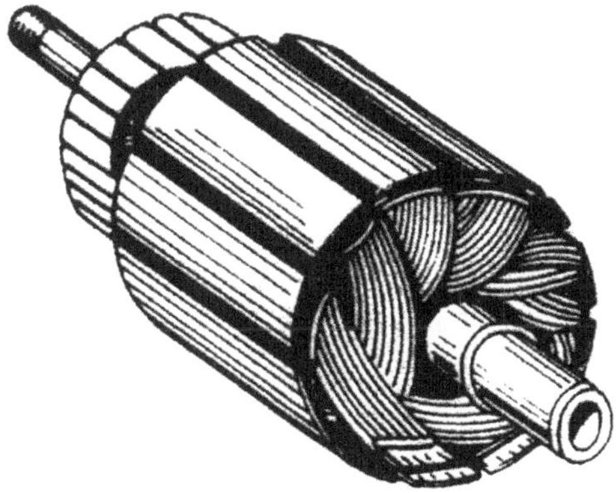

FIGURE 5.24 Use of high-strength enamel leads for the winding of electric motors. Modulus = MOD.

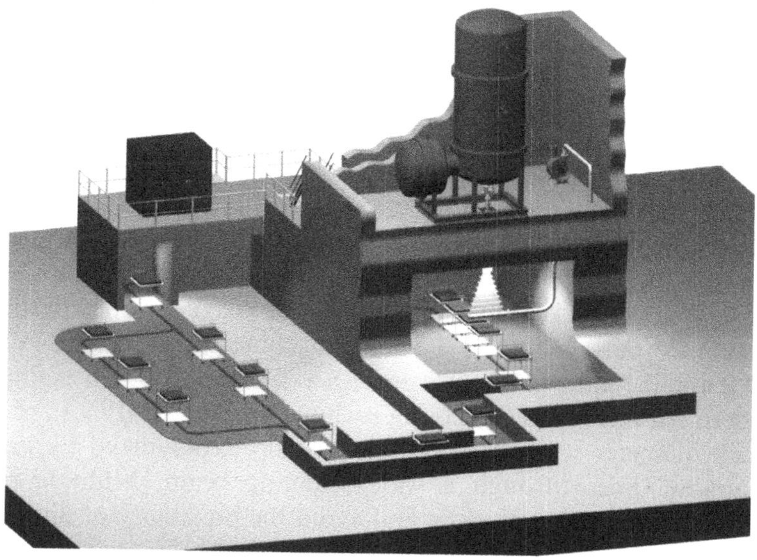

FIGURE 5.25 A 5-MeV Dynamitron facility. (Courtesy of IBA Industrial).

determined by the thickness, density, and atomic composition of the insulation and the diameter of the conductor. The EB current is usually in the range 25–100 mA. Facilities with electron energies <1.0 MeV can be shielded with steel panels or a combination of steel and lead panels to protect operating personnel from X-rays generated by the EB (see Section 5.2.3.3). Facilities with energies >1.0 MeV are usually shielded with thick concrete walls, which are less expensive than equivalent steel and lead panels for high energies (see Fig. 5.25) [70].

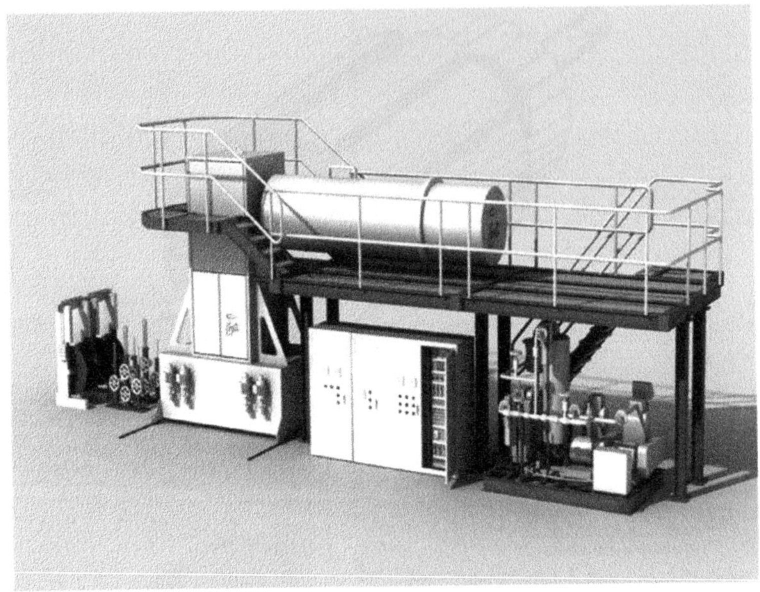

FIGURE 5.26 Example of a modern EB line for W&C curing. (Courtesy of IBA Industrial).

The larger machines (with higher accelerating voltages) are used for irradiating cables with thicker walls. As pointed out earlier, the penetration power of an accelerator is proportional to its accelerating voltage. For example, a 600-keV machine reaches its maximum penetration at ~1.3 mm (33 mils) [71] and a 1.5-MeV machine has its maximum point at ~3.2 mm (81 mils) in a substrate with a density of 1.0. An example of a modern EB line for W&C curing is shown in Fig. 5.26).

Small diameter wires or cables are usually cross-linked by passing them many times through a wide scanning beam. Multiple passes provide a sufficiently high dose and avoid the possibility of dissipating some of the heat generated by the EB processing. This is accomplished by using either the "figure eight" configuration (Fig. 5.27) for thin wires and a four drum fixture for larger wires and cables. The outer sheaths of larger cables can be treated by rotating the cables as they pass through the EB along the scanning direction (Fig. 5.28).

5.2.4 Cable Splices–Joints

Polymeric materials are used for completing a splice of W&Cs to assure an integral insulation for the joined conductor or conductor system. Some of the used methods will be described in Section 5.2.4.1–5.2.4.4.

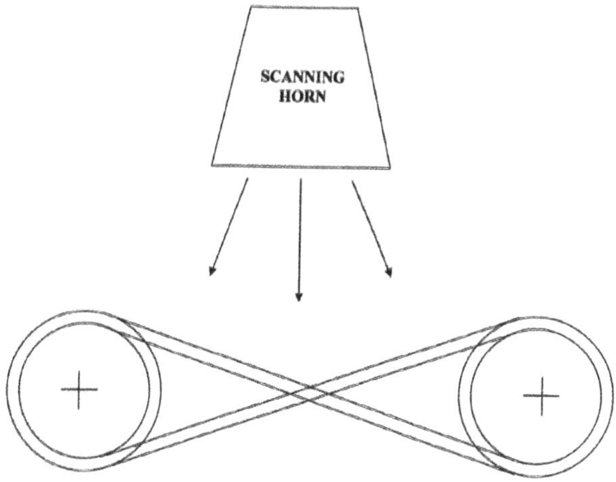

FIGURE 5.27 Two-sided irradiation of a W&C by EB (Figure Eight configuration).

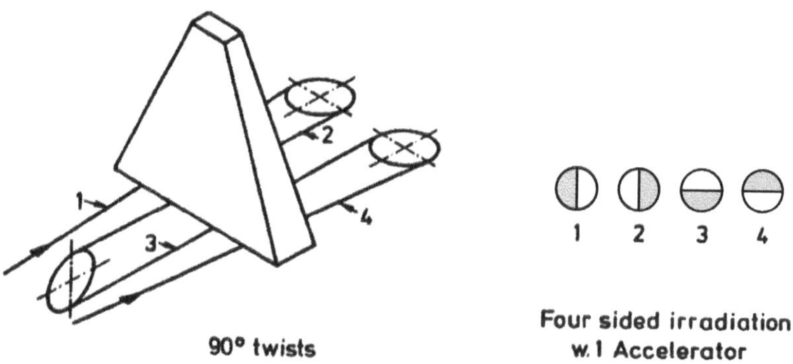

FIGURE 5.28 Four-sided irradiation of a W&C.

5.2.4.1 Tape Wrapped Joints Tape wrapped splices–joints generally are using EPR adhesive "self-amalgamating" tapes. The tape is pulled and stretched to about one-half of its original width at which point the adhesive property is activated. The tensile force applied to the tape allows the formation of a wrap of sufficient integrity. Tape-wrapped joints are suitable for service up to 300 kV [72].

5.2.4.2 Field Molded Joints In this method, cross-linkable PE or EPR tapes are wrapped around the cable ends. They are then heated and compressed to allow their cross-linking. A portable press with a split mold is used for that. After the elastomeric material is cross-linked, the joint is allowed to cool down in a controlled way to prevent contraction voids. These joints can be applied to cables for service up to 150 kV [72].

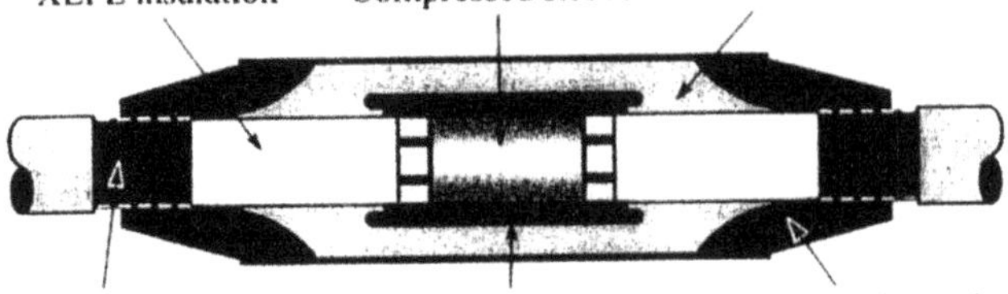

FIGURE 5.29 Slip-on elastomeric molding for a cable splice. (*Power and Communication Cables: Theory and Practice* (Bartnikas, R. and Srivastava, K.D., Eds.), Institute of Electrical and Electronic Engineers, New York, 2000, p. 238)

5.2.4.3 Premolded–Prefabricated Joints The simplest design is a one-piece rubber molding, usually from EPDM or silicone rubber, which incorporates an internal electrode of semiconducting rubber embedded into the joint insulation (see Fig. 5.29). This electrode serves to shield the conductor connection and the square-cut cable ends. A special tool is required to stretch and apply the molded unit [73]. Other, more complex prefabricated joints employ a rigid cast epoxy resin center portion with an embedded metal electrode [73]. Premolded–prefabricated joints are installed in a metal casing to provide mechanical protection and prevent water ingress after installation. This type of joint can be used for service up to 500 kV [74].

5.2.4.4 Heat Shrinkable Joints Heat shrinkable joins are an effective and economical method to protect splices of wires and cables with polymeric insulation. They have a practical preshrink to postshrink diameter of up to 4:1 [75]. This phenomenon results from the *elastic memory* of a polymer. Cross-linked PE is a well-known example. When cross-linked materials are heated above their crystalline melting point, they become rubbery because of the cross-linked polymeric chains. If the material is permitted to cool in an extended form, the crystalline segments will re-form in a lattice of the extended geometry (the material will be "frozen" in the extended form). When it is then heated, it will shrink to its natural relaxed state [76]. The tubing used for this process is frequently produced by cross-linking with an EB [77]. The individual steps of the process of producing heat-shrinkable tubing and its use for splicing a W&C is illustrated schematically by Fig. 5.30.

The joints are formed by applying heat-shrinkable tubing over the prepared cable. The tubing is then shrunk into place by the application of heat typically using a portable gas flame torch. The large diameter reduction allows for the use of a tube of a certain size for several different

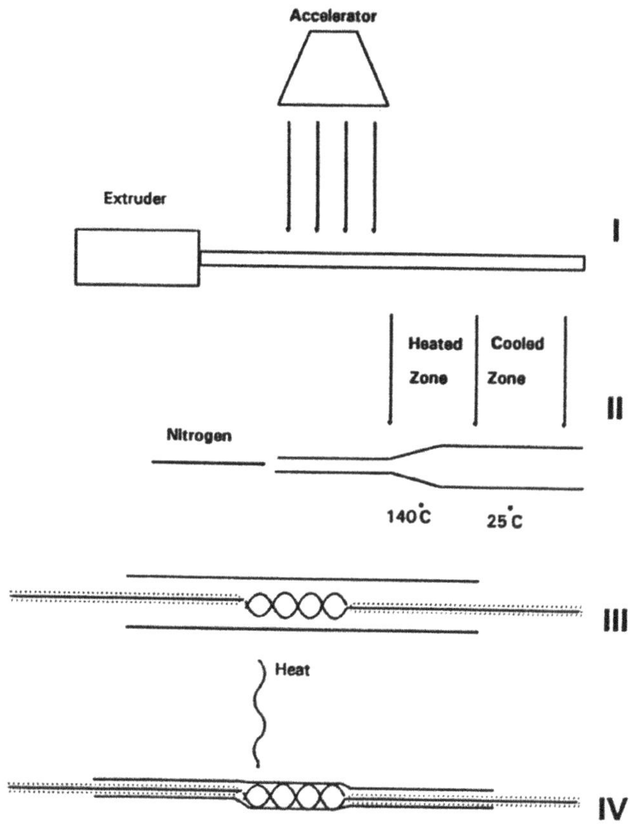

FIGURE 5.30 Production and application of heat-shrinkable tubing for W&C.

cable diameters. Heat shrinkable joints can be used for extruded cables up to 69 kV [72].

5.2.5 Cable Testing

All tests on a finished W&C represents some measure of both quality of the materials and the manufacturing techniques employed. Each of them can detect specific problems (poor dispersion of fillers, incorrect extrusion temperatures, speeds, cooling rate, degree of cross-linking, etc.).

A cable is subjected to multiple tests in its lifetime including a series of tests beginning at the factory and potentially continuing throughout the lifetime of the cable. Cable testing is performed in different phases including materials, qualification, production, and field testing (on reel, installation, acceptance, and/or maintenance testing). The types of tests performed in the various phases can depend on the environment, as well as the type of cable being tested, low or medium voltage, shielded, unshielded, and so on.

5.2.5.1 *Material Testing* Material testing pertains to all materials used, (i.e., metallic shields, armoring, and materials used as insulators). The requirements are set by industry standards, such as those from ICEA (Insulated Cable Engineers Association) and ASTM (American Society for Testing and Materials). The material testing requirements apply to both the physical and the electrical characteristics of the cable, but not necessarily to the specific ingredients of the materials. The applicable tests are covered in some detail in Chapter 7. Many of the cable test results are commonly summarized in a Certified Test Report [78].

5.2.5.2 *Qualification Testing* Qualification Testing (Type Testing) evaluates the quality of the overall design. This testing is performed on a particular cable design and some tests include accelerated aging as part of the testing protocol. The industry standard lists the following types of qualification tests:

- Core material qualification (including conductor–shield/insulation and insulation/insulation–shield).
- Thermomechanical qualification.
- Jacket material qualification.
- The CV extrusion qualification.
- Other qualification tests (includes insulation resistance, accelerated water absorption tests, etc.).

5.2.5.3 *Production Testing* Production tests are performed on a routine basis on various types of cables during and immediately following the manufacturing process. These include tensile strength and elongation at break of the extruded materials, hot creep and hot set, dimensional analysis, partial discharge (MV medium voltage, cables), and spark testing.

5.2.5.4 *Field Testing* Field testing of cables is commonly done to determine the as-received condition of the cable, the as-installed condition, and/or the operating condition of the cable. Field testing can be broken down into two broad categories, namely, *destructive* and *nondestructive* testing. Below are listed the usual tests in this category: high-potential testing (Hi-Pot), very low frequency testing (VLF), partial discharge testing (PD), dissipation factor/tan δ testing, megohm-meter testing of insulation resistance, capacitance-based fault location testing, arc-pulse fault location testing, and time domain reflectometry testing (TDR).

Maintenance tests are performed when a cable has been placed into service and is intended to detect in-service deterioration. Fault location tests, on the other hand, are used to pinpoint the exact failure site in a cable that has actually failed. Knowing exactly where the table has failed permits us to repair it or replaced it as necessary.

Details on the above and other specialized tests are beyond the scope of this book. The reader may wish to study [78] and/or Chapters 8–10 of [23].

5.3 OTHER APPLICATIONS OF POLYMERS AND POLYMERIC SYSTEMS

5.3.1 Potting and Encapsulation

Protection of the electronic system is of critical importance. An electronic circuit board houses various delicate components, (transistors, glass diodes, capacitors, semiconductors, silica chips, etc). Consequently, any damage to the components causes the device to fail. Reliable protection against the following environmental conditions is of importance: moisture, thermal shock, chemicals, and mechanical shock.

The solution to eliminate or considerably reduce these factors is *potting and encapsulation*, using liquid compounds based mainly on silicones, epoxies, silicones polyesters, and polyurethanes. Typical applications include modules, relays, power supplies, amplifiers, transformers, ferrite cores, connectors, circuit boards, capacitors, flyback transformers, ignition coils, thermostats, high-voltage transformers, and video game modules. The liquid systems are injected with precision equipment directly into the electronic component without harming delicate wiring or other sensitive parts.

In general, the materials used have to exhibit the following attributes:

- Surface adhesion: The material must adhere to the walls of the housing and to the individual components.
- Low shrinkage: The material must exhibit minimum shrinkage during the cure.
- Little or no volatile materials: Systems without solvents or other nonreactive dilutants are preferred.
- Tensile, flexural, and compressive strength.

- Chemical resistance.
- Low cost.

The amount of the liquid material depends on the dimensions of the part. Potting usually means using larger amounts of the protective liquid compound; typical examples are transformers, ignition coils, relays, and power supplies. More delicate parts (e.g., circuit boards) are protected by conformal coatings.

The systems for potting and conformal coating include the following: one- and two-part silicones, one- or two-part polyurethanes, epoxies, acrylics, and thermosetting esters. The curing of the potting compounds almost always is based on chemical reactions. Conformal coatings can be cured by a chemical reaction, as well as by UV radiation (polymerization and cross-linking). The latter assumes that the applied coat is very thin and transparent for UV radiation. The encapsulation materials are applied by spraying, brushing, dipping, and robotic application.

5.3.2 Electrically Conductive Adhesives, Inks, and Tapes

Electrically conductive adhesives (ECAs) are replacing the traditionally used tin–lead solder alloys. Most of them comprise usual commercial resins (e.g., epoxy, silicone, and polyimide resins), which provide necessary adhesion, mechanical strength, impact strength, and a metal filler (e.g., silver, gold, nickel, or copper) that conduct electricity. The ECAs offer many advantages over conventional solders, such as environmental friendliness, milder processing conditions (enabling the use of heat-sensitive and low-cost components and substrates), fewer processing steps (reducing processing cost), low stress on the substrates, and fine pitch interconnect capability (enabling further miniaturization of electronic devices) [79]. Typical applications are liquid-crystal displays, smart card applications as an interconnect material, and in flip–chip assembly, chip scale packages, and ball grid array applications. However, there are still some applications, where they cannot replace tin–lead solders, because of their lower electrical conductivity, conductivity fatigue (decrease of conductivity at elevated temperatures and humidity aging or normal use condition), limited current-carrying capacity, and poor impact strength when compared to tin–lead solders. These challenges appear mainly in high-power devices (e.g., microprocessors). Characteristics of tin–lead solders and electrically conductive adhesives are compared in Table 5.26.

The ECAs can be categorized with respect to conductive loading level into anisotropically conductive (with typically 3–5-μm sized conductive

TABLE 5.26 Comparison of Tin–Lead Solder and a Typical ECA

Characteristic	Tin–Lead Solder	ECA
Volume resistivity (Ω-cm)	0.000015	0.00035
Thermal conductivity (W/m K^{-1})	30	3.5
Shear strength [psi (MPa)]	2200 (15.2)	2000 (13.8)
Finest pitch (mil)	12	<6–8
Minimum processing temperature (°C)	215	150–170
Environmental impact	Negative	Very minor
Thermal fatigue	Yes	Minimal

TABLE 5.27 Conductive Adhesive Based on Two-Part Silver-Filled Epoxy Resin[a,b]

Mix Ratio, by Weight	1:1
Color	Silver
Mixed viscosity	Creamy paste
Pot Life, 100-g Mass @ 25°C	3 h
Specific gravity (25°C)	2.80
Tensile strength [psi (MPa)]	9,450 (65.1)
Hardness, Shore D	70
Compressive strength [psi (MPa)]	14,000 (96.5)
Heat distortion temperature (°C)	95
Volume resistivity (Ω-cm)	0.0001
Operating temperature range (°C)	−50 to +170

[a] 40-3900 Technical Bulletin, Epoxies Etc., Cranston, RI, 05/07.
[b] Curing times for this adhesive are either 24 h @ 25°C, 1 h @ 65°C, or 15 min @ 90°C.

TABLE 5.28 Typical Properties of a Conductive Heat Curable One-Part silicone Adhesive[a]

Color	Gray
Durometer, Shore A	82
Dynamic viscosity, cP (Pa.s)	27,000 (27)[b]
Heat cure, min @ 150°C	60
Specific gravity @ 25°C	3.7
Linear CTE	200 μm m^{-1} °C^{-1}
Shear [psi (MPa)]	246.5 (1.70)
Young's modulus [psi (MPa)]	1740 (12.0)
Volume Resistivity (Ω-cm)	0.00003

[a] Dow Corning® DA 6524, Data Sheet.
[b] Thixotropic.

fillers) and isotropically conductive types (with 1–10-μm sized fillers) [80]. Each of them has different properties and applications.

Examples of two commercially available types of ECAs and their properties are shown in Tables 5.27 and 5.28.

This adhesive is designed to be used for chip bonding in microelectronic and optoelectronic applications, for bonding of microwave electromagnetic interference (EMI) and radio-frequency interference (RFI) shielding. Other applications include assembly and repair of printed circuit board, wave guides, electronic modules, flat cable, high frequency shields, connectors, circuitry, and as a "cold solder". Both resin and hardener have silver powder dispersed in them.

The subject of conductive adhesives, inks, and tapes is covered in much detail in [78 and 79] and in the following publications: Gomatam, and Mittal, (2008), Yim and Paik (2006), and King (Feb. 18, 1997).

Other conductive products for the use in microelectronic and optoelectronic applications are conductive inks, films, and pressure adhesive films.

FURTHER READINGS

Yim, M.J. and Paik, K.W., Review of Electrically Conductive Adhesive Technologies, *Electron. Mater. Lett.*, **2**(3), 183–194 (2006).

King, D.R., U.S. Patent 5,604,026 (Feb. 18, 1997), to W.L. Gore Associates.

Drobny, J.G., *Radiation Technology for Polymers*. 2nd ed., CRC Press, Boca Raton, FL, 2010.

Rubber Technologist's Handbook, Vol **2** (White, J., De, S.K. and Naskar, K., Eds.), Smithers Rapra Technology Limited, Shawbury, Shrewsbury, Shropshire, 2009.

Rubber Technologist's Handbook, Vol. **1** (White, J.R., De, S.K., Eds.), Rapra Technology Limited, Shawbury, Shrewsbury, Shropshire, 2001.

Cleland, M.R. and Galloway, R.A, *Electron Beam Crosslinking of Wire and Cable Insulation*, Technical Information Series TIS 01812 IBA Industrial-White Paper, IBA Industrial, Edgewood, NY.

Power and Communication Cables: Theory and Practice (Bartnikas, R. and Srivastava, K.D., Eds.), Institute of Electrical and Electronic Engineers, New York, 2000.

Drobny, J.G., *Handbook of Thermoplastic Elastomers*, William Andrew Publishing, Norwich, NY, 2007.

Mehnert, R., Pincus, A., Janorsky, I., Stowe, R., and Berejka, A., *UV&EB Technology and Equipment*, Vol. **1**, John Wiley & Sons, Ltd., Chichester/SITA Technology, London, 1998.

Nagdi, K., *Rubber as an Engineering Material: Guideline for Users*, Hanser Publishers, Munich, 1993.

The Vanderbilt Rubber Handbook, 13th ed. (Ohm, R.F., ed.), R.T. Vanderbilt Company, Norwalk, CT, 1990.

Irradiation and Other Curing Techniques in the Wire and Cable Industry, Proceedings from the Regional Technical Conference of the Society of the Plastics Engineers, Inc. March 20–21, 1975, Newton, MA.

Gomatam, R. and Mittal, K.L., *Electrically Conductive Adhesives*, Brill, Leiden, The Nethertlands, 2008 (also available at Brill E-books, 2010).

REFERENCES

1. Available at Plastics News.com, January 7, 2008.
2. *World Rubber and Tire*. Freedonia Group, Inc., January 2010.
3. *World Thermoplastic Elastomers (Report)*, Freedonia Group, Inc., Cleaveland, OH, October 2009.
4. Wappel, K., Paper #1, *Polymers in Electronic 2007 Conference*, Munich, Germany, 30–31 January, 2007, iSmithers/Rapra Technology.
5. *Plastics for Electronics* (Goosey, M.T. Ed.), Elsevier Applied Science Publishers, London, 1985, p. 9.
6. *Thermoplastic Elastomers*, 3rd ed., (Holden, G., Kricheldorf, H.R., and Quirk, R.P., Eds.), Hanser Publishers, Munich, 2004, p. 115.
7. Drobny, J.G., *Handbook of Thermoplastic Elastomers*, Chapter 15, William Andrew Publishing, Norwich, NY, 2007.
8. *Handbook of Thermoplastic Elastomers* 2nd ed., Chapter 6 (Walker, B.M. and Rader, C.P., Eds.), Van Nostrand Reinhold, New York, 1988.
9. *Handbook of Thermoplastic Elastomers* 2nd ed., Chapter 4 (Walker, B.M. and Rader, C.P., Eds.), Van Nostrand Reinhold, New York, 1988.
10. *Handbook of Thermoplastic Elastomers* 2nd ed., (Walker, B.M. and Rader, C.P., Eds.), Van Nostrand Reinhold, New York, 1988 p. 212.
11. *SPI Plastics Engineering Handbook* (Berins, M.L., Ed.), Van Nostrand Reinhold, New York, 1991, p. 74.
12. *SPI Plastics Engineering Handbook* (Berins, M.L., Ed.), Van Nostrand Reinhold, New York, 1991, p. 525.
13. Chung, D.D.L., *Composite Materials: Science and Applications*, Chapter 4, Springer-Verlag, London, 2004.
14. Nagdi, K., *Rubber as an Engineering Material: Guideline for Users*, Hanser Publishers, Munich, 1993, p. 36.
15. *Power and Communication Cables: Theory and Practice* (Bartnikas, R. and Srivastava, K.D., Eds.), Institute of Electrical and Electronic Engineers, New York, 2000, p. 200.
16. Morshedian, J. and Hoseinpour, P.M., *Iranian Polym. J.*, **18**(2), 103–128 (2009).
17. *Power and Communication Cables: Theory and Practice* (Bartnikas, R and Srivastava, K.D., Eds.), Institute of Electrical and Electronic Engineers, New York, 2000, p. 201.

18. Mc. Cane, D.J., in *Encyclopedia of Polymer Science and Technology*, Vol. **13** (Bikales, N.M., Ed.), John Wiley & Sons, Inc., New York, 1970, p. 623.
19. Drobny, J.G., *Technology of Fluoropolymers*, 2nd ed., Chapter 4, CRC Press, Boca Raton, FL, 2009.
20. Hogan, J.D., in *The Vanderbilt Rubber Handbook*, 13th ed. (Ohm, R.F., Ed.), R.T. Vanderbilt Co., Inc., Norwalk, CT, 1660, p. 700.
21. Hogan, J.D., in *The Vanderbilt Rubber Handbook*, 13th ed. (Ohm, R.F., Ed.), R.T. Vanderbilt Co., Inc., Norwalk, CT, 1660, p. 701.
22. *Power and Communication Cables: Theory and Practice* (Bartnikas, R and Srivastava, K.D., Eds.), Institute of Electrical and Electronic Engineers, New York, 2000, p. 199.
23. *Power and Communication Cables: Theory and Practice* (Bartnikas, R and Srivastava, K.D., Eds.), Institute of Electrical and Electronic Engineers, New York, 2000, p. 225.
24. Drobny, J.G., *Radiation Technology for Polymers*, CRC Press, Boca Raton, FL, 2010, p. 4.
25. Hoffman, C.R., in *Irradiation and Other Curing Techniques in the Wire and Cable Industry*, Proceedings from the Regional Technical Conference of the Society of the Plastics Engineers, Inc. March 20–21, 1975, Newton, MA, p. 82.
26. Läuppi, U.V., *RadTtech Europe 97*, June 16–18, 1997, Lyon, France, Conference Proceedings, p. 96.
27. Mehnert, R., Pincus, A., Janorsky, I., Stowe, R., and Berejka, A., *UV&EB Technology and Equipment*, John Wiley & Sons, Ltd., Chichester/SITA Technology, London, 1998, p. 141.
28. Mehnert, R., Pincus, A., Janorsky, I., Stowe, R., and Berejka, A., *UV&EB Technology and Equipment*, John Wiley & Sons, Ltd., Chichester/SITA Technology, London, 1998, p. 143.
29. Hoffman, C.R., in *Irradiation and Other Curing Techniques in the Wire and Cable Industry*, Proceedings from the Regional Technical Conference of the Society of the Plastics Engineers, Inc., March 20–21, 1975, Newton, MA, p. 84.
30. Bernstein, B.R., in *Irradiation and Other Curing Techniques in the Wire and Cable Industry*, Proceedings from the Regional Technical Conference of the Society of the Plastics Engineers, Inc., March 20–21, 1975, Newton, MA, p. 120.
31. Bernstein, B.R., in *Irradiation and Other Curing Techniques in the Wire and Cable Industry*, Proceedings from the Regional Technical Conference of the Society of the Plastics Engineers, Inc., March 20–21, 1975, Newton, MA, p. 124.
32. Cleland, M.R. and Galloway, R.A., *Electron Beam Crosslinking of Wire and Cable Insulation*, Technical Information Series TIS 01812 IBA Industrial-White Paper, IBA Industrial, Edgewood, NY, p. 4.

33. Drobny, J.G., *Radiation Technology for Polymers*, CRC Press, Boca Raton, FL, 2010, p. 87.
34. Bernstein, B.R., in *Irradiation and Other Curing Techniques in the Wire and Cable Industry*, Proceedings from the Regional Technical Conference of the Society of the Plastics Engineers, Inc., March 20–21, 1975, Newton, MA, p. 122.
35. Böhm, G.G.A. and Tveekrem, J.O., *Rubber Chem. Technol.*, **55**, 607 (1982).
36. Böhm, G.G.A. and Tveekrem, J.O., *Rubber Chem. Technol.*, **55**, 612 (1982).
37. Bauman, R.G., *J. Appl. Polym. Sci.*, **2**, 328 (1959).
38. Turner, D., *J. Polym. Sci.*, **35**, 541 (1958).
39. Böhm, G.G.A. and Tveekrem, J.O., *Rubber Chem. Technol.*, **55**, 619 (1982).
40. Harmon, D.J., *Rubber World,* **138**, 585 (1959).
41. Harmon, D.J., *Rubber Age,* **86**, 251 (1959).
42. Kozlov, V.T. et al., *Vysokomol. Soedin.*, **12**, 592 (1970).
43. Böhm, G.G.A. and Tveekrem, J.O., *Rubber Chem. Technol.*, **55**, 627 (1982).
44. Nablo, S.V. and Makuunchi, K., *Meeting of the Rubber Division of American Chemical Society*, October 11–14, 1994 Pittsburg, PA, Paper 83.
45. Böhm, G.G.A. et al., *J. Appl. Polym. Sci.*, **21**, 3193 (1977).
46. Böhm, G.G.A. and Tveekrem, J.O., *Rubber Chem. Technol.*, **55**, 634 (1982).
47. Dogadkin, B.A., Mladenov, I., and Tutorskii, I.A., *Vysokomol. Soedin.*, **2**, 259 (1960).
48. Palalskii, B.K. et al., *Vysokomol. Soedin.*, **A16**, 2762 (1974).
49. Alexander, P. and Charlesby, A., *Proc. R. Soc. London,* **A230**, p. 136 (1955).
50. Loan, L.D., *J. Polym. Sci.*, **A3**, 2127 (1964).
51. Zapp, R.L. and Oswald, A.A., Rubber Division ACS Meeting in Cleveland, 1975. Paper #55.
52. Faucitano, A., Martinotti, F., and Buttafava, A., *Eur. Polym. J.*, **12**, 467 (1969).
53. Böhm, G.G.A. and Tveekrem, J.O., *Rubber Chem. Technol.*, **55**, 640 (1982).
54. Odian, G., Lamparella, D., and Canamare, J., *J. Polym. Sci.*, **C16**, 3619 (1967).
55. Eldred, R.J., *Rubber. Chem. Technol.*, **47**, 924 (1974).
56. Böhm, G.G.A. and Tveekrem, J.O, *Rubber Chem. Technol.*, **55**, 642 (1982).
57. El Milligy, A.A. et al., *Elastomerics*, **111**, 28 (1979).
58. Cleland, M.R. and Galloway, R.A., *Electron Beam Crosslinking of Wire and Cable Insulation*, Technical Information Series TIS 01812 IBA Industrial-White Paper, IBA Industrial, Edgewood, NY, p. 3.
59. Charlesby, A., *Proc. R. Soc. London Sec. A*, **230**, 120 (1955).
60. Miller, A.A., *J.Am. Chem. Soc.*, **82**, 3319 (1960).
61. Hilb, D.W., *J.Phys. Chem.*, **63**, 1838 (1959).
62. Delides, C.G. and Shephard, I.W., *Radiat. Phys. Chem.*, **10**, 379 (1977).

63. Miller, A.A., *J. Am. Chem. Soc.*, **83**, 31 (1961).
64. Drobny, J.G., *Technology of Fluoropolymers*, 2nd ed., CRC Publishers, Boca Raton, FL, 2009, p. 93.
65. Böhm, G.G.A. and Tveekrem, J.O., *Rubber Chem. Technol.*, **55**, 649 (1982).
66. Alexander, D.C., in *Irradiation and Other Curing Techniques in the Wire and Cable Industry,* Proceedings from the Regional Technical Conference of the Society of the Plastics Engineers, Inc., March 20–21, 1975, Newton, MA, p. 59.
67. Huyser, E.S., *Free Radical Chain Reactions*, John Wiley & Sons, New York, 1970.
68. Nablo, S.V., Uglum, J.R., and Quintal, B.S. in *Non-Poluting Coatings and Coating Processes* (Gardon, J.L. and Prane, J.W. Eds.), Plenum Press, New York, pp. 179–194) (1973).
69. Hipple, J.E. and Nablo, S.V., in *Irradiation and Other Curing Techniques in the Wire and Cable Industry*, Proceedings from the Regional Technical Conference of the Society of the Plastics Engineers, Inc., March 20–21, 1975, Newton, MA, p. 6.
70. Cleland, M.R. and Galloway, R.A., *Electron Beam Crosslinking of Wire and Cable Insulation*, Technical Information Series TIS 01812 IBA Industrial-White Paper, IBA Industrial, Edgewood, NY, p. 5.
71. Cleland, M.R., Galloway, R.A., and Berejka. A.J., *Nucl. Instrum. Methods Phyc. Res., Sect. B,* **261**(1–2), 94 (2007).
72. *Power and Communication Cables: Theory and Practice* (Bartnikas, R and Srivastava, K.D., Eds.), Institute of Electrical and Electronic Engineers, New York, 2000, p. 236.
73. *Power and Communication Cables: Theory and Practice* (Bartnikas, R and Srivastava, K.D., Eds.), Institute of Electrical and Electronic Engineers, New York, 2000, p. 237.
74. *Power and Communication Cables: Theory and Practice* (Bartnikas, R and Srivastava, K.D., Eds.), Institute of Electrical and Electronic Engineers, New York, 2000, p. 238.
75. *Power and Communication Cables: Theory and Practice* (Bartnikas, R and Srivastava, K.D., Eds.), Institute of Electrical and Electronic Engineers, New York, 2000, p. 444.
76. Drobny, J.G., *Radiation Technology for Polymers*, CRC Press, Boca Raton, FL, 2010, p. 196.
77. Drobny, J.G., *Radiation Technology for Polymers*, CRC Press, Boca Raton, FL, 2010, p. 197.
78. *Cable Testing, Excerpt from Prysmian Wire and Cable Engineering Guide*, Prysmian Cable Systems, Lexington, SC, October 2007.
79. Li, Y. and Wong, C.P., *Mater. Sci. Eng.*, **R 51**, 1–35, (2006).
80. *Conductive Adhesives in Electronic Packaging* (Liu, J., Ed.), Chapter 1, Electrochemical Publications Ltd., Isle of Man, British Isles, UK (1999).

CHAPTER 6

APPLICATIONS OF SPECIALTY POLYMERS

6.1 APPLICATIONS OF CONJUGATED POLYMERS

There are essentially two main groups of applications for the conjugated, inherently conductive polymers (ICPs). The first group of applications makes use of their conductivity property as its main property. The second group of applications makes extensive use of their electroactivity. The extended p-systems of conjugated polymers are susceptible to chemical or electrochemical oxidation or reduction, which alter the electrical and optical properties of these polymers. By controlling this oxidation and reduction, it is possible to systematically control the electrical properties of ICPs. Since these reactions are frequently reversible, it is possible to systematically control these electrical and optical properties with considerable precision, and even switch from conducting to an insulating state [1]. Typical applications of ICPs are summarized in Table 6.1 and characteristics of most common conjugated polymers are shown in Table 6.2.

Polymers for Electricity and Electronics: Materials, Properties, and Applications, First Edition. Jiri George Drobny.

TABLE 6.1 Typical Applications of ICPs

Conductive Properties	Electroactive Properties
Electrostatic materials	Polymer light-emitting diodes (LED)
Printed circuit boards	Rechargeable batteries
Shielding of electromagnetic interference	Solid electrolytes
Polymeric semiconductors	Polymer photovoltaic devices
Conductive adhesives	Polymer actuators/switches
Active electronics (diodes, transistors)	Corrosion protection
Polymer field-effect transistors	Supercapacitors
Controlled-release devices	Electrooptic devices
Conductive nanofibers and fabrics	Chemical and thermal sensors
Aircraft wiring	Biochemical sensors
Artificial nerves	Organic memory devices
Artificial muscles	Smart coatings and textiles

6.1.1 Polymer Light-Emitting Diodes

Polymer light-emitting diodes (PLEDs) are devices that transform electricity from conjugated polymers into light. Figure 6.1 illustrates the basic and simplified architecture of a PLED, which is largely builtup of conjugated polymers. Under the action of a driving voltage of several volts, electrons are injected from a metal cathode with a low work function into the electronic state corresponding to the *lowest unoccupied molecular orbital* (LUMO) of a conjugated polymer. Holes are injected from a bilayer anode (PEDOT: PSS) [poly(3,4-ethylene dioxythiophene): poly(styrene sulfonic acid)] on indium tin oxide [ITO]) into the electronic state corresponding to the *highest occupied molecular orbital* (HOMO) of a conjugated polymer. Both electrons and holes, coming from the different electrodes, move from opposite directions toward the recombination zone where they can combine to form *excitons* (tightly bound electron–hole pairs). This leads to a population of excited states of the emissive polymers that subsequently emits light [2].

6.1.2 Solar Cells

Solar cells generate power by absorbing sunlight. They are also referred to as *photovoltaic devices*. The commercial solar cells currently used for generating power are made from inorganic substances, (e.g., *single-crystal silicon* or so-called *multijunction cells*, based on Ga, GaAs, and GaInP are still costly and are mainly used on satellite–spaceships, in remote areas (e.g., on oil rigs, at homes on small islands, and on portable devices). The high cost of inorganic solar [photovoltaic (PV)] cells is due

TABLE 6.2 Main Characteristics of the Most Common Conjugated Polymers

Conjugated Polymer	Steady State	Method of Synthesis	Common State	Solubility	Applications
Polyacetylene	Unstable	Chemical polymerization	Film	Insoluble	Theoretical studies of the charge carriers and charge transport in conductive polymers
Polypyrrole	p-Doped conducting state	Electrochemical polymerization	Film	Insoluble	Conductive polymer films, electrode material for batteries, modified electrodes, biosensors, capacitors, electrochromics
Polyaniline	p-Doped conducting state	Electrochemical polymerization	Loosely bounded layer on electrode	Insoluble	Electrode materials for biosensors, batteries and electrochromics, modified electrodes
Polyaniline	p-Doped conducting state	Chemical oxidation polymerization	Powder	Soluble in *m*-cresol with camphor sulfonic acid or dodecylbenzene sulfonic acid as dopant	Electrode materials, anti-corrosion, microwave absorption material, antistatic, modification layer on ITO for the application in PLEDs
Polythiophene	p-Doped conducting state	Electrochemical polymerization	Film	Insoluble	Electrochromics
Poly(9,9′-dioxylfluorene (PFO)	Neutral state	Chemical polymerization	Powder	Soluble in THF[a], xylene, chlorobenzene	Blue luminescent polymer in PLEDs, host material for phosphorescent PLEDs and the white light PLEDs

[a] Tetrahydrofuran

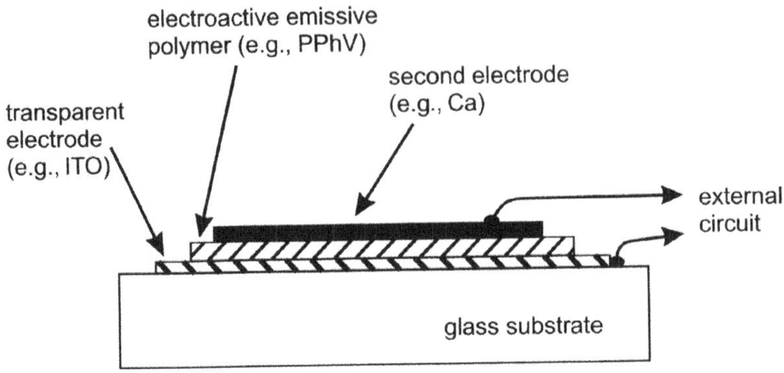

FIGURE 6.1 Schematic of a PLED. (Strobl, G., *The Physics of Polymers: Concepts for Understanding Their Structures and Behavior*, Springer-Verlag Berlin Heidelberg, 2007, p. 287)

to lengthy fabrication process, including crystal growth, wafer production, and high-temperature processing [3]. Organic, especially polymeric PV materials, allow development of low-cost solar cell technology due to the following advantages [4]:

- Low-cost solution-based thin-film fabrication via spin coating, spraying, coating, or painting.
- Flexibility, which allows low-cost and large area mass production and installation.
- Low processing temperatures allow easy integration of plastics cells with other products.
- The ability of optoelectronic and chemical properties of polymers to be tuned by molecular design and engineering.

The disadvantage of current PV devices is that they offer only one-third of the efficiency of the inorganic solar cells and are relatively unstable toward photochemical degradation. A simplified sketch of the design of a polymeric solar cell is shown in Fig. 6.2.

In an organic (polymeric) *p–n junction solar cell*, typically, a thin highly n-doped region is placed on top of a wide and lower doped p-region. Intrinsic regions are avoided to keep the series resistance small. Carrier diffusion creates a depletion region. This field separates electron–hole pairs generated by photon absorption. Electrons travel to the n-side and holes travel to the p-side, as in regular p–n diodes. However, with an open external circuit, the photo-generated carriers create a forward bias, called open circuit voltage (V_{oc}). When an external resistance is connected to the solar cell, the photogenerated bias leads to an external

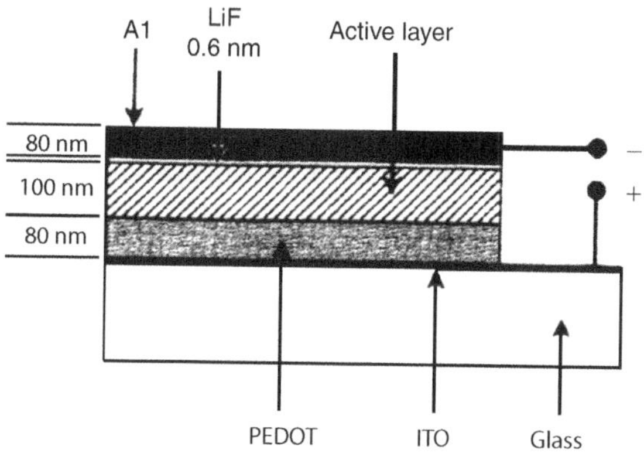

FIGURE 6.2 Schematic of a polymeric solar cell. (Shaheen, S.E. et al., *Appl. Phys.* Lett. 78, 841)

current that is given by the dark current of the diode minus the photocurrent. The key issue of the solar cell arises from the need to capture the solar spectrum over a large wavelength range.

In a donor–acceptor (D–A) binary-type organic PV cell, photoelectronic conversion is accomplished by the following five consecutive steps [5, 6]: (1) Adsorption of photons and the formation of excitons. (2) Exciton diffusion to a D–A interface. (3) Charge separation at the interface. (4) Charge transport to the anode (holes) and cathode (electrons). (5) Charge collection by electrodes.

Similar processes are involved in inorganic p–n junction PV cells, except that free charge carriers are generated upon absorption. Because of the special charge generation mechanism in organic devices, the efficiency-limiting factors are therefore distinct from those in conventional inorganic PV cells (e.g., silicon p–n junction PV cells). For example, while the V_{oc} achievable in silicon cells is generally limited to the magnitude of the builtin potential of the electrodes, it is common to observe experimentally V_{oc} greater than the builtin potential of electrodes in organic-based PV devices [4, 5]. It is believed the maximum V_{oc} of an organic binary cell is closely correlated with the gap between the HOMO of the donor and the LUMO of the acceptor [7].

The subjects of PLED and polymeric solar cells are covered in great detail in [3–7].

6.1.3 Schottky Barrier Diodes

The Schottky diode is a semiconductor diode with a low forward voltage drop and a very fast switching action desirable to switch mode power

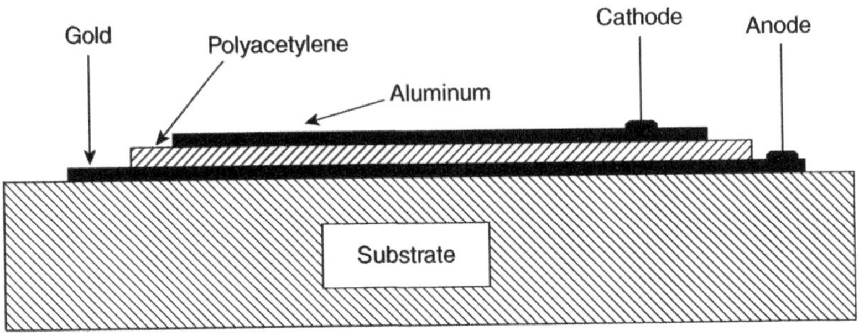

FIGURE 6.3 A typical Schottky barrier diode structure. (Burroughs, J.H., Jones, C.A. and Friend, R.H., *Nature* 335, 137 (1988))

supplies, discharge protection for solar cells connected to lead–acid batteries, prevention transistor saturation, and many others [8]. Instead of a semiconductor–semiconductor junction, a Schottky diode uses a metal–semiconductor junction as a Schottky barrier. A simplified structure of a Schottky barrier diode is shown in Fig. 6.3 with polyacetylene sandwiched between two metal contact layers, in which one of the layers (aluminum) forms the Schottky barrier with polyacetylene and the other (gold) provides an ohmic contact [9].

6.2 APPLICATIONS OF FERROELECTRIC POLYMERS

The advantage of PVDF providing light, rugged, and large-area flexible material with piezoelectric and pyroelectric properties opened innovative applications in the field of electromechanical and pyroelectric devices. Below lists some of these devices:

- *Acoustic transducers* [10–13] are based on the ability of PVDF to be used as a combined diaphragm and a transducer that both collects the sound and converts it to an electric signal without the need for armatures, coils, magnets, and so on.
- *Noise-canceling headphones* that are designed for the use in noisy environments (e.g., jet engine of commercial aircraft). The great simplicity of the PVDF element permits the construction of microphones that have a very symmetrical response to the sound field, an essential requirement for this application.
- *Mechanical applications* are based on the ability of the PVDF to generate an electric signal. These applications include telephone keypads [14], production of movement by an applied field to activate cooling fans [15], or small information displays [16].

6.3 APPLICATIONS OF LIQUID-CRYSTAL POLYMERS

Liquid-crystal polymers (LCPs) are widely used in electrical, electronic, optoelectronic, and fiberoptic applications because of their unique properties, including their high mechanical properties (thin-walled parts), high-impact resistance, low and easily adjustable coefficient of thermal expension (CTEs), low ionic contents, excellent dimensional stability, flame-retarding properties, excellent resistance to aggressive chemicals and solvents, and easy processing. Examples are displays, connectors, components of audio–video equipment, and medical equipment. Additional important commercial applications are discussed in Section 6.3.1.

6.3.1 Liquid-Crystal Displays

Liquid-crystal displays (LCDs) are currently the most common application of LCs. They are used in wristwatches, pocket calculators, high-resolutions computer screens, in advanced high definition (HD) LCD TVs and projectors. The main reason they are used in so many diverse applications is that they are compact and lightweight. The simplest case is a reflective LCD that consist of two glass plates (polarizers) coated with transparent electrodes made of indium tin oxide (ITO with some LC material between the electrodes and with thin-layer electronics behind it (see Fig. 6.4). The second important advantage is the low-power consumption. On the other hand, LCDs have some drawbacks, such as problems with viewing angle, contrast ratio, and response time [17].

6.3.2 Surface Mount Technology

Surface mount technology is based on the mounting of an electronic component directly onto the surface of a printed circuit board (PCB) rather than relying on anchoring leaded components using holes into the

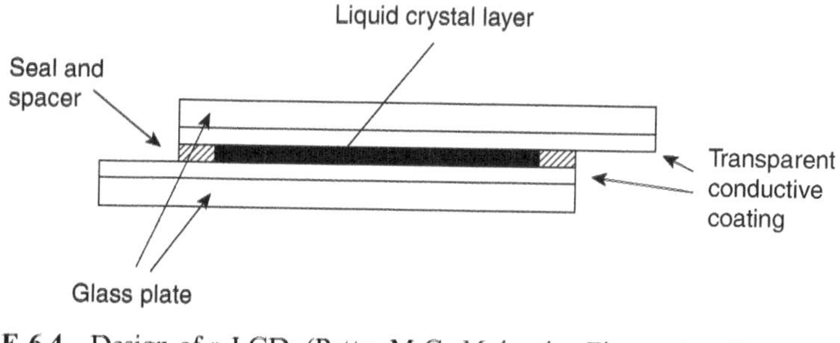

FIGURE 6.4 Design of a LCD. (Petty, M.C, *Molecular Electronics: From Principle to Practice* John Wiley & Sons, Ltd. 2007, p. 300)

PCB itself. Surface mounting reduces the overall size and spacing between electronic packages, and places greater performance demands on the plastics used in the production of components and circuits, particularly as quality and reliability are becoming increasingly important. It also enables components to be mounted on both sides of the board [18]. Surface-mount applications include high pincount, microminiature and precision–high-performance connectors, dual-in-line (DIL) packages' pin grid arrays, burn-in sockets, light emitting diode (LED) casings board-mounted switches, relays, programmable shunts, and capacitor covers [19].

6.3.3 Special Connectors

Currently, commercially available connectors include microminiature, D-subminiature, backplane, card-edge, and board to board. The board-to-board connector for flat-panel displays is an example of an effective way to solve a demanding problem. Unlike most connectors, the devices that interconnect flat panel displays for desktop and portable computers and data entry terminals are not permanently attached to their driver board or the glass-panel display. Permanent attachment is not used because of the difference in CTE between the glass substrate, the PCB, and connector itself [19].

6.3.4 Switches

The constant demand for miniaturization has influenced the market for special switches. The trend requires ever increasing accuracy and closer tolerances in parts. An example is a rotary switch for military radios. In this application, LCP replaced a diallyl phthalate (DAP) thermoset for the stator and resulted in a significant reduction of reject rates. This particular switch was formed by 11 terminals into the stator molding [20]

In most cases, it is necessary that the material has a sufficiently high resistance to withstand typical soldering temperature, typically in excess of 200°C (390°F) during the hard wiring of the switches. Most LCP have this resistance in addition to good electrical resistance properties and flame resistance that is frequently required.

6.3.5 Optoelectronic Components

Thermotropic LCPs are being used commercially to produce high-performance optoelectronic components. These include compact-disk components, where a high specific modulus and dimensional stability are important. Another application for LCP is in demanding fiber-optic couplers in light wave communications systems, test equipment, and sensors [21].

6.3.6 Fiber-Optic Cables

Because of their excellent mechanical properties, extremely low CTE, inherent flame retardance, good barrier properties, and ease of processing, LCPs meet the demanding requirements for fiber-optic cables. Their low-melt viscosity and high-melt strength allow the extrusion in the range of 0.2–10 mm (0.008 to 0.39 in) in diameter and different shapes using conventional extrusion equipment, such as found in typical fiber-optic and wire and cable production plants. Extrusion of LCP directly onto optical glass fibers as a reinforcing buffer coating have been used commercially [22, 23].

6.4 APPLICATIONS OF IONOMERS

6.4.1 Introduction

As pointed out in Section 4.4, the combination of hydrophilic and hydrophobic groups in the same polymeric molecule an ionomer results with unique properties and morphologies. Ionic groups may include a sulfonic or carboxylic group, as well as sulfonamides and sulfonimides. Examples of commercially available ionomers are a copolymer of ethylene and methacrylic acid and a group of fluorinated ionomers with several different ionic groups.

While the applications using certain commercially available polymers merely require stability in a sometimes hostile environment in terms of corrosive chemicals, high temperatures, strong mechanical stresses, or electric fields without failure, applications of some ionomers are based on their ability to actively engage the surrounding environment. As catalysts, they cause chemical reactions, as membranes, they allow the flow of an electric current under a small electric fields (the smaller the better). Most fluorinated ionomers, as well as some other ionomers, can discriminate between anions and cations or allow the passage of water while acting as a barrier for gases. As an actuator, they create mechanical stresses or movement as a result of an electrical stimulus. Also, they have found use as membranes, catalysts, protective clothing, and so on. From a commercial standpoint, the most important examples of the interaction with a chemical system are found in the use of fluorinated ionomers in membrane form as separators in *electrolytic cells* [24].

The above applications are interesting from both a theoretical and a practical point of view, but a detailed discussion of them would be outside the scope of this book. It appears that the subject of fuel cells based on specific ionomers deserves more attention and will be discussed at some length (see Section 6.4.2).

6.4.2 Fuel Cells

Fuel cells are electrochemical devices that convert chemical energy directly into electrical energy without combustion. In a way, they are similar to batteries: In a fuel cell or battery, the fuel releases electrons to an electrode (the anode). These electrons then travel through an external circuit (or load), doing useful work, until they reach the cathode of the fuel cell or battery, where they combine with the oxidizer. The movement of some ions through the electrolyte that connects the anode and the cathode completes the electrical circuit. In batteries, the fuel, as well as the oxidizer, is stored inside the battery. When all the fuel is consumed, the battery has to be recharged or replaced. In a fuel cell, the fuel and the oxidizer are stored external to the cell. Theoretically, the fuel cell can, therefore, operate indefinitely as long as fuel and oxidizer are supplied [25].

Fuel cells come in many varieties; however, they all work in the same general manner. They are made up of three segments that are sandwiched together: the anode, the electrolyte, and the cathode. Two chemical reactions occur at the interfaces of the three different segments. The net result of the two reactions is that fuel is consumed, water or carbon dioxide is created, and an electric current is created, which can be used to power electrical devices, normally referred to as the load.

The most important design features in a fuel cell are

- The electrolyte substance. The electrolyte substance usually defines the type of fuel cell.
- The fuel that is used.
- The anode catalyst, which breaks down the fuel into electrons and ions. The anode catalyst is usually made up of very fine platinum powder.
- The cathode catalyst, which turns the ions into the waste chemicals like water or carbon dioxide. The cathode catalyst is often made up of nickel.

A simplified block diagram of a fuel cell is shown in Fig. 6.5.

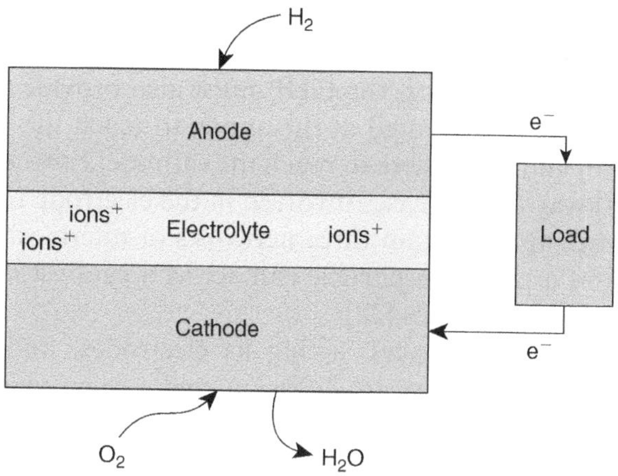

FIGURE 6.5 Simplified block diagram of a fuel cell.

There are different types of fuel cells based on the electrolyte used; this determines the operating temperature of the system. In this chapter the discussion is limited to fuel cells using a fluorinated ionomer as the electrolyte, referred to as, polymer electrolyte membrane (PEM) fuel cell, with a typical operating temperature ranging from 60 to 120°C (140 to 250°F).

There also are many combinations of fuels and oxidants possible. A hydrogen fuel cell uses hydrogen as its fuel and oxygen (usually from air) as its oxidant. Other fuels include hydrocarbons and alcohols. Other oxidants include chlorine and chlorine dioxide [26].

In the hydrogen fuel cell, hydrogen fuel is fed into the anode of the fuel cell. Oxygen (or air) enters the fuel cell through the cathode. The catalyst present in the cell splits the hydrogen atom into a proton and an electron, which take different paths to the cathode. The proton passes through the electrolyte. The electrons create a separate current that can be utilized before they return to the cathode, to be reunited with the hydrogen and oxygen in a molecule of water. In such a hydrogen–oxygen *proton-exchange membrane* fuel cell design, a proton-conducting polymer membrane (the electrolyte), separates the anode and cathode sides [27].

To maximize the surface area of the expensive catalyst, the catalyst is deposited on the surface of carbon black and distributed in a three-dimensional (3D) *gas-diffusion electrode* (GDE). This GDE, sometimes also called the catalyst layer, may be only a few microns thick, but still has an internal surface area many times larger than the geometric area of the electrode. Carbon black not only provides a high surface support for the platinum catalyst, but also a continuous electronically conducting phase. The ability of carbon black to form long strings of individual particles makes it uniquely suitable for this job. In addition to the continuous electronic conductor, the GDE must also provide a continuous pathway for the ions generated at the anode to reach the ionomer film (the electrolyte) and from there to reach the cathode. Furthermore, there must be a pathway for the gases involved in the electrode reactions. The GDE, therefore, must contain three networks of intermeshing continuous phases, and a platinum particle can act as a catalyst only if it is in contact with all these phases [28].

Between the catalyst layers, acting as electrodes, and the current collector–gas flow field, there are diffusion media to provide current distribution to those parts of the catalyst layer that are above a gas channel (the groves), and gas distribution to areas blocked by the ridges. In a typical cell, a felt or a paper made of carbon fibers is used as the gas-diffusion medium, and the current collectors are made of graphite. The two catalyst layers are usually bonded or laminated to the central membrane, creating a unit referred to as the *membrane electrode assembly* (MEA).

At the endplates, uniform clamping pressure is applied to the stack, not only to seal the individual components, but also to minimize the contact resistance between the current collectors, the diffusion media, and the catalyst layers. These three components rely on carbon black as the electronic conductor and the resistance of a carbon–carbon contact depends very much on the contact pressure. Some fuel cells use a pressurized gas bladder to distribute the clamping pressure uniformly over the entire cross-section of the stack. The contact resistance between the catalyst layers and the membrane is minimized by laminating these three components to form an MEA.

As pointed out above, an MEA consists of the following three layers. The first layer is a central ionomer film, typically 20–100 μm thick, is coated on both sides with the anode and cathode electrode (or catalyst) layers, each ~10 μm thick. The two electrode layers may not be identical: The cathode layer frequently contains higher platinum loading than the anode layer because of the more difficult cathode reaction.

The electrode layers are usually prepared by casting an "ink" on a support film or decal. The ink consists of a suspension of the catalyst (platinum on carbon black) in a liquid composition of the ionomer (the pigment and vehicle, respectively, in paint terms). A typical composition of the electrode layer after the removal of solvent is 50% ionomer, 25% platinum, and 25% carbon black. A typical fuel cell of current design produces a voltage from 0.6 V to 0.7 V at full rated load.

A fuel cell system that includes a *fuel reformer* can utilize the hydrogen from any hydrocarbon fuel, ranging from natural gas to methanol, and even gasoline. Since the fuel cell relies on chemistry and not combustion, emissions from this type of a system would still be much smaller than emissions from the cleanest fuel combustion processes.

The current market for PEM fuel cells can be divided into three main categories: portable power, stationary power, and transportation. Portable and the related small stationary power together account for ~85% of all the installed units, transportation account for ~12%, and large stationary power for only a small fraction. For large stationary power, other fuel cell systems, such as phosphoric acid, or different technologies, may be more economical [29]. As the development of fuel cells progresses, they will become a formidable competition to batteries, small internal combustion engines, and solar and wind power.

6.5 LITHIUM ION POLYMER BATTERIES

Lithium ion polymer batteries are rechargeable batteries (secondary cell batteries), which are essentially high-energy density devices. Although lithium with its high oxidation potential would be suited for high-energy density batteries, contact of metal lithium with a liquid electrolyte is hazardous. Therefore it had to be replaced with a safer lithium ion source. The current lithium ion polymer batteries are consisting of a cathode (usually lithium–carbon intercalation compound, e.g., $LiCoO_2$ or Li Mn_2O_4) an anode (Li_xC_8), and a separator (see Fig. 6.6).

The two electrodes must be separated by a medium, which allows easy transfer of metal ions during the charging and discharging process. In the usual arrangement, a separator frame of insulating plastic is filled with a

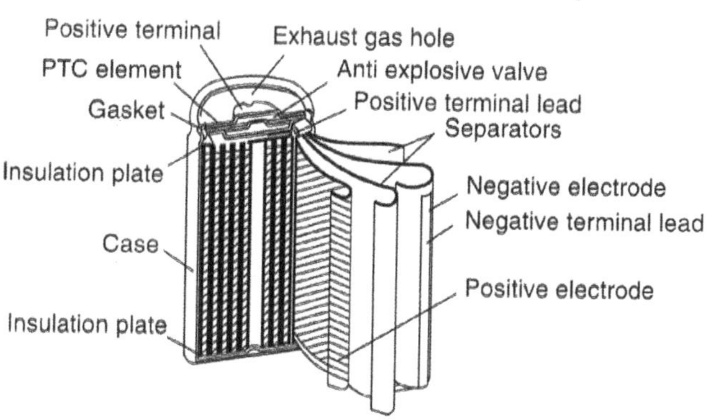

FIGURE 6.6 Schematic of a lithium ion polymer battery. (Courtesy of Panasonic Corporation). (*Lithium Ion Batteries Technical Handbook*, June 2007, Panasonic Corporation)

liquid electrolyte. At least two commercially successful designs are using a copolymer of vinylidene fluoride (VDF) and hexafluoropropylene (HFP) as a matrix for the composite electrodes and the electrolyte gelled with conventional solvents and salts. Below are the electrode reactions of the standard Li ion battery:

$$\text{The reaction on an anode:} \quad Li_xC_6 \rightarrow x\,Li^+ + x\,e^- + C_6$$

$$\text{The reaction on a cathode:} \quad xLi^+ + x\,e^- + Li_{1-x}MnO_4 \rightarrow LiMnO_4$$

The voltage of a Li ion polymer cell varies from ~2.7 V (discharged) to ~4.23 V (fully charged), and the cells have to be protected from overcharge by limiting the applied voltage to no >4.235 V cell used in a series combination. Overcharging a Li polymer battery can cause an explosion or fire. Currently, manufacturers have been declaring upward of 500 charge–discharge cycles before the capacity drops to 80%.

The applications of Li ion polymer batteries include personal digital assistants (PDAs), laptop computers, small digital music devices, and cell phones. They are desirable in applications where small form factors and energy density outweigh cost considerations. Current technology developments indicate that these batteries may also power the next generation of vehicles.

FURTHER READINGS

Conjugated Polymers: Processing and Applications (Skotheim, T. and Reynolds, J.R., Eds.), CRC Press, Boca Raton, FL, 2007.

Grot, W., *Fluorinated Ionomers*, William Andrew Publishing, Norwich, NY, 2008.

Introduction to Organic and Optoelectronic Materials and Devices (Sun, S.-S. and Dalton, L.R., Eds.), CRC Press, Boca Raton, FL, 2008.

Blythe, T. and Bloor, D., *Electrical Properties of Polymers*, 2nd ed., Cambridge University Press, Cambridge, UK, 2005.

REFERENCES

1. "Conductive Polymer Applications", *Plastics World Insight*, Special Issue, Vol. **2**, No.1, Momentum Press, New York, 1 May 2009.
2. *Introduction to Organic Electronic and Optoelectronic Devices* (Sun, S.-S. and Dalton, L.R., Eds.) Chapter 13, CRC Press, Boca Raton, FL, 2008, p. 374.

3. *Introduction to Organic Electronic and Optoelectronic Devices* (Sun, S.-S. and Dalton, L.R., Eds.) Chapter 14, CRC Press, Boca Raton, FL, 2008, p. 402.
4. *Introduction to Organic Electronic and Optoelectronic Devices* (Sun, S.-S. and Dalton, L.R., Eds.) Chapter 14, CRC Press, Boca Raton, FL, 2008, p. 403.
5. Organic Photovoltaics, Mechanisms, Materials and Devices (Sun, S.-S. and Sariciftci, N.S., Eds.), CRC Press, Boca Raton, FL, 2005.
6. Hoppe, H. and Sariciftci, N.S., *J. Mater. Sci.*, **19**, 1924–1945 (2004).
7. *Introduction to Organic Electronic and Optoelectronic Devices* (Sun, S.-S. and Dalton, L.R., Eds.) Chapter 14, CRC Press, Boca Raton, FL, 2008, p. 418.
8. *Introduction to Organic Electronic and Optoelectronic Devices* (Sun, S.-S. and Dalton, L.R., Eds.) Chapter 13, CRC Press, Boca Raton, FL, 2008, p. 241.
9. *Introduction to Organic Electronic and Optoelectronic Devices* (Sun, S.-S. and Dalton, L.R., Eds.) Chapter 13, CRC Press, Boca Raton, FL, 2008, p. 242.
10. Lerch, R., *J. Acous. Soc. Am.*, **67**, 1229 (1981).
11. Tamura, M., Yamaguchi, T., Oyaba, T., and Yoshimi, T., *J. Audio Engng. Soc.*, **23**, 71 (1975).
12. Lerch, R. and Sessler, G.M. *J. Acoust. Soc. Am.*, **64**, 1720 (1980).
13. Garner, G. M., *System Technol.*, **27**, 22 (1977).
14. Pearman, G.T., Hokanson, J.L., and Meeker, T.R., *Ferroelectrics*, **28**, 311 (1980).
15. Toda, M. and Osaka, S., *Proc. IEEE*, **67**, 1171 (1979).
16. Toda, M., *Ferroelectrics*, **32**, 127 (1981).
17. *Introduction to Organic Electronic and Optoelectronic Materials and Devices* (Sun, S.-S. and Dalton, L. R., Eds.), CRC Press, Boca Raton, FL, 2005, p. 784.
18. *Special Polymers for Electronics & Optoelectronics* (Chilton, J.A. and Goosey, M.T., Eds.), Chapman and Hall, London, 1995, p. 276.
19. *Special Polymers for Electronics & Optoelectronics* (Chilton, J.A. and Goosey, M.T., Eds.), Chapman and Hall, London, 1995, p. 277.
20. *Special Polymers for Electronics & Optoelectronics* (Chilton, J.A. and Goosey, M.T., Eds.), Chapman and Hall, London, 1995, p. 280.
21. Rogers, K, Mc Duffee, F.T., Holt, A., and Mekkaoui, A., *Proc. SPIE*, **574**, 29 (1985).
22. Shuto, Y., Yamamoto, F., Takeuchi, Y., and Yamakawa., *Proceedings of IWSF-85*, Hakone, Japan, p. 20, 1985.
23. Takeuchi, Y., Shuto, Y., and Yamamoto, F., Paper presented at Electronic Communications Society Meeting, 1985.
24. Grot, W., *Fluorinated Ionomers*, William Andrew Publishing, Norwich, NY, 2008, p. 75.
25. Grot, W., *Fluorinated Ionomers*, William Andrew Publishing, Norwich, NY, 2008, p. 137.
26. Meibuhr, S.G., *Electrochim. Acta*, **11**, 1301 (1966).

27. Du Puis, A. C., *Prog. Mater. Sci.*, **56**(3), 289–327, March 2011.
28. Grot, W., *Fluorinated Ionomers*, William Andrew Publishing, Norwich, NY, 2008, p. 139.
29. Grot, W., *Fluorinated Ionomers*, William Andrew Publishing, Norwich, NY, 2008, p. 161.

CHAPTER 7

TESTING OF ELECTRICAL PROPERTIES OF POLYMERS

7.1 INTRODUCTION

Polymers, (i.e., plastics, elastomers and their compounds) are used in many electrical applications, often as primary electrical insulation. Also, they may have primarily a structural function, and in some cases are combined with an electrical function. The electrical behavior of most insulators depends mainly on magnitude and frequency of the applied voltage, but also on temperature, time, moisture content, amount and type of contaminants, sizes and shapes of both the conductor and the insulation. Other severe ambient or design conditions of a widely different nature, which influence the electrical characteristics, also may be involved. At any rate, all these factors interact in a complex fashion.

To evaluate the capability of polymeric materials in electrical applications, they have to be subjected to a variety of tests. In addition, in most cases, mechanical and physical properties have to be evaluated as well, since they affect their overall performance. The test results can be used as tools for quality assurance, for research and development, as well as a means of identification, and serve as a research tool. In case of

Polymers for Electricity and Electronics: Materials, Properties, and Applications, First Edition. Jiri George Drobny.

TABLE 7.1 Selected ASTM Methods for Testing of Polymers as Insulators

ASTM Designation	Title
D149-09	Standard Test Method for Dielectric Breakdown Voltage and Dielectric Strength of Solid Electrical Insulating Materials at Commercial Power Frequencies
D150-98 (2004)	Standard Test Methods for ac Loss Characteristics and Permittivity (Dielectric Constant) of Solid Electrical Insulation
D257-07	Standard Test Methods for dc Resistance or Conductance of Insulating Materials
D495-99 (2004)	Standard Test Method for High-Voltage, Low-Current, Dry Arc Resistance of Solid Electrical Insulation
D1868-07	Standard Test Method for Detection and Measurement of Partial Discharge (Corona) Pulses in Evaluation of Insulation Systems
D2275-01 (2008) el	Standard Test Method for Voltage Endurance of Solid Electrical Insulating Materials Subjected to Partial Discharges (Corona) on the Surface

commercial products, certain test results provide a basis for the sales specification.

In the United States, ASTM Committee D09, on Insulating Materials, and Committee D20 on Plastics, has the primary responsibility of issuing standards for the evaluation and specification of plastics. Selected ASTM test methods pertaining to the electrical properties of polymers are listed in Table 7.1.

In addition to ASTM, several other groups, such as the National Electrical. Manufacturers Association (NEMA), the Underwriters Laboratories (UL), the Institute of Electrical and Electronics Engineers (IEEE), and several U.S. government agencies issue plastics specifications, as do German (VDE Association of German Electrical Engineers, and DIN, Deutsche Industrienorm) and British (BSI British Standards Institute) agencies. International standards are coordinated and written by the International Electrotechnical Commission (IEC). Despite such extensive and useful standardization activity, relatively little emphasis has been placed on the effect of electrical characteristics on the performance capability of plastics (i.e., on the engineering properties). Most standardized test methods emphasize speed and reproducibility, which are important in acceptance testing, but are not suitable for obtaining design values. Unfortunately, the electrical properties given in handbooks and reference works are generally based on such standardized tests. Such test values may have limited usefulness for screening and comparison. To determine the electrical properties of plastics in a fashion useful for engineering purposes, it is necessary to duplicate conditions of service as closely as possible. However, these conditions are not always known, and can never be completely duplicated in the laboratory. In fact, simple rather than complex and lengthy tests are frequently justified. Electrical properties should be determined as a function of variables (e.g., temperature and time). Time is an important variable in voltage degradation and moisture exposure. Properties should be given in the form of a curve where possible. Some plastics manufacturers supply such data [1].

This chapter deals with the measuring methods, calculation of pertinent quantities, and relationship between them. Additional test methods, which are more specific, are described in great detail in corresponding ASTM or other pertinent publications. These are listed in Appendix II.

7.2 INDIVIDUAL TESTS OF ELECTRICAL PROPERTIES

7.2.1 Dielectric Strength (ASTM D149)

Dielectric strength of an insulating material is the voltage gradient at which electric failure or breakdown occurs as a continuous arc (the electrical property analogous to tensile strength in mechanical tests). The dielectric strength of polymeric materials varies greatly with several conditions (e.g., humidity and geometry) and is not possible to directly apply the standard test values to field use unless all conditions, including specimen dimensions, are the same. Because of this, the dielectric strength test results are of relative rather then absolute value as a specification guide.

The specimen being tested is placed between two heavy cylindrical brass electrodes, which carry electric current during the test. There are two ways of running this test:

1. Short-Time Test: The voltage is increased from zero to breakdown at a uniform rate, typically 500 kV s^{-1} (in some cases 100, 1000, or 2000 kV s^{-1}). The precise voltage rate is specified in governing material specifications, which is the most common test.
2. Step-by-Step Test: The initial voltage applied is 50% of the breakdown voltage shown by the short-time test. It is increased at rates specified for each type of material and the breakdown level noted.

Breakdown by these tests means passage of a sudden excessive amount of current through the specimen and can be verified by instrument and

TABLE 7.2 Intrinsic Dielectric Strength (Peak Values) from Selected Polymeric Insulators Measured At 20°C

Polymer	Dielectric Strength (kV mm^{-1})
Polyisobutylene	100
Polystyrene	600
Plasticized poly(vinyl chloride)	650
Polyethylene	650
Chlorinated polyethylene	650

visible damage to the specimen. Dielectric strength is calculated by dividing the breakdown voltage by the thickness of the sample. The data is usually expressed in kV mm^{-1} or V mil^{-1}. The location of the failure is also recorded. A higher dielectric strength represents a better quality of insulator.

Intrinsic dielectric strength is defined as the value obtained at uniform voltage gradients, on thin test specimens, using direct voltage and short test times. Most insulating polymeric materials have intrinsic dielectric values ranging from 100 to 650 kV mm^{-1} [2].

The values of intrinsic dielectric strength measured at ambient temperature on selected polymers are shown in Table 7.2.

7.2.2 Dielectric Constant (ASTM D150)

The dielectric constant (relative permittivity) is the ratio of the capacity of a condenser (capacitor) containing a dielectric to the capacity of the same capacitor with air (or vacuum) as a dielectric (see Section 1.5). For a polymeric material used to support and insulate components of an electrical network from each other and the ground, generally it is desirable to have a low value of dielectric constant. For a material to function as the dielectric of a capacitor, it is desirable to have a high dielectric constant value, so the capacity may be physically as small as possible.

If the capacity is measured with alternating current (ac), then its frequency is an important factor for polar polymers because of the time taken to align the dipoles. At very low frequencies, the dipoles have sufficient time to align with the field before it changes direction and the measured value is high. At high frequencies, the dipoles do not have time to align before the field changes the direction and they must realign with the changed field. Polar polymer dielectrics at low frequencies (60 Hz) generally have dielectric constants between 3 and 9 and at higher frequencies (106 Hz) they generally have measured values between 3 and 5 [3]. For nonpolar polymers, the dielectric constant is independent of the ac frequency because the electron polarization is instantaneous. Nonpolar polymers have dielectric constants always <3 [3].

Methods for measuring capacitance and ac loss according to ASTM D150 can be divided into three groups: the null, resonance, and deflection methods. The choice of a method for any particular case will depend primarily on the operating frequency. The resistive- or inductive-ratio-arm capacitance bridge in its various forms can be used over the frequency range from <1 Hz to a few megahertz. For frequencies <1 Hz, special methods and instruments are required.

These test methods provide general information on a variety of electrodes, apparatus, and measurement techniques. The reader should also consult ASTM standards or other documents directly applicable to the material to be tested.

7.2.3 Power Factor (ASTM D150)

The power factor is a measure of the energy absorbed by the dielectric material as the ac constantly changes direction, and dipoles, if they are present, try to align themselves with the alternating field. As they do that they will always be slightly out of phase and will lag behind the field. The amount of lagging is measured by the phase angle θ and the power factor is defined as $\cos\theta$. The power factor can be thought of as a measure of the internal friction created by the ac and will define how much a material heats up when placed in an alternating field. It is defined as a ratio of real power P measured in watts and apparent power S measured in volt-amperes. For perfectly sinusoidal waveform $P/S = \cos\theta$.

For polar polymers, the power factor depends on the frequency of the ac [4]. At very low and very high frequencies, both the power factor and the amount of internal heating are low, since the dipoles either do or do not have time to align before the field changes direction. At intermediate frequencies, the power factor goes through a maximum and the internal friction is high with substantial heating of the dielectric taking place. For nonpolar polymers, the electronic polarization is effectively in phase with the external field (i.e., θ is ~ 0) thus they practically do not suffer from internal friction and internal heating.

Power factor in a single-phase circuit (or balanced three-phase circuit) can be measured with the wattmeter–ammeter–voltmeter method, where the power in watts is divided by the product of measured voltage and current. Digital instruments can be made that either directly measure the time lag between voltage and current waveforms and so calculate the power factor, or by measuring both true and apparent power in the circuit and calculating the quotient. The first method is only accurate if voltage and current are sinusoidal; loads (e.g., rectifiers) distort the waveforms from the sinusoidal shape, as mentioned above.

7.2.4 Dissipation Factor (ASTM D150)

Dissipation factor (D) is the ratio of the real power (in phase power) to the reactive power (power 90° out of phase) or the ratio of the loss index κ'' to the relative permittivity κ'. It is also defined as the ratio of parallel reactance X_p to the equivalent ac parallel resistance R_p. It is the tangent of the loss angle δ or cotangent of the phase angle θ [5]:

$$D = \frac{\kappa''}{\kappa'} = \frac{X_p}{R_p} = \text{cotg}\,\theta = \tan\delta \quad (7.1)$$

In principle, the dissipation factor is a measure of loss rate of power for a mode of oscillation that is not only electrical, or electromechanical, but is also mechanical in a dissipative system. Most polymers, particularly plastics, have a low dissipation factor, which is a desirable property because it minimizes the waste of electrical energy as heat.

7.2.5 Direct Current Resistance (ASTM D257)

In some applications, the resistance of a polymeric material or its reciprocal, the conductance is of functional significance, such as in electronic circuits for which a high resistance or impedance has operational importance. More often, resistance is used to measure indirectly the influence of other factors (e.g., moisture), which may affect electric strength [6]. Resistance measurements are also important in studying resin cure, determine the presence of impurities, and investigating the fundamental characteristics of polymeric materials. An extensive review of the theoretical aspects is found in [7].

Insulation resistance is defined as the direct current (dc) resistance of a material or device measured between electrodes or conductors without regard to geometry. Insulation resistance is a combination of surface and volume resistance [6].

Volume resistivity is the ratio of the potential gradient parallel to the current in the material to the current density unit in ohm-centimeters (Ω-cm) (the SI unit is Ω-m). It is the dc volume resistance multiplied by the function of specimen dimensions that transforms the resistance to that of a unit cube, usually expressed as Ω-cm ohm-centimeter (or Ω-m).

Surface resistivity is the ratio of the potential gradient parallel to the current along its surface to the current per unit width of the surface. It is the dc surface resistance multiplied by the function of the specimen dimensions that transform the resistance to that of a unit square on the surface. The unit of surface resistivity is the ohm(Ω).

The ASTM D 257 standard covers dc procedures for the measurement of dc insulation, volume, and surface resistance. From such measurements and the geometric dimensions of specimen and electrodes, both volume and surface resistivity of electrical insulating materials can be calculated, as well as the corresponding conductances and conductivities. Resistance measurements are usually made at 100 or 500 V dc, but in some cases the voltage may be increased to 5 kV or higher [8]. The instrument used to measure the resistance is an ohmmeter. The above techniques are described in detail in this standard. Knowing the volume and surface resistivity of an insulating material makes it possible to design an insulator for a specific application.

7.2.6 High-Voltage, Low-Current, Dry Arc Resistance (ASTM D495)

This test method covers, in a preliminary fashion, the differentiation among similar materials with respect to their resistance to the action of a high-voltage, low-current arc close to the surface of insulation. This arc tends to form a conducting path in that location or may cause the material to become conducting due to the localized thermal and chemical decomposition and erosion.

Because of its convenience and the short time required for testing, this test method is intended for preliminary screening of material, for detecting the effects of changes in formulation, and for quality control testing. In order to obtain meaningful results, it is necessary to established a correlation with other types of simulated service arc tests and field experience. Essentially, this test method is useful for preliminary evaluation of changes in structure and or composition without the complicating influence of environmental conditions, especially dirt and moisture.

Arc resistance, in accordance with ASTM D495, is expressed as the number of seconds that a material resists the formation of a surface-conducting path when subjected to an intermittently occurring arc of high-voltage, low-current characteristics. The results of testing the nominal 3-mm thickness are considered representative of the material's performance in any thickness. Performance level categories (PLC) were introduced by UL to avoid excessive implied precision and bias. The list of these categories is shown in Table 7.3.

TABLE 7.3 UL Performance Level Categories for Arc Resistance

Mean Time of Arc Resistance (s)	Assigned PLC
420 and longer	0
360–419	1
300–359	2
240–299	3
180–239	4
120–179	5
60–119	6
<60	7

7.2.7 Detection and Measurement of Partial Discharge Pulses (ASTM D1868)

As stated by ASTM D1868, this test method covers the detection and measurement of partial discharge (corona) pulses at the terminals of an insulation system under an applied test voltage, including the determination of partial discharge (corona) *inception* and *extinction* voltages as the test voltage is raised and lowered. The test method is also useful in determining quantities, such as apparent charge and pulse repetition rate together with such integrated quantities as average current, quadratic rate, and power. The test method is useful for test voltages ranging in frequency from 0 (direct voltage) to ~2000 Hz.

The test method is directly applicable to a simple insulation system that can be represented as a two-terminal capacitor. The test method is also applicable to (distributed parameter) insulation systems (e.g., high-voltage cable).

A partial discharge is defined as a type of localized discharge that results from transient gaseous ionization in an insulation system when the voltage stress exceeds a critical value. The ionization is localized over only a portion of the distance between the electrodes of the system. The resultant partial discharge signals appear as a very small magnitude, which are fast-rise pulses with irregular waveshapes superimposed on the high voltage at the terminals of the test sample. Partial discharges cause deterioration of insulation materials and are a primary cause of insulation failure at moderate and high voltages. The accepted unit for measurement of partial discharge magnitude is the picocoulomb (pc). The picocoulomb, a unit of charge, is $C \times 10^{-12}$

Corona inception voltage is the lowest voltage at which continuous corona of specified pulse amplitude occurs as the applied voltage is gradually increased. Corona inception voltage decreases as the frequency of the applied voltage increases. Corona can occur in applications as low as 300 V.

Corona extinction voltage is the highest voltage at which continuous corona of specified pulse amplitude no longer occurs as the applied voltage is gradually decreased from above the corona inception value. Thus, once corona starts, the voltage must be decreased to get it to stop.

The corona discharges in insulation systems result in voltage transients. These pulses are superimposed on the applied voltage and may be detected, by a detector. It is important that the voltage source and the coupling capacitor exhibit low noise so as not to obscure the corona. In its simplest form, the pulse detection network is a resistor monitored by an oscilloscope.

Commercially available corona detectors include electronic types (as above), as well as ultrasonic types. The classic corona measurement systems had been manufactured by the James G. Biddle Company (available at www.biddlemegger.com).

7.2.8 Surface Corona and Surface Discharge (ASTM D2275)

According to the description in ASTM D2275, this test method is used to compare the endurance of different materials to the action of the corona on the external surfaces. A poor test result does not indicate that the material is a poor selection for use at high voltage or at high-voltage stress in the absence of surface corona. Surface corona should be distinguished from corona that occurs in internal cavities for which no standardized test has been developed. Evaluation of endurance by comparison of data on specimens of different thickness is not valid.

The above standard describes a means of measuring degradation due to surface corona of sheets of materials between a 12.7-mm (0.5-in.) diameter bottom electrode and a top electrode, which is either a 6.35-mm (0.25-in.) diameter cylinder or a 12.7-mm (0.5-in) diameter steel ball. The assembly is kept very dry, while different voltage levels are applied between the electrodes in contact with the insulation [9], although tests are also made in open air, at 50% relative humidity. It may be important for some materials to make tests in circulating air at 20% relative humidity or less [10]. A curve is plotted of voltage versus time-to-failure by breakdown. In this test, the insulation surface is subjected directly to a chemical attack, bombardment, and resulting erosion from corona discharges, but only "volume failure takes place. Silicone resins and elastomers best resist such corona attack, followed in descending order by polyesters and PTFE [9].

Note that [11] (1) the term "corona" is used almost exclusively in this test method instead of "partial discharge", because it is a visible glow at the edge of the smaller electrode. This difference is in location, not in kind. Partial discharges also occur at the edges of electrodes, and in general corona describes an electrical discharge irrespective of its location. (2) This test method differentiates among solid electrical insulating materials for use at commercial power frequencies with respect to their voltage endurance under the action of corona. In general, this test method is more meaningful for rating materials with respect to their resistance to prolonged ac stress under corona conditions than is dielectric strength.

FURTHER READINGS

Blythe, T and Bloor, D., *Electrical Properties of Polymers*, 2nd ed., Chapter 5, Cambridge University Press, Cambridge, UK, 2005.

Riande, A. and Díaz-Calleja, R., *Electrical Properties of Polymers*, Chapter 4, Marcel Dekker, New York, 2004.

REFERENCES

1. Mathes, K.N., in *2010 Encyclopedia of Polymer Science and Technology*, John Wiley & Sons, Inc., Hoboken, NJ, Published Online September 15, 2010, p. 1.
2. Mathes, K.N., in *2010 Encyclopedia of Polymer Science and Technology*, John Wiley & Sons, Inc., Hoboken, NJ, Published Online September 15, 2010, p. 16.
3. *Zeus® Dielectric Properties of Polymers,* Technical Whitepaper, Zeus Industries Products, Inc., Orangeburg, SC, 2005, p. 2.
4. *Zeus® Dielectric Properties of Polymers,* Technical Whitepaper, Zeus Industries Products, Inc., 2005, Orangeburg, SC, p. 3.
5. ASTM D150–98 (2004), p. 1.
6. Mathes, K.N., in *2010 Encyclopedia of Polymer Science and Technology*, John Wiley & Sons, Inc., Hoboken, NJ, Published Online September 15, 2010, p. 31.
7. Wintle, H.J. in *Electrical Properties of Solid Insulating Materials* (Bartnikas, R. and Eichhorn, R.M., Eds.), ASTM STP 783, American Society or Testing and Materials, Philadelphia, PA, 1987, pp. 94–134.
8. Mathes, K.N., in *2010 Encyclopedia of Polymer Science and Technology*, John Wiley & Sons, Inc., Hoboken, NJ, Published Online September 15, 2010, p. 33.

9. Mathes, K.N. in *Engineering Dielectrics*, Vol. **IIB**, Chapter 4 (Bartnikas, R. Ed.), ASTM STP 926, American Society or Testing and Materials, Philadelphia, PA, 1987, p. 253.
10. ASTM D2275–01 (2008), p. 1.
11. *IHS Industry Standards and Regulations* (published March 10, 2001), available at www.global.ihs.com.

CHAPTER 8

CURRENT TRENDS AND DEVELOPMENTS

Our modern lifestyle increasingly demands more from polymers than their traditional role of insulators for electric current. The relevant characteristic of organic polymers is their ability to withstand high electric fields with negligible conduction. This coupled with favorable mechanical and processing properties make them the obvious choice as electric insulators. Today, however, organic polymers are used commercially not only as insulators, but also as conductors of electricity, alone or in the form of composite materials, conductive adhesives, and in electronic applications ranging from semiconductors to biomedicine. Thus their use in electrical and electronic applications has grown from simple insulation to sophisticated devices, such as semiconductors, fuel cells, light-emitting diodes (LEDs), to name a few. Organic polymers are used in many industries, including electronics, automotive, in medicine, and in the military. Emerging technologies (e.g., LEDs, soft electronics, and conductive fibers) are based predominantly on conductive and semiconductive polymers. Conductive fibers, for example, are intended for use in creating conductive yarns and realizing connections in smart clothing, or producing conductive fabrics that can be used as electromagnetic shields.

Polymers for Electricity and Electronics: Materials, Properties, and Applications, First Edition. Jiri George Drobny.

Thus it is a challenging task to summarize the new developments and draw some kind of conclusions as to current trends.

8.1 MOLECULAR ELECTRONICS

Molecular electronics is one of the new emerging technologies, relating to the exploitation of organic and biological materials in electronic and optoelectronic applications. It has its roots in material science and concerns the development of these devices by utilizing the unique macroscopic properties of organic compounds in general and polymers in particular. One of the most successful commercial products from this field of research are liquid-crystal displays (LCDs).

However, a more challenging field is that of molecular-scale electronics. Here the focus is on the behavior of individual organic molecules or groups of molecules and the precise three-dimensional (3D) positional control of individual atoms and molecules. Much of the research activity is directed toward computational architectures that may one day rival silicon microelectronics [1]. The explosive growth of nanotechnology is a factor here. The transition of materials from the microscale to nanoscale offers several advantages: The enormous increase in surface area may be achieved and bulk properties become governed by surface properties. A material can be made light absorbing by coating its constituting particles with a dye. The particle size reduction also induces both mechanical advantages and quantum effects [2]. Colors may be controlled and manipulated; an example is the finding that titanium dioxide in nanomaterial form absorbs much more ultraviolet (UV) light than its bulk counterpart.

The other field that may benefit from molecular-scale manipulation is medicine. Since all living organisms are composed of molecules, molecular biology has become the primary focus of biotechnology with the goal of finding cures for some diseases, which, to this day cannot yet be cured.

Given the growth of the applications of organic materials in semiconductor technology, (e.g., LCDs and certain chemical sensors), it can be expected that organic LEDs and polymeric circuitry for identification tags and smart cards will likely have a major impact on the marketplace in the near future.

8.2 INTELLIGENT MATERIAL SYSTEMS

One of the definitions of an *intelligent material* is that it is capable of recognizing appropriate environmental stimuli, processing that information arising from the stimuli, and responding to it in an appropriate manner and time frame. Another desirable feature of that material is that ideally it should be self-powered, having energy conversion–storage functions [3].

The appropriate responses are particularly important because they must result in desirable behavior that enhances the performance of the material itself or the system in which it operates. The simplest example is the response of the material responding to stress by increasing strength or it may release a drug by a specific stimulus (temperature, etc.). The main feature that differentiates intelligent materials from the conventional ones is their dynamic character [4]. As seen in Section 8.1, the process involves processes on the molecular level. The reason is that it is much easier to process and transmit information within the material. The most important task here is to identify the proper building blocks. Conducting electroactive polymers (CEPs) (e.g., polypyrrole, polythiophene, polyaniline, and sulfonated polyaniline) emerged as one of the promising groups of materials for this research.

These polymers have proven to be very useful even for practical. applications replacing metals. The following uses are already commercial or soon to be commercialized [5–14]: electrostatic discharge (ESD) protection, plastic chip, shielding of electromagnetic interference, "smart" label to replace bar codes for packaged products, lithium rechargeable batteries, supercapacitors, photovoltaic systems, electrochromics, changing colors in response to applied potential, electromechanical actuators, hollow fibers for membrane separation technologies, controlled-release devices, corrosion protection, chemical sensors, biomedical applications, and synthetic muscles.

8.3 OTHER DEVELOPMENTS AND TRENDS

8.3.1 Replacement of Poly(vinyl chloride) in Low-Voltage Cables

Halogen–free replacement of conventional poly(vinyl chloride) (PVC) based wire and cable insulation and jacketing materials is a continuing industry trend driven by green and sustainability concerns, particularly in the personal electronics and appliance market. Recently, brand owners have been aggressively looking to replace PVC with halogen-free flame-retardant (HFFR) compound for various wiring applications (e.g., computers, cell phones, and other portable devices). Specifically, the HFFR material will be used as part of assemblies for power cords, data cables including USB/mouse cord, and connectors, and so on.

Another driver for this trend has been the concern over the release of toxic chlorinated products when a PVC compound is exposed to fire, particularly in confined areas.

Replacement of PVC by cross-linked polyethylene (XLPE) has been practiced since 1990s because of the above concerns, but the trend has been gradual [15]. This replacement has established itself as an attractive alternative, both technically and economically [16]. As a variation of the use of polyethylene (PE) only, some manufacturers use a blend of PE and EPDM with aluminum trihydrate and cross-link the insulation by irradiating the insulation with an electron beam (EB) [17]. The use of highly effective HFFR in other polymeric compounds copolymer of ethylene and vinyl acitate ((EVA), polyvinylidene fluoride (PVDF) etc.) are reported [18].

8.3.2 Recent New Developments of Electrically Conductive Adhesives

The trend to replace eutectic tin–lead solder by electrically conductive adhesives is supported by a great amount of research and development. Adhesives based on epoxy resins, silicone, or polyimide, which provide adhesion, mechanical strength, and impact strength are filled by metal powders (e.g., silver, gold, nickel, or copper). These conductive adhesives have been used for several years, but recently their use has been growing rapidly [19]. A recent paper [20] reports the use of a silver-plated graphite nanosheet as a new effective and low-cost filler for conductive adhesives. The novel adhesive comprises acrylate resin and the silver-plated filler. The electrical conductivity of this adhesive containing the nanofillers was 2.6×10^{-2} S cm^{-1}. Conductivity, electrostatic behavior, as well as adhesion of silver-plated filler and short fibers in polyurethane and epoxy adhesives were investigated [21]. Both the adhesion to aluminum and impact strength decreased with increasing filler and fiber content, however, this disadvantage was offset by a significant improvement of conductivity and discharge characteristics (maximum of electrostatic potential and leakage half-time of the electrostatic charge). The best conductivity values were achieved with polyurethane filled with silver-plated polyurethane (2.5×10^3 S cm^{-1}) at a volume fraction of $\sim$0.5.

8.3.3 Field-Effect Transistors

Field-effect transistors are based on conjugated polymers. In this configuration, two metal electrodes, called the *source* and *drain* are deposited

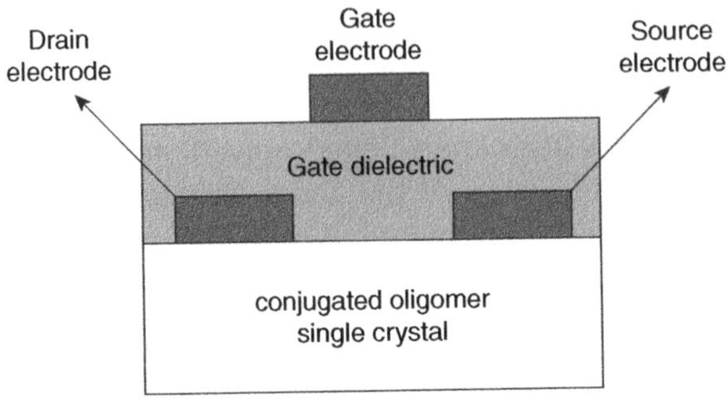

FIGURE 8.1 Schematic of a field-effect transistor. (Riande, E and Díaz-Calleja, R., *Electrical Properties of Polymers*, Marcel Dekker, 2004. p. 607)

on a semiconductor layer [22]. The two electrodes are separated from the third electrode, called the *gate* by a thin layer of dielectric, as shown in Fig. 8.1 [23]. By applying the voltage between the source and gate electrodes, carriers are created and flow between the source and drain. The gate voltage controls the carrier concentration. For practical applications, the ratio of conductivities with the gate voltage switches ON and OFF ratio should exceed 10^6. Therefore, it is important to use purified conjugated polymers without traces of impurities and doping [23].

8.3.4 Nanotechnology

Nanotechnology also touches electrical and electronic application. Besides the improved mechanical and insulating properties achieved by using nanofillers in elastomers and plastics, there are other unique applications, as shown in Section 5.3.2 by using special fillers in conductive adhesives. The use of carbon nanotubes in PVDF open the possibilities to develop various strain sensors or self-sensing composite structures and some mechanical damage self-monitoring materials utilizing this polymer's piezoelectric properties [24]. The possibilities of using different nanoparticles [e.g., Fe_2O_3 (magnetic)], other inorganic fillers, carbon nanotubes, and metallic nanoparticles were reviewed in [25]. Considering the opportunities the nanocomposites offer, one could even think that one day we might see a completely new types of batteries and an "all polymer cable" [26].

8.3.5 Plastic Optical Fibers

Plastic optical fibers (POFs) are manufactured by two primary methods. In the first method, the polymer is melted and extruded through a small orifice at high pressure and then coated with a lower index cladding.

TABLE 8.1 Typical Polymers used Currently for Plastic Optical Fibers

Polymer	Index of Refraction
Amorphous fluoropolymers	
Teflon® AF (DuPont)	1.31
Cytop® (AGC)	1.34
Poly(methyl methacrylate) (PMMA)	1.49
Polycarbonate	1.58
Polystyrene	1.59

A more common method is to make a rod of the fiber core–cladding structure first. Such a rod is called a fiber preform and is a few centimeters in diameter. One end of the preform is heated to the melting point and a thin fiber is drawn out of it, resulting in a fiber with the same aspect ratio and refraction index profile as the preform [27]. The preform can be made in a gradual decrease of refractive index from the center, yielding a graded-index fiber that offers a wider bandwidth than step-index fibers of the same diameter. Depending on the application, the diameters of plastic fibers range from tens of microns to over a millimeter. The most commonly used fibers are several hundred microns in diameter [28]. The most widely used plastics for optical fibers are listed in Table 8.1.

8.3.6 Conductive Textiles

Conducting polymers [e.g., polypyrrole (PPY), polyaniline (PAni), and polyethylenedioxythiophene (PEDOT)] have been deposited onto various textiles in the forms of woven fabrics, knitted fabrics, felt, other nonwoven structures, and fibers. For the given amount of coating add-on, it has been found that, of the three intrinsic conductive polymers (ICPs) mentioned, PPY tends to produce the most conductive products. The coatings are usually applied to full-width, long rolls to fabric or to fabricated articles. For example, electrostatic dissipative gloves, hook and loop bands, and 8 × 8-in. wipes, have been prepared by immersing the untreated fertile in the appropriate baths [29]. The following substrates are used commercially: glass and quartz fabrics, polyester, aramide (e.g., Kevlar®), polyamide (e.g., Nylon®), polyamide/spandex combinations.

An important property of any conductive fabric is its ability to shield against electromagnetic radiation. Conductive fabrics coated with PPY exhibit a high degree of absorption of electromagnetic radiation and can therefore be used in electromagnetic interference (EMI) suppression and cross-talk reduction. Another performance characteristic of conductive fabrics is their ability to reduce reflection of radar signals in various configurations [30]. Typical applications follow [29–31]:

- Microwave energy and radar absorption (artificial horizons at airports, camouflage netting, radar avoidance suits, antennas).
- Electrostatic discharge protection (clean room garments, bomb suits, X-ray machines).
- Heating (blankets, boots, aircraft wings deicing).
- Piezoresistive fabrics (pressure sensors).
- Biomedical instruments.
- Miscellaneous military applications.

Note that similar technology is used for applying conductive polymers (polyaniline and polypyrrole coatings) onto powdered carbonaceous substrates, onto PA12 powder, or titanium dioxide. These products are useful conductive additives for electronic packaging. The above coatings are also applied onto reticulated polyester mats or open-cell polyurethane foams. Such foams are used in damping cavity resonances, EMI suppression, and static dissipation [29–31].

FURTHER READINGS

Introduction to Organic Electronic and Optoelectronic Materials and Devices (Sun, S.-S and Dalton, L.R., Eds.), CRC Press, Boca Raton, FL. 2008.

Blythe, T. and Bloor, D., *Electrical Properties of Polymers*, Second Edition, Cambridge University Press, Cambridge, UK, 2005.

Special Polymers for Electronics & Optoelectronics (Chilton, J.A. and Goosey, M.T., Eds.) Chapman and Hall, London, UK, 1995.

Proceedings from the Conference *Polymers in Electronics*, 30–31 January, 2007, Rapra Technology, Ltd., Munich, Germany.

REFERENCES

1. Petty, M.C., *Molecular Electronics: From Principles to Practice*, John Wiley & Sons, Ltd., Chichester, UK, 2007, p. 1.
2. Petty, M.C., *Molecular Electronics: From Principles to Practice*, John Wiley & Sons, Ltd., Chichester, UK, 2007, p. 4.
3. Wallace, G.G., Spinks, G.M., Kane-Maguire, L.P., and Teasdale, P.R., *Conductive Electroactive Polymers: Intelligent Polymer Systems*, CRC Press, Boca Raton, FL, 2009, p. 2.

4. Wallace, G.G., Spinks, G.M., Kane-Maguire, L.P., and Teasdale, P.R., *Conductive Electroactive Polymers: Intelligent Polymer Systems*, CRC Press, Boca Raton, FL, 2009, p. 3.
5. Krumm, J. et al., *Elektron Lett. IEEE,* **25**(6), 399.
6. Wang, C.Y, et al., *J. Power Sources*, **163**, 1105 (2007).
7. *Conjugated Polymers, Vol. 2 Processing and Applications*, 3rd ed. (Skotheim, T. and Reynolds, J.R., Eds.), CRC Press, Boca Raton, FL, 2007, p.10–1.
8. Coskun, Y., Cirpan, A., and Toppare, L., *Polymer*, **45**, 4989 (2004).
9. Krishnamoorthy, K. et al, *Polymer*, **43**, 6465 (2002).
10. Somnez, G., Somnez, H.B., Shen, C.K.F., (and Wudl, F., *Adv. Mater.*, **16** (21), 1905 (2004).
11. Santos, M.J.L., et al., *J. Solid State Electrochem.*, **10**, 117 (2006).
12. Cutler, C.A. Bourguettaya, M., Kang, J-S., and Reynolds, J.R., *Macromolecules*, **135–136**, 139 (2003).
13. Lu, W., Fadeen, A.G., Qi, B., and Mates, B.R., *Macromolecules*, **38**, 3068 (2005).
14. Wallace, G.G., Spinks, G.M., Kane-Maguire, L.P., and Teasdale, P.R., *Conductive Electroactive Polymers: Intelligent Polymer Systems*, CRC Press, Boca Raton, FL, 2009, p. 24.
15. Moore, G.F., *Electric Cables Handbook*, Wiley Blackwell, Chittenden, UK, 1997, p. 349.
16. Laughton, M.A. and Warne, D.F., *Electrical Engineers Reference Book*, 16th ed., Elsevier Science, London, 2003, p. 31–20.
17. Cleland, M.R., and Galloway, R.A., *Electron Beam Crosslinking of Wire and Cable Insulation*, IBA Industrial-White Paper, Edgewood, NY.
18. Li, B., Guo, D.H., Bartlett, T., and Wasserman, S.H., Proceedings, 59th International Wire & Cable Symposium, November 7–10, 2010, Providence, RI, p. 170.
19. Li, Y. and Wong, C.P., *Mater. Sci. and Eng.*, **R 51**, 1–35 (2006).
20. Zhang, Y, Shuhua, Q, Wu, X., and Duan, G., *Synth. Metals*, **161**, 516–522 (2011).
21. Novák, I., Krupa, I., and Chodák, I., *Synthetic Metals*, **144**, 13 (2004).
22. Riande, E. and Díaz-Calleja, R., *Electrical Properties of Polymers*, Marcel Dekker, Inc., New York, 2004, p. 606.
23. Riande, E. and Díaz-Calleja, R., *Electrical Properties of Polymers*, Marcel Dekker, Inc., New York, 2004, p. 607.
24. Vidhate, S., Chung, J. and Souza, N.D., *Proceedings SPE ANTEC 2009,* June 22–24, 2009, Chicago, IL, p. 597.
25. Cizmic, H., *Wire & Cable Technology International* Vol. **XXXIII** No. 4, July 2005, p. 55.

26. G.C. Psarras, *Editorial Corner, EXPRESS Polymer Letters* Vol. **5**, No. 9 (2009), 533, available at www.expresspolymlett.com.
27. *Introduction to Organic Electronic and Optoelectronic Materials and Devices* (Sun, S.-S. and Dalton, L.R., Eds.), CRC Press, Boca Raton, FL, 2008, p. 829.
28. *Introduction to Organic Electronic and Optoelectronic Materials and Devices* (Sun, S.-S. and Dalton, L.R., Eds.), CRC Press, Boca Raton, FL, 2008, p. 830.
29. Avloni, J., Henn, A., and Lau, R., Proceedings from Conference *Polymers in Electronics*, Paper 3, p. 1, 30–31 January, 2007, Munich, Germany, Rapra Technology, Ltd.
30. Avloni, J., Henn, A., and Lau, R., Proceedings from Conference *Polymers in Electronics*, Paper 3, p.3, 30–31 January, 2007, Munich, Germany, Rapra Technology, Ltd.
31. Eeonyx Corporation, Pinole, CATechnical Data Sheets, available at www.eeonyx.com.

APPENDIX I

TYPICAL VALUES OF ELECTRICAL PROPERTIES OF SELECTED POLYMERS

Polymers for Electricity and Electronics: Materials, Properties, and Applications, First Edition. Jiri George Drobny.

A. Thermoplastics

Polymer	Acronym	Volume Resistivity (Ω-cm)	Surface Resistivity (Ω)	Dielectric Constant		Dissipation Factor	
				60 Hz	1 MHz	60 Hz	1 MHz
Low-density polyethylene	LDPE	$>10^{17}$	10^{14}	2.29	2.28	0.00015	0.00008
High-density polyethylene	HDPE	$>10^{17}$	10^{14}	2.35	2.34	0.00024	0.0002
Ethylene–vinyl acetate	EVA	$>10^{15}$	10^{13}	2.5–3.2	2.6–3.2	0.01	0.03
Polypropylene	PP	$>10^{17}$	10^{13}	2.27	2.25	0.0004	0.0005
Polyisobutylene	PIB	$>10^{15}$	10^{13}	2.3		0.0004	
Poly(vinyl chloride), rigid	PVC	$>10^{15}$	10^{13}	3.5	3.0	0.011	0.05
Poly(vinyl chloride), plasticized		$>10^{11}$	10^{11}	4–8	4–4.5	0.05	0.12
Polystyrene	PS	$>10^{16}$	$>10^{13}$	2.5	2.5	0.0005	0.00007
Acrylonitrile–butadiene–styrene	ABS	$>10^{15}$	$>10^{13}$	2.4–4.7	2.4–3.8	0.001	0.001
Poly(methyl methacrylate)	PMMA	$>10^{15}$	10^{15}	3.4–3.9	2.2–3.2	0.05	0.01
Polytetrafluoroethylene	PTFE	$>10^{18}$	10^{17}	2.1	2.1	0.0002	0.0002
Copolymer of TFE and HFP	FEP	$>10^{18}$	10^{16}	2.1	2.1	0.0002	0.0007
Polychlorotrifluoroethylene	PCTFE	$>10^{18}$	10^{16}	2.3–2.8	2.3–2.5	0.001	0.02
Copolymer of ethylene and TFE	ETFE	$>10^{16}$	10^{13}	2.6	2.6	0.0008	0.005
Polyamide 6	PA 6	10^{12}	10^{10}	3.8	3.4	0.12	0.03
Polyamide 66	PA 6/6	10^{12}	10^{10}	10.0	4.0	0.02	0.03
Polyamide 11	PA 11	10^{13}	10^{11}	3.7	3.5	0.04	0.03
Polyamide 12	PA 12	10^{13}	10^{11}	4.2	3.1	0.04	0.03
Polycarbonate	PC	10^{17}	10^{15}	3.0	2.9	0.0007	0.01
Polybutylene terephthalate	PBT	10^{16}	10^{15}	3.8	3.2	0.002	0.02
Polyphenylene sulfide	PPS	$>10^{16}$		3.1	3.2	0.0004	0.0007
Polyimide	PI	10^{16}	10^{15}	3.5	3.4	0.002	0.005

B. Thermoplastic Elastomers (TPEs)

Polymer	Acronym	Volume Resistivity (Ω-cm)	Surface Resistivity (Ω)	Dielectric Constant		Dissipation Factor	
				50 Hz	1 MHz	50 Hz	1 MHz
Styrene–butadiene–styrene	SBS	10^{16}		2.51	2.50	0.0004	0.0008
Styrene–ethylenebutylene–styrene	SEBS	10^{16}	10^{17}	2.30	2.30	0.0001	0.0002
Thermoplastic vulcanizates	TPV	10^{14}	10^{15}	3.7	3.0	0.027	0.073
Polyolefin blends elastomers	TPO	10^{16}	$>10^{15}$	2.23	2.20	0.0003	0.021
Thermoplastic polyurethane elastomers	TPU	10^{12}	10^{11}	6.5	5.6	0.03	0.06
Polyamide TPEs	COPA	10^{11}	10^{12}	10.3	5.7.0	0.092	0.010
Copolyester TPEs	COPE	10^{13}	10^{14}	4.6	4.4	0.007	0.023

APPENDIX II

ELECTRICAL PROPERTIES, ASTM TEST METHODS, AND SPECIFICATIONS

Polymers for Electricity and Electronics: Materials, Properties, and Applications, First Edition. Jiri George Drobny.

Designation	Title
D257-07	Test Methods for dc Resistance or Conductance of Insulating Materials
D150-98 (2004)	Test Methods for ac Loss Characteristics and Permittivity
D669-03	Test Method for Dissipation Factor and Permittivity Parallel with Laminations of Laminated Sheet and Plate Insulating Materials
D1531-06	Test Methods for Relative Permittivity (Dielectric Constant) and Dissipation Factor of Polyethylene by Liquid Displacement Procedure
D1673-94 (1998)	Test Methods for Relative Permittivity and Dissipation Factor of Expanded Cellular Plastics Used for Electrical Insulation
D2520-95	Test Method for Complex Permittivity (Dielectric Constant) of Solid Electrical Insulating Materials at Microwave Frequencies and Temperatures Up to 1650°C
D-3380-10	Test Method for Permittivity (Dielectric Constant) and Dissipation Factor of Plastic-Based Microwave Substrates
D149-09	Test Method for Dielectric Breakdown and Dielectric Strength of Solid Electrical Insulating Materials at Commercial Power Frequencies
D495-99 (2004)	Test Method for High-Voltage, Low Current Dry Arc Resistance of Solid Electrical Insulation
D1868-07	Test Method for Detection and Measurement of Partial Discharge (Corona) Pulses in Evaluation of Insulation Systems
D3382-07	Methods for Measurement of Energy and Integrated Charges Transfer Due to Partial Discharge (Corona) Using Bridge Techniques
D2275-01 (2008e)	Test Method for Voltage Endurance of Solid Electrical Insulating Materials Subjected to Partial Discharge (Corona) on the Surface
D2132-03	Test Method for Dust-and-Fog Tracking and Erosion Resistance of Electrical Insulating Materials
D2303-97	Test Method for Liquid-Contaminant, Inclined-Plane Tracking and Erosion of Insulating Materials
Partial List: Tests and Specifications for Specific Materials	
D229-09b	Method of Testing Rigid Sheet and Plate Materials Used for Electrical Insulation
D348-07	Methods of Testing Rigid Tubes Used for Electrical Insulation
D349-07	Methods of Testing Laminated Round Rods Used for Electrical Insulation
D709-01 (2007)	Specification for Laminated Thermosetting Materials
D1532-88 (2007)	Specification for Polyester Glass-Mat Sheet Laminate
D619-99 (2004)	Methods for Testing Vulcanized Fiber Used for Electrical Insulation
D710-97 (2009e1)	Specification for Vulcanized Fiber Sheet, Rods and Tubes Used for Electrical Insulation
D2305-02	Methods of Testing Polymeric Films Used for Electrical Insulation
D1351	Specification for Polyethylene Insulated Wire and Cable

APPENDIX III

BASIC ELECTRICAL TESTS: ASTM AND CORRESPONDING INTERNATIONAL STANDARDS

Property	ASTM[a]	ISO[b]	IEC[c]	SI Unit
Arc resistance	D495	1325	62039	S
Dielectric constant	D150	1325	60250	Dimensionless
Dielectric strength	D149	3915	60245	kV mm^{-1}
Dissipation factor	D150		60243-1	Dimensionless
Surface resistivity	D257	1325	60093	Ω
Volume resistivity	D257	1325	60093	Ω-m

[a] American Society for Testing and Materials
[b] International Organization for Standardization
[c] International Electrotechnical Commission

Polymers for Electricity and Electronics: Materials, Properties, and Applications, First Edition. Jiri George Drobny.

APPENDIX IV

NOBEL PRIZE 2000 FOR CHEMISTRY, TITLE PAGE

Polymers for Electricity and Electronics: Materials, Properties, and Applications, First Edition. Jiri George Drobny.

KUNGL.
VETENSKAPSAKADEMIEN
THE ROYAL SWEDISH ACADEMY OF SCIENCES

Information Department, P.O. Box 50005, SE-104 05 Stockholm, Sweden
Phone: +46 8 673 95 00, Fax: +46 8 15 56 70, E-mail: info@kva.se, Web site: www.kva.se

---

(Advanced Information)

The Nobel Prize in Chemistry, 2000:
Conductive polymers

Contents

Prize motivation

The Royal Swedish Academy of Sciences has decided to award the Nobel Prize in Chemistry for 2000 to three scientists who have revolutionised the development of electrically conductive polymers.

Professor Alan J. Heeger at the University of California at Santa Barbara, USA
Professor Alan G. MacDiarmid at the University of Pennsylvania, USA and
Professor Hideki Shirakawa at the University of Tsukuba, Japan

are rewarded *"for the discovery and development of electrically conductive polymers"*.

The choice is motivated by the important scientific position that the field has achieved and the consequences in terms of practical applications and of interdisciplinary development between chemistry and physics.

Conductive polymers – a surprising discovery

We are used to polymers – that is, plastics – being somehow the opposite of metals. They insulate, they do *not* conduct electricity. Electric wires are coated with polymers to protect them – and us – from short-circuits. Yet Alan J. Heeger, Alan G. MacDiarmid and Hideki Shirakawa have changed this view with their discovery that a polymer, polyacetylene, can be made conductive almost like a metal.

Polyacetylene was already known as a black powder when in 1974 it was prepared as a silvery film by Shirakawa and co-workers from acetylene, using a *Ziegler-Natta* catalyst (K. Ziegler and G. Natta, Nobel Prize in Chemistry 1966). But despite its metallic appearance it was not a conductor. In 1977, however, Shirakawa, MacDiarmid and Heeger discovered that oxidation with chlorine, bromine or iodine vapour made polyacetylene films 10^9 times more conductive than they were originally.[1] Treatment with halogen was called "doping" by analogy with the doping of semiconductors. The "doped" form of polyacetylene had a conductivity of 10^5 Siemens per meter, which was higher than that of any previously known polymer. As a comparison, teflon has a conductivity of 10^{-16} S $m^{-1}$and silver and copper 10^8 S m^{-1}.
A key property of a conductive polymer is the presence of conjugated double bonds along the backbone of the polymer. In conjugation, the bonds between the carbon atoms are alternately single and double. Every bond

ACRONYMS AND ABBREVIATIONS

ABR	Acrylate–butadiene rubber
ABS	Acrylonitrile–Butadiene–Styrene
ac	Alternating Current
ACN	Acrylonitrile
ACS	American Chemical Society
AEIC	Association of Edison Illuminating Companies
AEM	Ethylene–acrylic elastomer
ANSI	American National Standards Institute
API	American Petroleum Institute
AREMA	American Railway Engineering and Maintenance of Way Association
ASTM	American Society for Testing and Materials
AU	Polyester urethane
AWG	American Wire Gauge
BIIR	Brominated isobutylene–isoprene (also bromobutyl rubber)
BMC	Bulk molding compound
BMI	Bismaleimide
BR	Polybutadiene rubber
BSI	British Standards Institute
CCV	Catenary continuous vulcanization
CED	Cohesive energy density
CEP	Conducting electroactive polymers (homopolymers) (chlorobutyl rubber)
CIIR	Chlorinated isobutylene–isoprene
CMP	Plenum-rated communication wire
CO	Epichlorohydrin rubber (homopolymer)
COPA	Polyamide thermoplastic elastomer
COPE	Copolyester thermoplastic elastomer

Polymers for Electricity and Electronics: Materials, Properties, and Applications,
First Edition. Jiri George Drobny.

CPE	Chlorinated polyethylene
CR	Polychloroprene rubber
CSM	Chlorosulfonated polyethylene (hypalon), also Cure site monomer
CTE	Coefficient of thermal expansion
CTFE	Chlorotrifluoroethylene
CV	Continuous vulcanization
DAP	Dialyl phthalate
DAIP	Dialyl isophthalate
3D	Three dimensional
dc	Direct current
DIL	Dual in line
DIDP	Diisodecyl phthalate
DIV	Discharge inception voltage
DIN	Deutsche Industrienorm
DOP	Dioctyl phthalate
DOTG	Diorthotolyl guanidine (rubber accelerator)
DSC	Differential scanning calorimetry
DTA	Differential thermal analysis
E	Elastic modulus
E^*	Complex modulus
EAM	Ethylene–vinylacetate copolymer (also EVA)
EB	Electron beam
ECA	Electrically conductive adhesives
ECO	Copolymer of ethylene oxide and epichlorohydrin
ECTFE	Copolymer of ethylene–chlorotrifluoroethylene
EMI	Electromagnetic interference
EOL	End-of-life
EPDM	Terpolymer of ethylene, propylene, and a diene (cure site monomer)
EPM	Copolymer of ethylene–propylene rubber
EPR	Copolymer of ethylene-propylene rubber
EPS	Expanded polystyrene
ETFE	Copolymer of ethylene–tetrafluoroethylene
ESD	Electrostatic discharge
ENB	Ethylene norbornene
ETPV	Engineering thermoplastic vulcanizates
EU	Polyether urethane
EW	Equivalent weight
EVA	Copolymer of ethylene–vinyl acetate
F	Force

FCM	Farrell Continuous Mixer
FEF	Fast extrusion furnace carbon black
FEP	Copolymer of tetrafluoroethylene (TFE) and hexafluoroethylene (HFP), also known as *fluorinated ethylene–propylene*
FFKM	Perfluoro rubber (with a CSM cure site monomer)
FKM	Fluorocarbon rubber (ASTM abbreviation) (also refered to as FPM)
FPM	Fluorocarbon rubber (European abbreviaton) (also refer to as FKM)
FRPE	Flame retardant polyethylene
FVMQ	Flurosilicone rubber
FZ	Fluorinated polyphosphazene rubber
G	Gibbs free energy
GDE	Mass-diffusion electrode
H	Enthalpy
HAF	High-abrasion furnace carbon black
HD	High diffusion
HDPE	High-density polyethylene
HFFR	Halogen-fire flame-retardant
HFP	Hexafluoropropylene
HFPO	Hexafluoropropylene epoxide
HNBR	Hydrogenated nitrile rubber (NBR)
HOMO	Highest occupied molecular orbital
ICEA	Insulated Cable Engineers Association
ICP	Intrinsically conducting polymer
IEC	International Electrotechnical Commission
IEEE	Institute of Electrical and Electronic Engineers
IIR	Isobutene–isoprene rubber (butyl rubber)
IM	Polysobutylene
IR	Polyisoprene rubber (synthetic NR)
ISO	International Standard Association
ITO	Indium tin oxide
LCD	Liquid-crystal display
LCP	Liquid-crystal polymer
LDPE	Low-density polyethylene
LED	Light emitting diode
LIM	Liquid injection molding
LLDPE	Linear low-density polyethylene
LSR	Liquid silicone rubber
LUMO	Lowest unoccupied molecular orbital

M	Molar weight of a polymer
MBTS	2-Mercaptobenzothiazole (rubber accelerator)
MD	Machine direction to follow
MDCV	Long land die process of continuous curing
MEA	Membrane electrode assembly
MFA	Copolymer of tetrafluroethylene and perfluoromethylvinyl ether (PMVE)
MFI	Melt Flow Index
MPR	Melt processable rubber
MQ	Silicone rubber with methyl substituent groups on the polymer chain
MT	Medium thermal carbon black
NBR	Copolymer of acrylonitrile–butadiene rubber (nitrile rubber)
NEC	National Electrical Code
NEMA	National Electrical Manufacturers Association
NFPA	National Fire Protection Agency
NP	Nonpolar
NR	Natural rubber (polyisoprene)
ODP	Ozone depletion potential
OLED	Organic (polymeric) light-emitting diodes
PAA	Poly (acrylic acid)
PA	Polyamide
PB	Polybutadiene
PAc	Polyacetylene
PAE	Polyarylether
PAEK	Poly(arylether ketone)
PAI	Poly(amide-imide)
PAN	Poly(acrylonitrile)
PAni	Polyaniline
PAS	Poly(aryl sulfone)
PBT	Polybutylene terephthalate
PC	Polycarbonate
PCB	Printed circuit board
PCTFE	Poly(chlorotrifluoroethylene)
PD	Partial discharge testing
PDA	Personal digital assestant
PDMS	Polydimethyl siloxane
PE	Polyethylene
PEBA	Polyether block amide
PEDOT	Polyethylenedioxythiophene or poly(3,4-ethylenedioxythiophene)

PEEK	Poly(ether–ether ketone)
PEI	Poly(ether imide)
PEM	Polymer electrolyte membrane
PEO	Poly (ethyleneoxide)
PET	Poly(ethylene terephthalate)
PFA	Copolymer of tetrafluoroethylene (TFE) and perfluoropropylvinyl ether (PPVE)
PFOA	Perfluorooctanoic acid
PFP	Perfluoropolymer
phr	Part per hundred parts of rubber
PI	Polyimide
PIB	Polyisobutylene
PLC	Performance level categories
PLCV	Pressured liquid salt continuous vulcanization
PLED	Polymer light emitting diode
PLLA	Poly(L-Lactic acid)
PMMA	Poly(methyl methacrylate)
PMVE	Perfluoromethylvinyl ether
PMTFPS	polymethyltrifluoropropylsiloxane
PMQ	Silicone rubber with both methyl and phenyl substituent groups
PAni	Polyaniline
POM	Poly(oxymethylene)
PNR	Polynorbornene rubber
PP	Polypropylene
PPME	Perfluoromethylvinyl ether
PPO	Poly (propylene oxide)
PPP	Poly(*p*-phenylene)
PPS	Polyphenylene sulfide
PPV	Poly(*p*-phenylene vinylene)
PS	Polystyrene
PPVE	Perfluoropropylvinyl ether
PPY	Polypyrrole
PSF	Polysulfone
PSS	Poly(styrene sulfonic acid)
PTFE	Poly(tetrafluoroethylene)
PTH	Polythiophene
PV	Photovoltaic
PVAc	Poly(vinylacetate)
PVC	Poly(vinyl chloride)
PVDC	Polyvinylidene chloride
PVDF	Poly(vinylidene fluoride)

PVF	Poly(vinyl fluoride)
PVME	Poly(vinylmethylether)
PZ	Polyphosphazine rubber
RAM	Random access memory
RCP	Radiation-curing process
RF	Radio frequency
RHC	Rubber hydrocarbon content
RFI	Radio-frequency interference
RFID	Radio-frequency identification
RIM	Reaction injection molding
RRIM	Reinforced reaction injection molding
RT	Room temperature
S	Entropy
SBC	Styrenic block copolymer
SBR	Copolymer of styrene–butadiene
SBS	Styrene–butadiene–styrene block copolymer
SEBS	Styrene-ethylenebutylene–styrene block copolymer
SN	Poly(sulfur nitride)
SRF	Semireinforcing furnace carbon black
SRIM	Structural reaction injection molding
ST	Systemé International
T	Polysulfide rubber
TAC	Triallyl cyanurate
TAIC	Triallyisocyanurate
TD	Transverse direction of flow
TDR	Time domain reflectivity testing
TFE	Tetrafluoroethylene
T_g	Glass transition temperature
TMTM	Tetramethylthiuram monosulfide
T_m	Crystalline melting temperature
TMTD	Tetramethyl thiuram disulfide (rubber accelerator)
TMPS	Poly(tetramethyl-*p*-silphenylenesiloxane)
TMPT	1,2-dihydroxy-2,2,4-tremethylquinoline (polymerized)
TMPTA	Trimethylolpropane triacylate
TPE	Thermoplastic elastomer
TPO	Polyolefin blend thermoplastic elastomer
TPU	Thermoplastic polyurethane (elastomer)
TPV	Thermoplastic vulcanizate
TrFE	Vinylidene fluoride with trifluoroethylene
UHF	Ultrahigh frequency
UHM	Ultrahigh molecular weight

UHMWPE	Ultrahigh molecular weight polyethylene
UL	Underwriters Laboratories, Inc.
UPS	Uninterruptible power supplies
UTP	Unshielded twisted pair
URD	Underground residential
UV	Ultraviolet
VDE	Association for German Electrical Energies
VDF	Vinyledene fluoride
VF	Vinylfluoride
VLF	Very low frequency testing
VMQ	Silicone rubber with both vinyl and methyl substituent groups
VOC	Volatile organic compounds
W&C	Wire and cable
XLPE	Cross-linked polyethylene
XNBR	Carboxylic–acrylonitrile–butadiene rubber

GLOSSARY

A

Absorbed Dose: Mean value of the energy of ionizing radiation absorbed by a unit of mass. The unit of absorbed dose is 1 gray (Gy), which is equal to 1 joule per kilogram (J kg^{-1}). The commonly used practical quantity is 1 kGy (10^3 Gy) and 1 kGy is equal to one-tenth of a megarad (Mrad) a unit used until 1986 and still is being used occasionally.

Acceleration Voltage: Potential difference between cathode and anode of the electron beam (EB) accelerator, usually expressed in kilovolts (kV). Typical values of acceleration voltages of commonly used industrial equipment are in the range 100 kV–5 MV, although accelerators with voltages $<$100 kV and $>$5 MV are available.

Accelerated Aging: A test, in which voltage, temperature, and so on, are increased above normal operating values in order to obtain observable deterioration in a relatively short period of time.

Admittance: The measure of ease with which an ac flows in a circuit; the reciprocal of impedance.

Alternating Current (ac): Electric current that continuously reverses its direction. It is expressed in cycles per second (unit hertz, abbreviated as Hz).

Ambient Temperature: The temperature of a medium surrounding an object.

American Wire Gauge (AWG): A standard system for designating wire diameter. Primarily used in the United States.

Ampere (A): The unit of current. One ampere is the current flowing through one 1 Ω at 1-V potential.

Analogue: Representation of data by continuously variable quantities.

Polymers for Electricity and Electronics: Materials, Properties, and Applications, First Edition. Jiri George Drobny.

Antioxidant: A chemical substance, which prevents or slows down oxidation of polymeric materials.

Antiozonant: A chemical substance, which prevents or slows down the effect of ozone on polymeric materials.

Arc Resistance: Time required for an arc to establish a conductive path in a material. Breakdown between electrodes usually occurs as a conducting path and is burned on the surface of the dielectric material.

Armor: A mechanical protection usually accomplished by a metallic layer of tape, braid, or served wires

Attenuation: Power loss in an electrical system in cables, generally expressed in decibels (dB) per unit length.

Audio Frequency: Those frequencies audible to the human ear, are generally considered to be in the range of 32–16,000 Hz.

B

Bandwidth: The difference between the upper and lower limits of a given band of frequencies, expressed in hertz (Hz).

Balanced Circuit: A circuit so arranged that the impressed voltages on each conductor of the pair are equal in magnitude, but opposite in polarity, with respect to the ground.

Baud: Unit of data transmission speed representing bits per second; 9600 baud = 9600 bits per second.

Bend Radius: The radius of curvature that a wire can bend without causing any damaging effects.

Binder: A polymeric substance, a resin together with the ingredients of a coating or adhesive.

Bit: One binary digit.

Bond Strength: Amount of adhesion between two surfaces.

Braid: A fibrous or metallic group of filaments interwoven in cylindrical form to cover one or more wires.

Breakdown Voltage: The voltage at which the insulation between two conductors is destroyed.

Bus: Wire used to connect two terminals inside of an electrical unit. A common point for electrical circuits to return to. It can be bare, tinned, or insulated.

Byte: A group of eight binary digits.

C

Capacitance: Storage of electrically separated charges between two plates having different potentials. The value depends largely on the surface area of the plates and the distance between them.

Circuit: A complete path over which electrons can flow from the negative terminals of a voltage source through parts and wires to the positive terminals of the same voltage source.

Cold Flow: Permanent deformation of material due to mechanical force or pressure (not due to heat softening).

Conductance: The ability of a conductor to carry an electrical charge. The ratio of the current flow to the potential difference causing the flow. The reciprocal of resistance.

Conductivity: The capability of a material to carry electrical current, which is usually expressed as a percentage of copper conductivity (copper being 100%).

Conductor: An uninsulated wire suitable for carrying electrical current.

Conduit: A tube or trough in which insulated wires and cables are passed.

Connector: A device used to physically and electrically join two or more conductors.

Continuous Vulcanization (CV): Simultaneous extrusion and curing of elastomeric wire coating materials.

Copolymer: A polymeric material resulting from polymerization of two different monomers.

Cord: A small, very flexible insulated cable.

Corona: A luminous discharge due to ionization of the gas surrounding a conductor, around which a voltage gradient exceeding a certain critical value exists.

Corona Resistance: The time that insulation will withstand a specified level of field-intensified ionization that does not result in the immediate complete breakdown of the insulation.

Coulomb (C): Unit quantity of electricity, that is, the quantity transferred by 1 A in 1 s.

Crazing: Minute cracks on or near the surface of plastics.

Creep: The dimensional change of a material under load.

Cross-linking: A process introducing intermolecular bonds between long polymeric chains resulting from chemical or irradiation techniques.

Cure: Change of physical and mechanical properties of polymeric materials by chemical reaction, by the action of heat and catalysts, alone or in combination with or without pressure. Radiation cure is done by irradiation by EB, gamma rays, or UV–visible radiation.

Curing Cycle: The time, temperature, and pressure required for curing; mostly applicable to molding of thermosets.

Current: The rate of transfer of electricity; practical unit is the ampere (A) that represents the transfer of 1 C s^{-1}.

Cycle: The complete sequence, including reversal of flow of an alternating electric current.

D

Decibel (dB): A unit to express differences of power levels. A term that expresses two power levels used to indicate gains or losses in a system. *Decibel loss* is the loss of signal in a conductor or passive circuits or cables expressed in decibels. *Decibel gain* is the power gain in amplifiers expressed in decibels.

Dielectric: (1) Any insulating material between two conductors, which permits electrostatic attraction and repulsion to take place across it. (2) A material having the property that energy required to establish an electric field is recoverable in whole or in part, as electric energy.

Dielectric Constant: The ratio of capacitance using the material in question as the dielectric, to the capacitance resulting when the material is replaced by air.

Dielectric Loss: The time rate at which electric energy is transformed into heat in a dielectric when it is subjected to a changing electric field.

Dielectric Loss Angle: The difference between 90° and the dielectric phase angle.

Dielectric Loss Factor: The product of the dielectric constant and the tangent of the dielectric loss angle of the dielectric material.

Dielectric Strength: The voltage, which an insulation material can withstand before breakdown occurs.

Digital: Representation of data by discrete characters.

Direct Current (dc): An electric current that flows in one direction.

Dissipation Factor (Loss Tangent, tan δ): The tangent of the loss angle of the insulating material; the ratio of the power loss to the circulating kilovolt ampere (kVA).

Durometer: An instrument for measurement of hardness of materials, also the value denoting the hardness (e.g., Durometer 50 Shore A).

E

Elastomer: A polymer that can be extended into a multiple of its original length and returns to the approximate original length after the applied force is removed. Most of the elastomers possess different degrees of viscoelasticity, which reduces the amount of return to the original length.

Electromagnet: A coil of wire, usually wound on an iron core, which produces a strong magnet field when current is sent through the coil.

Electromagnetic: Pertaining to the combined electric and magnetic fields associated with movements of electrons through conductors.

Electromotive Force: Pressure or voltage; the force that causes current to flow in the circuit.

Electrostatic: Pertaining to static electricity or electricity at rest.

F

Farad (F): Unit of capacitance; the capacitance of a capacitor which, when charged with 1 C, gives a potential of 1 V.

Fiber Optics: Light transmission through optical fibers for communication, including voice, video, and data.

Film: Sheeting having a nominal thickness not >0.0010 in. (0.25 mm).

Flame Resistance: The ability of a material not to propagate flame once the flame is removed.

Flammability: The measure of the material's ability to support combustion.

Foamed Plastics: Resins in flexible or rigid sponge form with the cells closed or interconnected. Foamed insulations provide low dielectric constants and weight savings.

Foaming Agents: Chemicals added to plastics and rubbers that cause them to assume a cellular structure.

Frequency: The number of times an ac repeats its cycle in 1 s.

G

Gauge: A term used to denote the physical size of a wire.

Gray (Gy): Unit of absorbed dose of ionizing radiation; 1 Gy equals to 1 J kg^{-1}. One kilogray (kGy) (1000 Gy) is a commonly used practical unit.

Ground: The connection between an electrical circuit and earth or other large conducting body to serve as earth, thus making a complete electrical circuit.

H

Henry (H): Unit of conductance when the induced electromotive force of 1 V is produced by the inducing current, changing at the rate of 1 A s^{-1}.

Hertz (Hz): A term replacing cycles per second as an indication of frequency.

Hi-pot: A test to determine the highest voltage that can be applied to a conductor without electrically breaking down the insulation.

Homopolymer: A polymer composed of one type of monomer.

Hook-Up Wire: A single insulated conductor used for low-current, low-voltage (usually under 1000 V) applications within enclosed electronic equipment.

I

Impedance: The total opposition the circuit offers to the flow of ac or any other varying current at a particular frequency. It is a combination of resistance R and reactance X, measured in ohms.

Inductance: The property of a circuit or circuit element that opposes a change in current flow, thus causing current changes to lag behind voltage changes. It is measured in henrys.

Interference: Electrical or electromagnetic disturbances that introduce undesirable responses into other electronic equipment.

Ionized radiation: High-energy radiation, which includes electron beam (accelerated electrons), gamma rays, and X-rays.

J

Jacket: An outer covering, usually nonmetallic, mainly used for protection against environmental effects.

Junction: A point where two or more conductors are connected.

K

Kilowatt (kW): A unit equal to 1000 W.

Kilovolt Ampere (kVA): 1000 V × A.

Kirchhoff's Laws: (1) The algebraic sum of the currents, which meet at any point, is zero. (2) In any closed circuit, the algebraic sum of the products of the current and the resistance in each conductor in the circuit is equal to the electromotive force in the circuit.

L

Longitudinal Shrinkage: A term generally applied to shrink products denoting the discrete axial length lost through heating in order to obtain the recovered diameter.

Loss Index: The product of the power factor and the dielectric constant.

Low-Loss Dielectric: An insulating material that has a relatively low dielectric loss, such as polyethylene (PE) or PTFE.

M

Magnetic Field: The region within which a body or current experiences magnetic forces.

Megarad (Mrad): Commonly used unit of radiation dose (10^6 rad); replaced by gray (Gy).

Micrometer (μm): One-millionth of a meter (Micron (μm) is a non-SI unit, which is discouraged).

Microwave: A short electrical wave, usually <30 cm.

Microwave Frequency: Frequency usually >1000 MHz.

mil: A unit used in measuring thickness of film, thickness of wire insulation, or wire diameter. One thousandth of an inch.

Mining Cable: A flame-retardant heavy duty portable cable for use with portable power supply systems and on mobile mining equipment.

Modulus of Elasticity: The ratio of stress to strain in an elastic material.

Monomer: The basic chemical unit used in building a polymer.

N

Nanosecond (ns): One-thousandth of a millionth of a second (10^{-9} s).

Nanometer (nm): One-thousandth of a millionth of a meter (10^{-9} m).

O

OEM: Original Equipment Manufacturer(s).

Ohm: Unit of resistance such that a constant current of 1 A produces a force of 1 V; also denoted as Ω.

Ohm's Law: Current I is equal to the ratio of electromotive force E to resistance R:

$$I = \frac{E}{R}$$

Oligomer: A polymer with a very low molecular weight, usually a liquid substance.

P

Permittivity: Preferred term for dielectric constant.

Plastic Deformation: Change in dimensions under load that is not recovered.

Plasticizer: A chemical agent added to polymers to make them softer, easier processed, and pliable.

Plenum: The air return path of a central air handling system, either ductwork or open space over a dropped ceiling.

Plenum Cable: Cable listed by Underwriters Laboratories for installation in plenums without a need for conduit.

Potting: Sealing of a cable termination or other electronic component with a liquid that hardens into an elastomer or a plastic.

Power Factor: The ratio of resistance to impedance; mathematically the cosine of the angle of phase difference between current and voltage applied.

Primary Insulation: The first layer of nonconductive material applied to a conductor, whose prime function is to act as an electrical barrier.

R

Reactance: The opposition offered by the flow of ac by inductance or capacitance of a component or a circuit.

Recovered Diameter: Diameter of shrinkable products after heating has caused it to return to its original extruded diameter.

Refractive Index: Ratio of light velocity in a vacuum to its velocity in the transmitting medium.

Resistance: A measure of the difficulty in moving electrical current through a medium when voltage is applied. The unit of resistance is the ohm (Ω).

Resistivity: The ability of a material to resist passage of electrical current, either through the bulk or on the surface. The unit of volume resistivity is the Ω-cm, of surface resistivity, the ohm (Ω).

rem: The amount of radiation absorbed that will produce one electrostatic unit of ions per cubic centimeter volume. The SI unit is 1 sievert (Sv); 1 Sv = 100 rem.

S

Sheath: The outer covering or jacket of a multiconductor cable.

Shield: In cables, a metallic layer placed around a conductor or group of conductors to prevent electrostatic interference between the enclosed wires and external fields.

Shrinkage Ratio: The ratio between the expanded diameter and recovered diameter of heat shrinkable products.

Shrink Temperature: The temperature that effects a complete recovery of a shrinkable product from the expanded state.

Shrink Tubing: Tubing that has been extruded, cross-linked, and mechanically expanded, which when reheated will return to its original diameter.

Sintering: Fusion of formed articles, (tapes, rods, films, etc.) at temperatures above the melting point of the material used; commonly used for PTFE.

Skived Tape: Tape shaved in thin layer from a cylindrical block of material (e.g., PTFE).

Surface Resistivity: The resistance of a material between two opposite sides of a unit square of its surface. It is usually expressed by ohms (Ω).

Surge: A temporary, large increase in the voltage or current in an electric circuit or cable.

T

Tear Strength: Force required to initiate or continue to tear in a material under specific conditions.

Tensile Strength (T.S.): The stress in tension required to break a given sample. The SI unit of tensile strength is MPa (alternative unit pounds per square inch-psi is often used in the United States.

Thermoplastic: A material, that softens when heated or reheated, and becomes firm on cooling.

Thermoplastic Elastomers (TPEs): Polymers, copolymers, and blends of polymers that have elastic properties, yet can be processed by melt-processing techniques in the essentially same way as plastics.

Thermoset: A material that hardens or sets by heat, chemicals, or radiation cross-linking techniques, and which once set, cannot be resoftened by heating.

Twisted Pair: A cable composed of two small-insulated conductors, twisted together without a common covering. (*Note*: the two conductors of a twisted pair are usually substantially insulated so that the combination is a special case of a cord).

U

Ultraviolet Degradation: Deterioration caused by a long-time exposure of a material to sunlight, or other ultraviolet (UV) rays containing radiation.

UV/Vis Cure: A method of curing coatings, varnishes, and adhesives using the source of ultraviolet (UV) or vis (visible) radiation. The result is polymerization and/or cross-linking of mostly liquid monomers or oligomers. The reaction is practically instantaneous.

V

VHF: Very high frequency, typically 30–300 MHz.

Volt (V): A unit of electromotive force.

Voltage: The term used in place of electromotive force, potential, potential difference, or voltage drop to designate the electric pressure that exists between two points and is capable of producing a current when a closed circuit is connected between two points.

Volume Resistivity: The electrical resistance between opposite faces of 1 $cm^3$of insulating material; commonly expressed in Ω-cm.

W

Watt (W): A unit of electric power. One watt is the power required to do work of 1 J s^{-1}.

Yield Strength: The minimum stress at which a material will start to physically deform (yield).

BIBLIOGRAPHY

Goosey, M.T., *Plastics for Electronics*, 2nd ed., Kluwer Academic Publishers, Dordrecht, The Netherlands, 2010.

Blythe, T. and Bloor, D., *Electrical Properties of Polymers*, 2nd ed., Cambridge University Press, Cambridge, UK, 2005.

Riande, E. and Diaz-Calleja, R., *Electrical Properties of Polymers*, Marcel Dekker, New York, 2004.

Goosey, M.T., *Electronics Application of Polymers II*, Report 120, Rapra Publishing, 2000.

Ku, C.C. and Liepins, R., *Electrical Properties of Polymers-Chemical Principles*, Hanser Publishers, Munich, 1987.

Introduction to Organic Electronic and Optoelectronic Materials and Devices (Sun, S.-S. and Dalton, L.R., Eds.), CRC Press, Boca Raton, FL, 2008.

So, F., *Organic Electronics: Materials, Processing, Devices and Applications*, CRC Press, Boca Raton, FL, 2010.

Wallace, G.G., Spinks, G.M., Kane-Maguire, L.P., and Teasdale, P.R., *Conductive Electroactive Polymers: Intelligent Polymer Systems*, 3rd ed., CRC Press, Boca Raton, FL, 2009.

Special Polymers for Electronics & Optoelectronics (Chilton, J.A. and Goosey, M.T., Eds.), Chapman & Hall, London, 1995.

Held, G., *Introduction to Light Emitting Diode Technology and Applications*, CRC Press, Boca Raton, FL, 2010.

Plastics for Electronics (Goosey, M.T., Ed.), Elsevier, London, 1985.

Organic Photovoltaic: Mechanisms, Materials, and Devices (Sun, S.-S. and Sariciftci, N.S., Eds.), CRC Press, Boca Raton, FL, 2005.

Polymers for Electricity and Electronics: Materials, Properties, and Applications, First Edition. Jiri George Drobny.

Power and Communication Cables: Theory and Practice (Bartnikas, R. and Srivastava, K.D., Eds.), Institute of Electrical and Electronic Engineers, New York, 2000.

Electrical, Optical Polymer Systems: Fundamentals, Methods, and Applications (Wise, D.L., Wnek, G.E., Trantolo, D.J., Cooper, T.M., and Griesser, J.D., Eds.), Marcel Dekker, New York, 1998.

Polymers in Electronics (Davidson, T., Ed.), ACS Symposium Series 242, American Chemical Society, Washington, DC, 1984.

Conjugated Polymers: Theory, Synthesis, Properties and Characterization, 3rd ed (Skotheim, T. and Reynolds, J.R., Eds.), CRC Press, Boca Raton, FL, 2007.

Conjugated Polymers: Processing and Applications, 3rd ed., (Skotheim, T. and Reynolds, J.R., Eds.), CRC Press, Boca Raton, FL, 2007.

Petty, M.C., *Molecular Electronics: From Principle to Practice*, John Wiley & Sons, Ltd., Chichester, UK, 2007.

Uchino, K., *Ferroelectric Devices*, 2nd ed., CRC Press, Boca Raton, FL, 2010.

Hainich, R.R. and Bimber, O., *Displays: Fundamentals and Applications*, CRC Press, Boca Raton, FL, 2011.

Nanotechnology for Photovoltaics (Tsakalacos, L., Ed.), CRC Press, Boca Raton, 2011.

Thin Film Solar Cells (Poortmans, J. and Arkhipov, V., Eds.), John Wiley & Sons, Ltd., Chichester, UK, 2006.

Dielectric Materials for Electrical Engineering (Martinez-Vega, J., Ed.), Wiley-ISTE, Hoboken, NJ, 2010.

Arora, R. and Mosch, W., *High Voltage and Electrical Insulation Engineering*, Wiley-IEEE Press, Hoboken, NJ, 2011.

Organic Light-Emitting Materials and Devices (Li, Z. and Meng, H., Eds.), CRC Press, Boca Raton, FL, 2007.

Morrison, R., *The Fields of Electronics: Understanding Electronics Using Basic Physics*, John Wiley & Sons, Inc., Hoboken, NY, 2002.

Fuel Cells-Clean Energy for Today's World, Proceedings from 1998 Fuel Cell Seminar, November 16–19, 1998, Palm Springs, CA.

Mayofis, I.M., *Plastics Insulating Materials*, London Life Books Ltd, Princeton, NJ, 1966.

Black, R.M., *The History of Wires and Cables*, Peter Peregrinus Ltd., London, 1983.

Nussbaum, A., *Electronic and Magnetic Behavior of Materials*, Prentice-Hall, Englewood Cliffs, NJ, 1967.

INDEX

Polymers for Electricity and Electronics: Materials, Properties, and Applications, First Edition. Jiri George Drobny.

Printed and bound by CPI Group (UK) Ltd, Croydon, CR0 4YY
16/09/2026
14963412-0004